Higher Engineering Mathematics

Higher Engineering Mathematics

Dr H. S. Govinda Rao

Viva Books

New Delhi | Mumbai | Chennai | Kolkata | Bangalore | Hyderabad | Kochi | Guwahati

First Published 2007
Reprinted 2009

VIVA BOOKS PRIVATE LIMITED

- 4737/23 Ansari Road, Daryaganj, New Delhi 110 002
  E-mail: vivadelhi@vivagroupindia.net, Tel. 42242200
- 76, Service Industries, Shirvane, Sector 1, Nerul, Navi Mumbai 400 706
  E-mail: vivamumbai@vivagroupindia.net, Tel. 27721273, 27721274
- Jamals Fazal Chambers, 26 Greams Road, Chennai 600 006
  E-mail: vivachennai@vivagroupindia.net, Tel. 28290304, 28294241
- B-103, Jindal Towers, 21/1A/3 Darga Road, Kolkata 700 017
  E-mail: vivakolkata@vivagroupindia.net, Tel. 22836381, 22816713
- 7, Sovereign Park Apartments, 56-58, K. R. Road, Basavanagudi, Bangalore 560 004
  E-mail: vivabangalore@vivagroupindia.net, Tel. 26607409, 26607410
- 101-102 Mughal Marc Apartments, 3-4-637 to 641, Narayanguda, Hyderabad 500 029
  E-mail: vivahyderabad@vivagroupindia.net, Tel. 27564481, 27564482
- First Floor, Beevi Towers, SRM Road, Kaloor, Kochi 682 018, Kerala
  E-mail: vivakochi@vivagroupindia.net, Tel: 0484-2403055, 2403056
- 232, GNB Road, Beside UCO Bank, Silpukhuri, Guwahati - 781 003
  E-mail: vivaguwahati@vivagroupindia.net, Tel: 0361-2666386

www.vivagroupindia.com

ISBN: 978-81-309-0618-8

Published by Vinod Vasishtha for Viva Books Private Limited, 4737/23, Ansari Road, Daryaganj, New Delhi - 110 002.

Printed and bound by Sanat Printers, Kundli - 131 028.

Contents

Preface

The book *Higher Engineering Mathematics* is intended to serve as a reference book for students of Mathematics, Science and Engineering who have acquired some knowledge of advanced calculus and elementary integral transformations. Chapters 1 and 2 provide an introduction to *Differential Difference Equations* and *Transform Techniques*. Chapter 3 presents a detailed treatment of *Partial Differential Equations*. Chapter 4 gives the derivation and implementation of *Discrete Fourier Transforms*, *Fast Fourier Transforms* and standard problems. Chapter 5 includes a brief discussion on *Orthogonal Set of Functions*.

I cannot find adequate words to express my gratitude to Dr P. Martin Jebaraj, Principal, Dr Ambedkar Institute of Technology, Bangalore, and Dr V. Seenappa, former Principal and Director of Technical Education, Karnataka for their support and encouragement. I am also grateful to the management trustees of Dr Ambedkar Institute of Technology for their constant encouragement which helped me complete this book. My sincere thanks are due to K.R. Krishnamurthy, Administrative Officer, Dr Ambedkar Institute of Technology, and Viva Books (P) Ltd.

Author

CHAPTER 1

Ordinary Differential Difference Equations

Introduction

In the field of Engineering and Science, we come across many physical and natural phenomena which, when represented by mathematical models, happen to be **differential equations.** Ordinary differential equations occur in a natural way while studying certain problems of physics and engineering. The objective of this chapter is to describe an elementary theory of such differential equations, along with some important techniques of solving the equations of most frequent occurrence.

1.1 Ordinary differential equations of second order with constant coefficients

(i) Homogenous

We start with the homogenous equation, i.e., equations of the form:

$$\frac{d^2y}{dx^2}+a\frac{dy}{dx}+by=0 \qquad [1]$$

where the coefficients a and b are constants. We assume that a and b are real and the range of x considered is the entire x-axis.

Let us consider the function:

$$y=e^{mx} \qquad [2]$$

then, $\dfrac{dy}{dx}=me^{mx};\ \dfrac{d^2y}{dx^2}=m^2e^{mx}$

Substituting the values of $y, \dfrac{dy}{dx}$ and $\dfrac{d^2y}{dx^2}$ in [1], we get

$$(m^2+am+b)e^{mx}=0$$

Since $e^{mx} \neq 0$; $y = e^{mx}$ is a solution of [1] if m satisfies the equation:

$$m^2 + am + b = 0 \qquad [3]$$

Equation [3] is known as auxiliary equation or characteristic equation of [1]. Thus, Equation [3] is obtained from differential Equation [1] by replacing 'y' by '1', $\frac{dy}{dx}$ by 'm' and $\frac{d^2y}{dx^2}$ by 'm^2'.

The complete solution or general solution of the homogeneous differential Equation [1] is obtained depending on the nature of the two roots of the auxiliary Equation [3] as follows.

Case 1: If m_1 and m_2 are the two distinct roots of the quadratic equation:

$$m^2 + am + b = 0 \qquad [1]'$$

then $y = c_1 e^{m_1 x} + c_2 e^{m_2 x}$ [2]'

(where c_1 and c_2 are arbitrary constants) is the general solution of Equation [1].

Proof: For $y = e^{m_1 x}$, we find that

$$\frac{d^2y}{dx^2} + a\frac{dy}{dx} + by = \left[\frac{d^2}{dx^2}(e^{m_1 x})\right] + a\left[\frac{d}{dx}(e^{m_1 x})\right] + be^{m_1 x}$$

$$= m_1^2 e^{m_1 x} + am_1 e^{m_1 x} + be^{m_1 x}$$

$$= (m_1^2 + am_1 + b)e^{m_1 x}$$

$$= 0$$

because m_1 is a root of Equation [1]'

$\Rightarrow \quad y = e^{m_1 x}$ satisfies $\frac{d^2y}{dx^2} + a\frac{dy}{dx} + by = 0.$

Similarly, $y = e^{m_2 x}$ also satisfies $\frac{d^2y}{dx^2} + a\frac{dy}{dx} + by = 0.$

We find that $e^{m_1 x}$ is not a constant multiple of $e^{m_2 x}$ (and vice versa). Hence $y = e^{m_1 x}$ and $y = e^{m_2 x}$ are linearly independent solutions of equation.

Case 2: If m_1 is a repeated root (real and double) of the auxiliary equation: $m^2 + am + b = 0$ given by Equation [1]'

then $y = (c_1 + c_2 x)e^{m_1 x}$

(where c_1 and c_2 are arbitrary constants) in the general solution of $\frac{d^2y}{dx^2}+a\frac{dy}{dx}+by=0.$

Proof: If m_1 is a repeated root of $m^2+am+b=0$, we have

$$m^2+am+b=(m-m_1)^2$$

$$\Rightarrow \quad a=-2m_1 \text{ and } b=m_1^2$$

Since m_1 is a root of $y''+ay'+by=0$,

$y=e^{m_1x}$ is a solution of $y''+ay'+by=0$

(from case 1)

For $y=xe^{m_1x}$, we find that

$$\begin{aligned} y''+ay'+by &= \frac{d^2}{dx^2}(xe^{m_1x})+a\frac{d}{dx}(xe^{m_1x})+b(xe^{m_1x}) \\ &= e^{m_1x}\left[x(m_1^2+am_1+b)+(2m_1+a)\right] \\ &= 0 \ (\because a=-2m, \text{ and } b=m_1^2) \end{aligned}$$

$\Rightarrow$ $y=xe^{m_1x}$ is a solution of $y''+ay'+by=0$ and e^{m_1x} and xe^{m_1x} are linearly independent.

Hence, $y=(c_1+c_2x)e^{m_1x}$ is the general solution of $y''+ay'+by=0$.

Case 3: If one of the roots of the equation $m^2+am+b=0$, say m_1, is complex, then the other root m_2 is also complex and $m_2=\bar{m}_1$, the complex conjugate of m_1. Hence if $m_1=\alpha+i\beta$ and $m_2=\alpha-i\beta$, then the general solution $y=c_1e^{m_1x}+c_2e^{m_2x}$ becomes $y=e^{\alpha x}[c_1\cos\beta x+c_2\sin\beta x]$.

Proof: For the complex conjugate roots, $e^{\alpha+i\beta}$ and $e^{\alpha-i\beta}$ are linearly independent solutions of $m^2+am+b=0$, therefore the general solution of $y''+ay'+by=0$ is

$$\begin{aligned} y &= Ae^{(\alpha+i\beta)x}+Be^{(\alpha-i\beta)x} \\ &= Ae^{\alpha x}(e^{i\beta x}+Be^{\alpha x}(e^{-i\beta x}) \\ &= e^{\alpha x}[A(\cos\beta x+i\sin\beta x)+B(\cos\beta x-i\sin\beta x)] \\ &= e^{\alpha x}[(A+B)\cos\beta x+(A-B)i\cdot\sin\beta x] \end{aligned}$$

$$y = e^{\alpha x}[c_1 \cos \beta x + c_2 \sin \beta x]$$

where $c_1 = (A + B)$ and $c_2 = (A - B)$

The above results are listed in Table 1.1.

Table 1.1

Nature of roots of auxiliary equation	Set of linearly independent solution	General solution
1. Distinct real roots (m_1, m_2)	$e^{m_1 x}, e^{m_2 x}$	$y = c_1 e^{m_1 x} + c_2 x e^{m_2 x}$
2. Real double root (two equal roots) (repeated roots) $m = m_1 = m_2$	$e^{m_1 x}, x e^{m_1 x}$	$y = (c_1 + c_2 x) e^{m_1 x}$
3. Complex conjugate $m_1 = \alpha + i\beta$ and $m_2 = \alpha - i\beta$	$e^{\alpha x} \cos \beta x, e^{\alpha x} \sin \beta x$	$y = e^{\alpha x}(c_1 \cos \beta x + c_2 \sin \beta x)$

(A) EXAMPLES (real and distinct roots)

1. Solve: $\dfrac{d^2 y}{dx^2} - 7\dfrac{dy}{dx} - 44y = 0$

Solution: First we express the equation in terms of the operators operating on y i.e., we put

$$\left(\frac{d}{dx}\right) \equiv D, \left(\frac{d^2}{dx^2}\right) \equiv D^2, \left(\frac{d^3}{dx^3}\right) \equiv D^3, \text{ etc.}$$

Equation: $\left(\dfrac{d^2}{dx^2} - 7\dfrac{d}{dx} - 44\right)y = 0;\ (D^2 - 7D - 44)y = 0$

Next, we form the auxiliary equation by replacing D by m in the operator $(D^2 - 7D - 44)$,

i.e., $(m^2 - 7m - 44) = 0;$

$\Rightarrow$ $(m - 11)(m + 4) = 0;$

we get $m = 11, -4$

Complementary function $\equiv y = Ae^{3x}$

$\therefore$ Complementary solution is $y = c_1 e^{11x} + c_2 e^{-4x}$

2. Solve the equation: $y'' + y - 2y = 0$, given that $y = 0$ and $y' = 3$ for $x = 0$.

Solution: The auxiliary equation is $(m^2 + m - 2) = 0$

$\Rightarrow \quad (m-1)(m+2) = 0$

$\Rightarrow \quad m_1 = 1$ and $m_2 = -2$ are the roots which are real and distinct.

Hence, the general solution is

$$y = c_1 e^x + c_2 e^{-2x} \qquad [1]$$

It is given that $y = 0$ and $y' = 3$ for $x = 0$.

Hence, applying this condition in [1], we get

$$0 = c_1 + c_2 \qquad [2]$$

therefore [1] becomes

$$y' = c_1 e^x - c_2 e^{-2x}$$

Using the condition $y' = 3$ for $x = 0$ in this we get,

$$3 = c_1 - 2c_2 \qquad [3]$$

Solving [2] and [3], we get $c_1 = 1, c_2 = -1$.

Substituting these value of c_1 and c_2 in [1], we get

$$y = e^x - e^{-2x}$$

which is the required particular solution.

3. Solve $y'' - y' - 12y = 0$,

Solution: The auxiliary equation is $(m^2 - m - 12) = 0$

$\Rightarrow \quad (m+3)(m-4) = 0$

Thus, there are two real distinct roots -3 and 4. The general solution is

$$y(x) = c_1 e^{-3x} + c_2 e^{4x}$$

where c_1 and c_2 are arbitrary constants.

4. Solve $y'' - 3y' + 2y = 0,\ y(0) = -1,\ y'(0) = 0$

Solution: The auxiliary equation is $(m^2 - 3m + 2) = 0$

$\Rightarrow \quad (m-2)(m-1) = 0$

$\Rightarrow \quad m = 1, 2$ are real and distinct roots.

The general solution is: $y = c_1 e^x + c_2 e^{2x}$.

Particular solution: Given $y(0) = -1$, so general solution, we get

$$-1 = y(0) = c_1 + c_2 \qquad [1]$$

Using $y'(0) = 0$ is the derivative of general solution so,

$$y' = c_1 e^x + 2c_2 e^{2x}$$

or $\quad 0 = y'(0) = c_1 + 2c_2 \qquad [2]$

Solving [1] and [2], we get $c_1 = -2,\ c_2 = 1$

Hence, the particular solution is $y = -2e^x + e^{2x}$.

EXERCISES

Solve the following.

1. $y'' + 2y' - 8 = 0$ $\left[Ans.\ y = c_1 e^{2x} + c_2 e^{-4x}\right]$

2. $y'' - 3y' + 2y = 0$ $\left[Ans.\ y = c_1 e^{x} + c_2 e^{2x}\right]$

3. $2y'' - 2y' + y = 0$ $\left[Ans.\ y = c_1 e^{(2+\sqrt{3})x} + c_2 e^{2-\sqrt{3})x}\right]$

4. $4y'' - 12y' + 5y = 0$ $\left[Ans.\ y = c_1 e^{\frac{x}{2}} + c_2 e^{\frac{5x}{2}}\right]$

5. $y'' - 6y' + 8y = 0,\ y(0) = 1,\ y'(0) = 6$ $\left[Ans.\ y = -e^{2x} + 2e^{4x}\right]$

6. $2y'' + y' - 6y = 0$ $\left[Ans.\ y = c_1 e^{\frac{3x}{2}} + c_2 e^{-2x}\right]$

7. $y'' + y' - 2y = 0,\ y(0) = 4,\ y'(0) = -5$ $\left[\textit{Ans.}\ y = e^{x} + 3e^{-2x}\right]$

8. $y'' - y' - 12y = 0,\ y(0) = 3,\ y'(0) = -5$ $\left[\textit{Ans.}\ y = 2e^{4x} + e^{-3x}\right]$

(B) EXAMPLES (equal roots, repeated roots)

Solve the following equations.

1. $y'' - 6y' + 9y = 0$

Solution: The auxiliary equation is given by $(m^2 - 6m + 9) = 0$

$\Rightarrow \quad (m-3)^2 = 0$

$\Rightarrow \quad m = 3, 3$, a double (repeated) root

General solution: $y = (c_1 + c_2 x)e^{3x}$.

2. $y'' + 10y' + 25y = 0$

Solution: The auxiliary equation is $(m^2 + 10m + 25) = 0$

$\Rightarrow \quad (m+5)^2 = 0$

$\Rightarrow \quad m_1 = m_2 = -5$

Therefore, the general solution is $y = (c_1 + c_2 x)e^{-5x}$.

3. $y'' + 3y' + 2.25y = 0$

Solution: The auxiliary equation is $(m^2 + 3m + 2.25) = 0$

$$\Rightarrow \quad \left(m + \frac{3}{2}\right)^2 = 0$$

$\Rightarrow \quad m = -\frac{3}{2}, -\frac{3}{2}$, double (repeated) roots

General solution is $y = (c_1 + c_2 x)e^{-\frac{3}{2}x}$.

EXERCISES

Solve the following.

1. $y'' - 8y' + 16y = 0$ $\left[Ans.\ y = (c_1 + c_2 x)e^{4x}\right]$

2. $9y'' - 30y' + 25y = 0$ $\left[Ans.\ y = (c_1 + c_2 x)e^{\frac{5x}{3}}\right]$

3. $y'' - 4y' + 4y = 0,\ y(0) = 3,\ y'(0) = 1$ $\left[Ans.\ y = (3 - 5x)e^{2x}\right]$

4. $y'' + 2ky' + k^2 y = 0$ $\left[Ans.\ y = (c_1 + c_2 x)e^{-kx}\right]$

5. $y'' + 6y' + 9y = 0,\ y(0) = 2,\ y'(0) = -3$ $\left[Ans.\ y = (3x + 2)e^{-3x}\right]$

6. $y'' + 4y' + 4y = 0,\ y(0) = 3,\ y'(0) = 7$ $\left[Ans.\ y = (13x + 3)e^{-2x}\right]$

7. $y'' + 8y' + 16y = 0$ $\left[Ans.\ y = (c_1 + c_2 x)e^{-4x}\right]$

8. $4y'' - 4y' + y = 0$ $\left[Ans.\ y = (c_1 + c_2 x)e^{\frac{x}{2}}\right]$

9. $16y'' - 8y' + y = 0$ $\left[Ans.\ y = (c_1 + c_2 x)e^{\frac{x}{4}}\right]$

(C) EXAMPLES (Complex Conjugate roots)

Solve the following.

1. $y'' + 4y = 0$

Solution: The auxiliary equation is $(m^2 + 4) = 0$

$\Rightarrow \quad m^2 = -4;\ m = \pm 2i = 0 \pm 2i = \alpha \pm i\beta$, (where $\alpha = 0, \beta = 2$)

$\therefore \quad m = 2i, -2i$ imaginary roots

General solution is: $y = (c_1 \cos 2x + c_2 \sin 2x)e^{0x} = (c_1 \cos 2x + c_2 \sin 2x)$

2. $y'' - 4y' + 13y = 0$

Solution: The auxiliary equation is $(m^2 - 4m + 13) = 0$

$$\Rightarrow \quad m = \frac{4 \pm \sqrt{(16-52)}}{2}$$

$$\therefore \quad m = \frac{4 + \sqrt{-36}}{2}$$

$$= \frac{4 \pm 6i}{2} = 2 \pm 3i$$

therefore, there are two complex roots,

$$m = (2 + 3i), (2 - 3i)$$

General solution is $y = e^{2x}(c_1 \cos 3x + c_2 \sin 3x)$.

3. $y'' + 5y' + 12.5y = 0$

Solution: The auxiliary equation is $m^2 + 5m + 12.5 = 0$

$$\Rightarrow \quad m = \frac{-5 \pm 5i}{2}$$

Therefore, there are two complex roots,

$$m = \frac{(-5+5i)}{2}, \frac{(-5-5i)}{2}$$

General solution is $y = e^{-\frac{5}{2}x}\left(c_1 \cos\frac{5}{2}x + c_2 \sin\frac{5}{2}x\right)$.

4. $y'' - 6y' + 13y = 0$

Solution: The auxiliary equation is $(m^2 - 6m + 13) = 0$

$$\Rightarrow \quad m = 3 \pm 2i$$

$$\Rightarrow \quad m = (3 + 2i), (3 - 2i)$$

Therefore, there are two complex roots

The general solution is $y = e^{3x}(c_1 \cos 2x + c_2 \sin 2x)$.

Note: $m = \dfrac{-(-6) \pm \sqrt{(-6)^2 - 4 \cdot 1 \cdot 13}}{2 \cdot 1}$

$= \dfrac{6 \pm \sqrt{36 - 52}}{2}$

$= \dfrac{6 \pm \sqrt{-16}}{2}$

$= 3 \pm i4$

5. $y'' - 2y' + 5y = 0$, $y(0) = -3$, $y'(0) = 1$

Solution: The auxiliary equation is $(m^2 - 2m + 5) = 0$

$\Rightarrow \quad m = 1 \pm 2i$

$\Rightarrow \quad m = (1 + 2i), (1 - 2i)$

There are two complex roots

$\therefore$ general solution is $y = e^x(c_1 \cos 2x + c_2 \sin 2x)$

using $y(0) = -3$ in general solution we get $-3 = e^0(c_1 + 0) \Rightarrow c_1 = -3$

Differentiating general solution. w.r.t. x,

$$y' = e^x(c_1 \cos 2x + c_2 \sin 2x) + e^x(-2c_1 \sin 2x + c_2 \cos 2x)$$

Using $y'(0) = 1$ in the above equation, we have,

$$1 = y'(0) = e^0[c_1 \cdot 1 + 0] + e^0[0 + 2c_2 \cdot 1]$$

$$1 = c_1 + 2c_2, \text{ thus } c_2 = 2.$$

therefore the particular solution is $y' = e^x(-3\cos 2x + 2\sin 2x)$.

EXERCISES

Solve the following.

1. $y'' + 2y' + 5y = 0$ $\left[Ans.\ y = e^{-x}(c_1 \cos 2x + c_2 \sin 2x)\right]$

2. $y''+8y=0,\ y(0)=1,\ y'(0)=2\sqrt{2}$ $\left[Ans.\ \ y=(\cos 2\sqrt{2}x+\sin 2\sqrt{2}x)\right]$

3. $y''-4y'+29y=0,\ y(0)=0,\ y'(0)=5$ $\left[Ans.\ \ y=e^{2x}\sin 5x\right]$

4. $y''+y'+y=0$ $\left[Ans.\ \ y=e^{-\frac{x}{2}}\left(c_1\cos\frac{\sqrt{3}}{2}x+c_2\sin\frac{\sqrt{3}}{2}x\right)\right]$

5. $9y''+6y'+5y=0,\ y(0)=6,\ y'(0)=0$ $\left[Ans.\ \ y=3e^{-\frac{x}{3}}\left(\sin\frac{2}{3}+2\cos\frac{2}{3}x\right)\right]$

6. $y''+0.2y'+4.01y=0,\ y(0)=0,\ y'(0)=2$ $\left[Ans.\ \ y=e^{-0.1x}\sin 2x)\right]$

1.2 Linear equations of higher order

A linear homogeneous equation of nth order with constant coefficients is of the form:

$$a_0\frac{d^n y}{dx^n}+a_1\frac{d^{n-1}y}{dx^{n-1}}+a_2\frac{d^{n-2}y}{dx^{n-2}}+\cdots+a_{n-1}\frac{dy}{dx}+a_n y=0 \qquad [1]$$

Here $a_0, a_1, a_2,\ldots,a_n$ are all constants.

Introducing the notation of differential operator $D\equiv\frac{d}{dx}$ and higher order operators as $D^2=\frac{d^2}{dx^2},\ldots,D^n=\frac{d^n}{dx^n}$, etc., the given equation can be rewritten with this notations as:

$$a_0D^n y+a_1D^{n-1}y+a_2D^{n-2}y+\ldots+a_{n-1}Dy+a_n y=0$$

or $$(a_0D^n+a_1D^{n-1}+a_2D^{n-2}+\ldots+a_{n-1}D+a_n)y=0 \qquad [2]$$

or $$f(D)y=0 \qquad [3]$$

where $f(D)=a_0D^n+a_1D^{n-1}+\ldots+a_{n-1}D+a_n$ is function of D.

Here, the auxiliary equation of [1] is obtained by replacing y by 1, y' by m, y'' by m^2,... $y^{(n)}$ by m^n in [1]. The auxiliary equation is

$$a_0m^n+a_1m^{n-1}+a_2m^{n-2}+\ldots+a_{n-1}m+a_n=0 \qquad [4]$$

is the nth degree polynomial in m, i.e.,

$$f(m)=0 \qquad [5]$$

The general solution of [1] containing n arbitrary constants are obtained according to the nature of the roots of auxiliary Equation (4), as given in Table 1.2.

Table 1.2

Nature of roots of auxiliary equation	Linearly independent solution set	General solutions
1. n distinct and real roots $m_1, m_2, m_3 \dots$	$e^{m_1x}, e^{m_2x}, e^{m_3x} \dots$	$y = c_1e^{m_1x} + c_2e^{m_2x} + c_3e^{m_3x} + \dots$
2. Two equal and real roots, $n - 2$ distinct roots $m_1, m_2, m_3, m_4 \dots$	$e^{m_1x}, xe^{m_1x}, e^{m_3x}, e^{m_4x} \dots$	$y = (c_1 + c_2x)e^{m_1x} + c_3e^{m_3x} + c_4e^{m_4x} + \dots$
3. Three equal real roots, $n - 3$ distinct roots $m_1, m_2, m_3, m_4, m_5 \dots$	$e^{m_1x}, xe^{m_1x}, x^2e^{m_1x}, e^{m_4x} \dots$	$y = (c_1 + c_2x + c_3x^2)e^{m_1x} + c_4e^{m_4x} + c_5e^{m_5x} + \dots$
4. Two complex conjugate roots, $n - 2$ distinct real roots $m_1 = \alpha + i\beta$ $m_2 = \alpha - i\beta$, $m_3, m_4 \dots$	$e^{\alpha x}\cos\beta x$, $e^{\alpha x}\sin\beta x$ $e^{m_3x}, e^{m_4x}, \dots$	$y = e^{\alpha x}(c_1\cos\beta x + c_2\sin\beta x) + c_3e^{m_3x} + c_4e^{m_4x} + \dots$
5. Two equal complex conjugate roots, $n - 4$ distinct real roots, $m_1 = m_2 = \alpha + i\beta$ $m_3 = m_4 = \alpha - i\beta$ $m_5, m_6 \dots$	$e^{\alpha x}\cos\beta x$, $xe^{\alpha x}\cos\beta x$ $e^{\alpha x}\sin\beta x$, $xe^{\alpha x}\sin\beta x$ $e^{m_5x}, e^{m_6x}, \dots$	$y = e^{\alpha x}[(c_1 + c_2x)\cos\beta x + (c_3 + c_4x)\sin\beta x] + c_5e^{m_5x} + \dots$

Synthetic Division Scheme

For finding the quotient and remainder on dividing a polynomial $f(x) = a_0x^n + a_1x^{n-1}y + a_2x^{n-2}y + \dots + a_{n-1}x + a_n$ by $(x - h)$, we apply a method known as the *synthetic division* method which is outlined below.

$$\begin{array}{c|l} x = h & a_0 + a_1 + a_2 + \dots + a_{n-1} + a_n \\ & \quad +a_0h + (a_1h + a_0h^2)\dots \\ \hline & a_0 + (a_1 + a_0h) + (a_2 + a_1h + a_0h^2) + \dots + R \end{array}$$

If we denote $a_0, (a_1 + a_0h), (a_2 + a_1h + a_0h^2), \dots$ by $b_0, b_1, b_2, \dots$ respectively, then the quotient will be, $b_0x^{n-1} + b_1x^{n-2} + a_2x^{n-3} + \dots + b_{n-1}$, and the remainder is

$R = b_n$ and $R = 0$, if $(x-h)$ is a factor of $f(x)$ (i.e., h is a root of $f(x) = 0$. This method is called the synthetic division.

Examples (i): Find the quotient and remainder on dividing
$x^4 - 5x^3 + 6x^2 + 7x - 9$ by $(x-1)$.

Solution:

$$\begin{array}{r|l} x=1 & 1-5-6+7-9 \\ & \quad +1-4+2+9 \\ \hline & 1-4+2+9\,(+0) \end{array}$$

$\therefore$ quotient $= x^3 - 4x^2 + 2x + 9$ and remainder $= 0$.

(ii): Find the quotient and the remainder in the division of
$x^5 - 3x^3 + 4x - 1$ by $x + 2$.

Solution:

$$\begin{array}{r|l} x=-2 & 1+0-3+0+4-1 \\ & \quad -2+4-2+4-16 \\ \hline & 1-2+1-2+8\,(-17) \end{array}$$

$\therefore$ quotient $= x^4 - 2x^3 + x^2 - 2x + 8$ and remainder $= -17$.

(D) EXAMPLES

Solve the following equations.

1. $(D^3 - 9D^2 + 23D - 15)y = 0$

Solution: The auxiliary equation is $m^3 - 9m^2 + 23m - 15 = 0$

$\Rightarrow \quad (m-1)(m-5)(m-3) = 0$

$\Rightarrow \quad m = 1, 3, 5.$

$\therefore$ General solution is $y = c_1 e^x + c_2 e^{3x} + c_3 e^{5x}$.

2. $y''' - 2y'' + y' = 0$

Solution: The auxiliary equation is $m^3 - 2m^2 + m = 0$

$\Rightarrow \quad m(m^2 - 2m + 1) = 0$

$$\Rightarrow \quad m(m-1)^2 = 0$$

$$\Rightarrow \quad m = 0, 1, 1.$$

There is a pair of real and coincident roots and another distinct roots.

$\therefore$ General solution is $y = c_1 e^{0x} + (c_2 + c_3 x)e^x$

$$y = c_1 + (c_2 + c_3 x)e^x.$$

3. $y''' - 4y'' + y' + 6y = 0$

***Solution*:** The auxiliary equation is $m^3 - 4m^2 + m + 6 = 0$

By inspection -1 is a root.

$\therefore$ by synthetic division:

$$\begin{array}{r|rrrl} -1 & 1 & -4 & 1 & 6 \\ & 0 & -1 & 5 & -6 \\ \hline & 1 & -5 & 6 & \lfloor 0 = R \end{array}$$

$$\Rightarrow \quad m = 0, 1, 1.$$

We obtain the factorization as

$$(m - (-1))(m^2 - 5m + 6) = 0$$

$$\Rightarrow \quad (m+1)(m-2)(m-3) = 0$$

Thus, there are three real distinct roots $-1, 2, 3$.

$\therefore$ General solution is $y = c_1 e^{-x} + c_2 e^{+2x} + c_3 e^{3x}$.

4. $4y^{(4)} - 8y''' - 7y'' + 11y' + 6y = 0$

***Solution*:** The auxiliary equation is $4m^4 - 8m^3 + 7m^2 + 11m + 6 = 0$

Observe that -1 is a root $(4 + 8 - 7 - 11 + 6 = 0)$

[Note that 2 is also a root, $(64 - 64 - 28 + 22 + 6) = 0$

By synthetic division

−1	4	−8	−7	11	6
		−4	12	−5	−6
2	4	−12	5	6	0
		8	−8	−6	
	4	−4	−3	0	

We obtain the factorization as

$$(m+1)(m-2)(4m^2-4m-3)=0$$

$$(m+1)(m-2)\left(m+\frac{1}{2}\right)\left(m-\frac{3}{2}\right)=0$$

with four distinct real roots, $-\frac{1}{2}, -1, \frac{3}{2}, 2.$

Therefore the general solution is $y=c_1e^{-\frac{x}{2}}+c_2e^{-x}+c_3e^{\frac{3x}{2}}+c_4e^{2x}.$

5. $\left(\frac{d^3y}{dx^3}\right)-3\left(\frac{d^2y}{dx^2}\right)+3\left(\frac{dy}{dx}\right)-y=0$

***Solution*:** Equation $(D^3-3D^2+3D-1)y=0$

Auxiliary equation: $(m^3-3m^2+3m-1)=0;$

or, $(m-1)^3=0;$

$\therefore$ $m=1, 1, 1$ (three equal roots)

$\therefore$ General solution is $y=(c_1+c_2x+c_3x^2)e^x$

6. $\left(\frac{d^3y}{dx^3}\right)-y=0$

***Solution*:** Equation $(D^3-1)y=0$

Auxiliary equation $(m^3-1)\ y=0$

$\therefore$ $(m-1)(m^2+m+1)=0;$

$m=1$, or $(m^2+m+1)=0$

$$m=\frac{-1\pm\sqrt{(1-4)}}{2}=\frac{-1\pm i\sqrt{3}}{2}=\left(-\frac{1}{2}\right)\pm i\left(\frac{\sqrt{3}}{2}\right)$$

$$\therefore \text{ roots are: } 1, \left[\left(-\frac{1}{2}\right)+i\left(\frac{\sqrt{3}}{2}\right), \left(-\frac{1}{2}\right)-i\left(\frac{\sqrt{3}}{2}\right)\right]$$

$$\therefore \text{ general solution } y=c_1e^x+e^{-\frac{x}{2}}\left[c_2\cos\left(x\left(\frac{\sqrt{3}}{2}\right)\right)+c_3\sin\left(x\left(\frac{\sqrt{3}}{2}\right)\right)\right]$$

7. $\left(\dfrac{d^4y}{dx^4}\right)-13\left(\dfrac{d^2y}{dx^2}\right)+36y=0$

***Solution*:** Here, the equation is apparently of the fourth order; but as you can see it is reducible to quadratics easily,

$$\left[\left(\frac{d^4}{dx^4}\right)-13\left(\frac{d^2}{dx^2}\right)+36\right]y=0;$$

$$(D^4-13D^2+36)y=0$$

Auxiliary equation $(m^4-13m^2+36)-0$

or, $\quad (m^2-9)(m^2-4)=0$

$\therefore \quad (m+3)(m-3)(m+2)(m-2)=0$

$\therefore \quad m=-3,3,-2,2$

$\therefore$ General solution $y=c_1e^{-3x}+c_2e^{3x}+c_3e^{-2x}+c_4e^{2x}$.

8. $\left(\dfrac{d^4y}{dx^4}\right)+\left(\dfrac{d^2y}{dx^2}\right)-20y=0$

***Solution*:** Auxiliary equation is

$$(D^4+D^2-20)y=0; \Rightarrow \left[(D^2)^4+5D^2-4D^2-20\right]y=0$$

$$(m^2+5)(m^2-4)=0$$

$\Rightarrow \quad m^2=-5, \text{ or } 4;$

$m = \pm i\sqrt{5};\ \pm 2$

$\therefore$ General solution $y = \left[c_1 \cos\left(x\sqrt{5}\right) + c_2 \sin\left(x\sqrt{5}\right)\right] + \left(c_3 e^{2x} + c_4 e^{-2x}\right).$

9. $\left(\dfrac{d^4 y}{dx^4}\right) + 13\left(\dfrac{d^2 y}{dx^2}\right) + 36y = 0$

Solution: Auxiliary equation is $(D^4 + 13D^2 + 36)y = 0$

$\Rightarrow \quad m^4 + 13m^2 + 36 = 0;$

$\Rightarrow \quad (m^2 + 9)(m^2 + 4) = 0$

$\therefore \quad m^2 = -9;$ or $m^2 = -4;$

so, $\quad m = \pm 3i, \quad m = \pm 2i$

There are two pairs of distinct imaginary roots

$\therefore \quad y = (c_1 \cos 3x + c_2 \sin 3x) + (c_3 \cos 2x + c_4 \sin 2x).$

10. $\left(\dfrac{d^4 y}{dx^4}\right) - 8\left(\dfrac{d^2 y}{dx^2}\right) + 16y = 0$

Solution: Auxiliary equation is $(m^4 - 8m^2 + 16) = 0;$

$\Rightarrow \quad (m^2 - 4)^2 = 0; \quad m^2 = 4;$

$\Rightarrow \quad m = (+2, -2); \quad (+2, -2)$

There are two pairs of real coincident roots.

$\therefore \quad y = (c_1 + c_2 x)e^{2x} + (c_3 + c_4 x)e^{-2x}.$

11. $\left(\dfrac{d^4 y}{dx^4}\right) + 8\left(\dfrac{d^2 y}{dx^2}\right) + 16y = 0$

Solution: Auxiliary equation is: $m^4 + 8m^2 + 16 = 0$

$(m^2 + 4)^2 = 0;$

$m^2 = -4, -4;$

$$m = \pm 2i, \pm 2i$$

so, there are two pairs of coincident imaginary roots.

$\therefore$ General solution is $y = (c_1 + c_2 x)\cos 2x + (c_3 + c_4 x)\sin 2x.$

12. $y'''' + 6y''' + 9y'' = 0$

Solution: Auxiliary equation is: $m^4 + 6m^3 + 9m^2 = 0$

$$m^2(m^2 + 6m + 9) = m^2(m+3)^2 = 0$$

so $\quad m = 0, 0, -3, -3$ are the roots

i.e., $\quad m = 0$ and $m = -3$ each is a double root.

$\therefore$ General solution is $y = (c_1 + c_2 x)e^{0 \cdot x} + (c_3 + c_4 x)e^{-3x}.$

13. $4y''' + 4y'' + y' = 0$

Solution: Auxiliary equation is: $4m^3 + 4m^2 + m = 0$

$$m(4m^2 + 4m + 1) = m\left(m + \frac{1}{2}\right)^2 = 0$$

so $\quad m = 0, -\frac{1}{2}, -\frac{1}{2}$ are the roots of which

$m = -\frac{1}{2}$ is a repeated (double) root.

$\therefore$ General solution is $y = (c_1 \cdot e^{0 \cdot x} + (c_2 + xc_3)e^{-\frac{x}{2}}.$

14. $\left(\frac{d^3y}{dx^3}\right) - 2\left(\frac{d^2y}{dx^2}\right) + 4\left(\frac{dy}{dx}\right) - 8y = 0$

Solution: Equation: $(D^3 - 2D^2 + 4D - 8)y = 0$

Auxiliary equation: $m^3 - 2m^2 + 4m - 8$

$\Rightarrow \quad m^2(m-2) + 4(m-2) = 0$

$\Rightarrow \quad (m-2)(m^2+4) = 0$

$\Rightarrow \quad m = 2,\ m = \sqrt{-4} = \sqrt{4i^2} = \pm 2i$

$\therefore$ General solution is $y = c_1 e^{2x} + c_2 \cos 2x + c_3 \sin 2x.$

15. $\left(\dfrac{d^3 y}{dx^3}\right) + y = 0$

Solution: Equation: $(D^3 + 1)y = 0$

Auxiliary equation: $m^3 + 1 = 0 = f(m)$

If $\quad m = -1,\ f(-1) = -1 + 1 = 0$

$\Rightarrow \quad -1$ is a root by inspection.

The other two roots are found by synthetic division.

$$\begin{array}{r|rrrr} -1 & 1 & 0 & 0 & 1 \\ & 0 & -1 & 1 & -1 \\ \hline & 1 & -1 & 1 & \underline{|0} \end{array}$$

$\Rightarrow \quad m^2 - m + 1 = 0$

$\Rightarrow \quad m = \dfrac{1 \pm \sqrt{1-4}}{2} = \dfrac{1 \pm \sqrt{-3}}{2} = \dfrac{1 \pm i\sqrt{3}}{2}$

Thus, $m = -1,\ \left(\dfrac{1}{2}\right) \pm i\left(\dfrac{\sqrt{3}}{2}\right)$ are the root of auxiliary equation.

Therefore the general solution is

$$y = c_1 e^{-x} + e^{\frac{x}{2}}\left[c_2 \cos\left(\frac{\sqrt{3}}{2}\right)x + c_3 \sin\left(\frac{\sqrt{3}}{2}\right)x\right].$$

16. $y''' - 3y'' - y' + 3y = 0$

Solution: Auxiliary equation: $m^3 - 3m^2 - m + 3 = 0$

Observe that $m = 1$ is a root that becomes $(1 - 3 - 1 + 3 = 0)$

By synthetic division; we have

$$\begin{array}{r|rrrr} 1 & 1 & -3 & -1 & 3 \\ & & -1 & -2 & -3 \\ \hline & 1 & -2 & -3 & 0 \end{array}$$

we now factorize the above as

$$(m-1)(m^2-2m-3)$$
$$=(m-1)(m+1)(m-3)=0$$

Three real and distinct roots are $-1, 1, 3$

General solution is $y=c_1e^{-x}+c_2e^{x}+c_3e^{3x}$.

17. If $\left(\dfrac{d^4x}{dt^4}\right)=m^4x$ show that $x=c_1\cos mt+c_2\sin mt+c_3\cosh mt+c_4\sinh mt$

Solution: Given equation is $(D^4-m^4)x=0$

$$\left(D=\frac{d}{dt}\right)$$

It is not appropriate to use 'm' to denote 'D' for writing the auxiliary equation, so the auxiliary equation is $(D^4-m^4)=0$

$$\Rightarrow \quad (D^2+m^2)(d^2-m^2)=0$$
$$D=\pm im,\ D=m,-m$$

Therefore the complete solution is: $x=c_1\cos mt+c_2\sin mt+Ae^{mt}+Be^{-mt}$

$$=c_1\cos mt+c_2\sin mt+A(\cosh mt+\sinh mt)+B(\cosh mt-\sinh mt)$$

$$\left[\because \cosh mt=\frac{e^{mt}+e^{-mt}}{2};\ \sinh mt=\frac{e^{mt}-e^{-mt}}{2}\right]$$

$$\left(\therefore \begin{array}{l} e^{mt}=\cosh mt+\sinh mt \\ e^{-mt}=\cosh mt-\sinh mt \end{array}\right)$$

$$=c_1\cos mt+c_2\sin mt+(A+B)\cosh mt+(A-B)\sinh mt$$
$$=c_1\cos mt+c_2\sin mt+c_3\cosh mt+c_4\sinh mt$$

on replacing $(A+B)$ and $(A-B)$ by c_3 and c_4 respectively.

18. $y''''' - y''' = 0$

Solution: Auxiliary equation: $m^5-m^3=0$

$$m^3(m^2-1)=0$$

so $m=0$ is a triple (repeated three times) root and $m=-1, 1$ are distinct real roots.

General solution is $y = c_1e^{-x} + c_2e^x + (c_3 + xc_4 + x^2c_5)e^{0\cdot x}$

$$y = c_1e^{-x} + c_2e^x + (c_3 + xc_4 + x^2c_5).$$

EXERCISES

Solve the following equations.

1. $y''' - 3y'' - y' + 3y = 0$ $\quad\left[Ans.\ \ c_1e^x + c_2e^{-x} + c_3e^{3x}\right]$

2. $y''' - 6y'' + 11y' - 6y = 0$ $\quad\left[Ans.\ \ c_1e^x + c_2e^{2x} + c_3e^{3x}\right]$

3. $y''' + y'' + y' + y = 0$ $\quad\left[Ans.\ \ y = c_1e^{-x} + c_2\cos x + c_3\sin x\right]$

4. $y''' + 3y'' + 3y' + y = 0$ $\quad\left[Ans.\ \ y = (c_1 + c_2x + c_3x^2)e^{-x}\right]$

5. $y''' - 2y'' + y' = 0$ $\quad\left[Ans.\ \ y = c_1e^{0x} + (c_2 + c_3x)e^x = c_1 + (c_2 + c_3x)e^x\right]$

6. $y'''' + y'' - 20y = 0$ $\quad\left[Ans.\ \ y = \left[c_1\cos\left(\dfrac{x}{\sqrt{5}}\right) + c_2\sin\left(\dfrac{x}{\sqrt{5}}\right)\right] + c_3e^{2x} + c_4e^{-2x}\right]$

7. $y'''' + 8y'' + 16y = 0$ $\quad\left[Ans.\ \ y = (c_1 + c_2x)\cos 2x + (c_3 + c_4x)\sin 2x\right]$

8. $y'''' + a^4y = 0$ $\quad\left[Ans.\ \ e^{-\left(\frac{a}{\sqrt{2}}\right)x}\left(c_1\cos\dfrac{a}{\sqrt{2}}x + c_2\sin\dfrac{a}{\sqrt{2}}x\right)\right.$

$$\left. + e^{-\left(\frac{a}{\sqrt{2}}\right)x}\left(c_3\cos\frac{a}{\sqrt{2}}x + c_4\sin\frac{a}{\sqrt{2}}x\right)\right]$$

9. $y''' - 4y'' + y' + 6y = 0$ $\quad\left[Ans.\ \ c_1e^{-x} + c_2e^{2x} + c_3e^{3x}\right]$

10. $y''' - 6y'' + 11y' - 6y = 0,\ y(0) = 0,\ y''(0) = 2$ $\quad\left[Ans.\ \ y = e^x - 2e^{2x} + e^{3x}\right]$

11. $y'''' - 3y''' + 2y'' = 0,\ y(0) = 2,\ y'(0) = 0,\ y''(0) = 2,\ y'''(0) = 2$

$\left[Ans.\ \ y = 2(e^x - x)\right]$

12. $y'''' + 4y = 0$ $\quad \left[Ans. \quad y = e^{-x}(c_1 \cos x + c_2 \sin x) + e^{x}(c_3 \cos x + c_4 \sin x)\right]$

13. $y^{(6)} + 3y^{(4)} + 3y^{(2)} + y = 0$ $\left[Ans. \quad y = (c_1 + c_2 x + c_3 x^2)\sin x + (c_4 + c_5 x + c_6 x^2)\cos x\right]$

(ii) Non-Homogenous

A linear non-homogenous equations of second order with constant coefficients is of the form:

$$\frac{d^2 y}{dx^2} + a\frac{dy}{dx} + \text{by} = X = F(x) \qquad [1]$$

where a and b are constants and X is a specified function of x. A linear non-homogenous equation of nth order with constant coefficients is of the form:

$$(a_0 D^n + a_1 D^{n-1} + a_2 D^{n-2} + \ldots + a_n)y = X = F(x) \qquad [2]$$

where $D = \dfrac{d}{dx}, D^2 = \dfrac{d^2}{dx^2}, \ldots, D^n = \dfrac{d^n}{dx^n}$.

The general solution of [2] is given by $y(x) = y_c(x) + y_p(x)$

where, $y(x) = c_1 y_1(x) + c_2 y_2(x) + \ldots c_n y_n(x)$ containing n arbitrary constants and $y_p(x)$ is any (particular) solution of the non-homogenous Equation [2] containing no arbitrary constants.

Inverse differential operator methods of finding particular integrals (P.I)

Inverse operator: $D^{-1}[F(x)] = \dfrac{1}{D}F(x) = \int F(x)dx$. If $Du(x) = v(x)$ then $u(x) = D^{-1}v(x)$. Here D^{-1} is known as inverse operator of D such that $DD^{-1}(F(x)) = F(x)$ i.e., $DD^{-1} = 1$ when D is differential operator then D^{-1} also denoted by $\dfrac{1}{D}$ represents integral operator $\int$. Thus in a similar way $\dfrac{1}{D^2}, \dfrac{1}{D^3}, \ldots$ etc., denotes integration twice, thrice etc., w.r.t. x.

Equation [2] can now be rewritten as: $f(D)y = F(x)$ $\qquad [3]$

where $f(D) = a_0 D^n + a_0 D^{n-1} + \ldots + a_{n-1}D + a_n$ $\qquad [4]$

is a function of the differential operator D. Then the particular integral (P.I.) of [3] is $f(D)y = F(x)$

is $$y_p = \frac{1}{f(D)} F(x) \qquad [5]$$

where $\frac{1}{f(D)}$ is the inverse operator of $f(D)$, such that

$$f(D)\left\{\frac{1}{f(D)} F(x)\right\} = F(x) = \frac{1}{f(D)}\{f(D)F(x)\}.$$

General method of finding the particular integral of any function *F*(*x*).

To Prove that if $(D-a)y = F(x)$, [1]

then $$y = \frac{1}{(D-a)} F(x) = e^{ax}\int F(x)e^{-ax}dx \qquad [2]$$

Proof: Given $(D-a)y = F(x)$ [1]

$$\Rightarrow \quad \frac{dy}{dx} - ay = F(x); \qquad \left(D = \frac{d}{dx}\right)$$

which is a first order linear differential equation with integrating factor as:

$$e^{\int(-a)dx} = e^{-ax},$$

Solution: $y(\text{Integral factor}) = \int e^{-ax} \cdot F(x)dx + c$

Thus, a particular solution of [1] (with $c = 0$) is $y = e^{ax}\int F(x)e^{-ax}dx.$

Note: (i) In case $F(x) = b = \text{constant}$, then $(D-a)y = b$

Integrating, $ye^{-ax} = \int be^{-ax}dx = \frac{be^{-ax}}{-a}$, $a \neq 0$

or $$y = \frac{1}{(D-a)} b = \frac{b}{-a} \text{ with } a \neq 0 \qquad [2]$$

(ii) In RHS of $y_p = \frac{1}{f(D)} F(x)$

if $F(x) = e^{ax}$, then

$$y = \frac{1}{(D-a)} F(x) = e^{ax}\int F(x)e^{-ax}dx \text{ reduces to}$$

$$y = e^{ax}\int F(x)e^{-ax}dx = e^{ax}\int e^{ax}e^{-ax}dx = xe^{ax}.$$

thus, $y_p = \frac{1}{(D-a)}e^{ax} = xe^{ax}$ [3]

(iii) If $(D-a)^r y = F(x) = e^{ax}$, then by the repeated application of the above result, by integrating 'r' times successively, we get

$$y_p = \frac{1}{(D-a)^r}e^{ax} = \frac{x^r}{r!}e^{ax}$$

I. Here are some methods of finding $\left[\frac{1}{f(D)}\right]F(x)$, where $F(x)$, the RHS function in the non-homogenous differential equation $f(D)y = F(x)$, has the following forms:

$$\left\{\begin{array}{l} F(x) = e^{ax+b} \\ F(x) = \sin(ax+b) \text{ or } \cos(ax+b) \\ F(x) = x^m \text{ or polynomial in } x \\ F(x) = e^{ax}V(x) \text{ (exponential shift)} \\ F(x) = xV(x) \end{array}\right\}$$

(i) Particular integral when $f(a) \neq 0$, then

$$y_p = \frac{1}{f(D)}e^{ax+b} = \frac{1}{f(a)}\cdot e^{ax+b} \qquad \text{(put } D = a\text{)}$$

(ii) Let $f(a) = 0$

If $f(a) = 0$ then $(m-a)$ is a factor of $f(m)$ so that $(D-a)$ is a factor of $f(D)$ i.e.,

$$f(D) = (D-a)\phi(D)$$

such that $\phi(a) \neq 0$, then

$$\frac{1}{f(D)}e^{ax+b} = \frac{1}{(D-a)\phi(D)}e^{ax+b}$$

$$= \frac{1}{\phi(a)}\frac{1}{D-a}e^{ax}$$

$$= \frac{1}{\phi(a)} \cdot xe^{ax+b}$$

Suppose $f(D) = (D-a)^r \psi(D)$ with $\psi(a) \neq 0$ Then

$$\frac{1}{f(D)} e^{ax+b} = \frac{1}{(D-a)^r \psi(D)} e^{ax+b}$$

$$= \frac{1}{\psi(a)} \frac{1}{(D-a)^r} e^{ax+b}$$

$$= \frac{1}{\psi(a)} \cdot \frac{x^r}{r!} e^{ax+b}$$

(E) EXAMPLES

1. $(D-2)^2 y = e^{2x}$

Solution: The auxiliary equation is $(m-2)^2 = 0$

$\Rightarrow \quad m = 2, 2$

$\therefore$ Complementary function $= y_c = e^{2x}(c_1 + c_2 x)$

$$y_p = \text{particular integral} = \frac{1}{(D-2)^2} e^{2x} = \frac{x^2 e^{2x}}{2!}$$

$$\therefore \quad y = y_c + y_p = (c_1 + c_2 x)e^{2x} + \frac{x^2 e^{2x}}{2!}$$

2. $y'' - 4y = e^{2x} + e^{-4x}$ or $(D^2 - 4)y = e^{2x} + e^{-4x}$

Solution: The auxiliary equation is $m^2 - 4 = 0$

$\Rightarrow \quad m = 2, -2$

$\therefore$ Complementary function $= y_c = c_1 e^{2x} + c_2 e^{-2x}$

$$\text{Particular integral} = \left(\frac{1}{D^2 - 4}\right)\left(e^{2x} + e^{-4x}\right)$$

$$= \left(\frac{1}{D^2 - 4}\right) e^{2x} + \left(\frac{1}{D^2 - 4}\right) e^{-4x}$$

$$=\left(\frac{1}{(2+2)(D-2)}\right)e^{2x}+\left(\frac{e^{-4x}}{4^2-4}\right)$$

$$=\frac{1}{4}\left(\frac{1}{D-2}\right)e^{2x}+\frac{e^{-4x}}{12}=\frac{1}{4}xe^{2x}+\frac{1}{12}e^{-4x}$$

$$\therefore \quad y=c_1e^{2x}+c_2e^{-2x}+\frac{1}{4}xe^{2x}+\frac{1}{12}e^{-4x}$$

3. $(D^2+2D+1)y=2e^{3x}$

***Solution*:** Complementary function: Here auxiliary equation $m^2+2m+1=0$ i.e., $(m+1)^2=0$ so $m=-1$ is a double root. Thus the complementary function is

$$y_c=(c_1+c_2x)e^{-x}$$

Particular integral $=\; y_p=\dfrac{1}{D^2+2D+1}2e^{3x}$

$$=2\cdot\frac{1}{32+2\cdot 3+1}e^{3x}=\frac{e^{3x}}{8}$$

Hence, the general solution is $y=y_c+y_p=(c_1+c_2x)e^{-x}+\dfrac{e^{3x}}{8}$.

4. $y''+4y'+13y=2e^{-x},\; y(0)=0,\; y'(0)=-1$

***Solution*:** The auxiliary equation is: $m^2+4m+13=0$

$\therefore \quad m=-2\pm 3i$

Hence complementary function $=y_c=e^{-2x}(c_1\cos 3x+c_2\sin 3x)$

$$y_p=\text{particular integral}=\left(\frac{1}{D^2+4D+13}\right)2e^{-x}=\frac{2e^{-x}}{1-4+13}=\frac{1}{5}e^{-x}$$

$\therefore$ General solution is $y=y_c+y_p$

$$=e^{-2x}(c_1\cos 3x+c_2\sin 3x)=\frac{1}{5}e^{-x}$$

Now, $y(0)=0\Rightarrow c_1=-\dfrac{1}{5}$

$$y' = e^{-2x}[-3c_1 \sin 3x + 3c_2 \cos 3x] - 2e^{-2x}[c_1 \cos 3x + c_2 \sin 3x] - \frac{1}{5}e^{-x}$$

$$\therefore \quad y'(0) = -1 \Rightarrow -1 = 3c_2 - 2c_1 - \frac{1}{5}$$

Hence $c_2 = -\frac{2}{5}$

$$\therefore \quad y = -\frac{1}{5}e^{-2x}(\cos 3x + 2\sin 3x) + \frac{1}{5}e^{-x}$$

5. $(D^3 - 2D^2 - 5D + 6)y = 2e^x + 4e^{3x} + 7e^{-2x} + 8e^{2x} + 15.$

Solution:

Complementary function: Here auxiliary equation is $m^3 - 2m^2 - 5m + 6 = 0$ having roots $m = 1, 3, -2$. Thus the complementary function is

$$y_c = c_1 e^x + c_2 e^{3x} + c_3 e^{-2x}$$

Particular integral

$$= y_p = \frac{1}{D^3 - 2D^2 - 5D + 6 = 0} \times (2e^x + 4e^{3x} + 7e^{-2x} + 8e^{2x} + 15)$$

$$= \frac{2}{(D+2)(D-3)(D-1)}e^x + \frac{4}{(D-1)(D+2)(D-3)}e^{3x}$$

$$+ \frac{7}{(D-1)(D-3)(D+2)}e^{-2x} + \frac{8}{D^3 - 2D^2 - 5D + 6}e^{2x}$$

$$+ \frac{15}{D^3 - 2D^2 - 5D + 6} \times 1$$

$$y_p = \frac{2}{-6}\frac{1}{D-1}e^x + \frac{4}{10}\frac{1}{D-3}e^{3x} + \frac{7}{15}\frac{1}{D+2}e^{-2x} + \frac{8}{2^3 - 2\cdot 2^2 - 5\cdot 2 + 6}e^{2x}$$

$$+ \frac{15}{0 - 0 - 0 + 6}$$

$$= -\frac{1}{3}xe^x + \frac{2}{5}xe^{3x} + \frac{7}{15}xe^{-2x} - 2e^{2x} + \frac{15}{6}$$

$\therefore$ General solution is $y = y_c + y_p$

$$y = c_1 e^x + c_2 e^{3x} + c_3 e^{-2x} - \frac{1}{3}xe^x + \frac{2}{5}xe^{3x} + \frac{7}{15}xe^{-2x} - 2e^{2x} + \frac{15}{6}$$

6. $y'' + 4y' + 3y = e^{e^x}$

Solution: The auxiliary equation is: $D^2 + 4D + 3 = 0$

$\Rightarrow \quad m^2 + 3m + m + 3 = 0$

$\Rightarrow \quad m(m+3) + 1(m+3) = 0$

$\Rightarrow \quad (m+3)(m+1) = 0$

$\Rightarrow \quad m = -3, -1$

$\therefore$ Complementary function $= c_1 e^{-3x} + c_2 e^{-x}$

Particular integral: $\dfrac{1}{D^2 + 4D + 3} e^{e^x}$

$$= \frac{1}{(D+3)(D+1)} \cdot e^{e^x}$$

$$= \frac{1}{2}\left[\frac{1}{(D+1)} - \frac{1}{(D+3)}\right] \cdot e^{e^x}$$

$$= \frac{1}{2}\left[\frac{1}{(D+1)} e^{e^x} - \frac{1}{(D+3)} e^{e^x}\right]$$

$$= \frac{1}{2}\left[e^{-x}\int e^{e^x} . e^x dx - e^{-3x}\int e^{e^x} \cdot e^{3x} dx\right]$$

Put $\quad e^x = t; e^x dx = dt$ in the integrals.

$\therefore$ Particular integral $= \dfrac{1}{2}\left[e^{-x}\int e^t dt - e^{-3x}\int e^t \cdot t^3 dt\right]$

$$= \frac{1}{2}\left[e^{-x} \cdot e^t - e^{-3x}\left(t^3 \cdot e^t - 3\int t^2 \cdot e^t dt\right)\right]$$

$$= \frac{1}{2}\left[e^{-x} \cdot e^{e^x} - e^{-3x}\left(t^3 \cdot e^t - 3\int t^2 \cdot e^t - 2\int te^t dt\right)\right]$$

$$= \frac{1}{2}\left[e^{-x} \cdot e^{e^x} - e^{-3x}\left(e^{3x} \cdot e^{e^x}\right) + \left(e^{-3x} \cdot e^{2x} e^{e^x}\right) - 6e^{-3x} . \left(te^t - et\right)\right]$$

$$= \frac{1}{2}\left[e^{e^x}\right]\left[e^{-x} - 1 + e^{-x} - 6e^{-2x} + 6e^{-3x}\right].$$

$\therefore$ General solution is $y = c_1 e^{-3x} + c_2 e^{-x} + \frac{e^{e^x}}{2}\left[2e^{-x} - 1\right] - 3\left(e^{-2x} - e^{-3x}\right)$.

7. $\frac{d^2 y}{dx^2} - 4y = \cosh(2x-1) + 3^x$,

Solution: Given $(D^2 - 4)y = \cosh(2x-1) + 3^x$,

The auxiliary equation is $(m^2 - 4) = 0$

$\Rightarrow \quad (m-2)(m+2) = 0$

$\Rightarrow \quad m = 2, -2$

$\therefore$ complementary function $y_c = c_1 e^{2x} + c_2 e^{-2x}$

Particular integral: $y_p = \frac{\cosh(2x-1) + 3^x}{D^2 - 4} = \frac{1}{2}\left[\frac{e^{2x-1}}{D^2 - 4} + \frac{e^{-(2x-1)}}{D^2 - 4}\right] + \frac{3^x}{D^2 - 4}$

$= p_1 + p_2 + p_3$, (say)

$$p_1 = \frac{1}{2} \cdot \frac{e^{2x-1}}{D^2 - 4} = \frac{1}{2} \cdot \frac{e^{2x-1}}{2^2 - 4} = \frac{1}{2} \cdot \frac{e^{2x-1}}{0} = ? \quad \text{(a case of failure)}$$

$$= \frac{1}{2} x \cdot \frac{e^{2x-1}}{2D} = \frac{1}{2} \cdot \frac{x e^{2x-1}}{4} = \frac{x}{8} \cdot e^{2x-1}$$

$$p_2 = \frac{1}{2} \cdot \frac{e^{-(2x-1)}}{D^2 - 4} = \frac{1}{2} \cdot \frac{e^{-(2x-1)}}{(-2)^2 - 4} = \frac{1}{2} \cdot \frac{e^{-(2x-1)}}{0} = ?$$

$$= \frac{1}{2} \cdot x \cdot \frac{e^{-(2x-1)}}{2D} = \frac{1}{2} \cdot x \cdot \frac{e^{-(2x-1)}}{-4} = -\frac{x}{8} \cdot e^{-(2x-1)}$$

$$p_3 = \frac{3^x}{D^2 - 4} = \frac{\left(e^{\log^3}\right)^x}{D^2 - 4} = \frac{e^{\log^{3 \cdot x}}}{D^2 - 4} = \frac{e^{\log^{3 \cdot x}}}{\left(\log^3\right)^2 - 4} = \frac{3^x}{\left(\log^3\right)^2 - 4}$$

$\therefore \quad y = y_c + y_p$ i.e., $y = y_c + p_1 + p_2 + p_3$

$$= c_1 e^{2x} + c_2 e^{-2x} + \frac{x}{8} e^{2x-1} - \frac{x}{8} e^{-(2x-1)} + \frac{3^x}{\left(\log^3\right)^2 - 4}$$

$$y = c_1 e^{2x} + c_2 e^{-2x} + \frac{x}{4} \sinh 2x - 1 + \frac{3^x}{\left(\log^3\right)^2 - 4}$$

8. $\dfrac{d^2y}{dt^2}+4\dfrac{dy}{dx}+13y=e^{3t}\cosh 2t+2^t$

Solution: Given equation is $(D^2-4D+13)y=e^{3t}\cosh 2t+2^t$

Auxiliary equation is $m^2-4m+13=0$

$$\therefore \quad m=\frac{4\pm\sqrt{16-52}}{2}=\frac{4\pm 6i}{2}=2\pm 3i$$

$$y_c=e^{2t}(c_1\cos 3t+c_2\sin 3t)$$

$$y_p=\frac{e^{3t}\cosh 2t}{(D^2-4D+13)}+\frac{2^t}{(D^2-4D+13)}$$

$$=\frac{1}{2}\cdot\frac{e^{3t}(e^{2t}+e^{-2t})}{(D^2-4D+13)}+\frac{2^t}{(D^2-4D+13)}$$

$$=\frac{1}{2}\cdot\frac{e^{5t}}{(D^2-4D+13)}+\frac{1}{2}\cdot\frac{e^{t}}{(D^2-4D+13)}+\frac{2^t}{(D^2-4D+13)}$$

$$=p_1+p_2+p_3$$

$$p_1=\frac{1}{2}\cdot\frac{e^{5t}}{(D^2-4D+13)}=\frac{1}{2}\cdot\frac{e^{5t}}{(5)^2-4(5)+13}=\frac{1}{2}\cdot\frac{e^{5t}}{18}=\frac{e^{5t}}{36}$$

$$p_2=\frac{1}{2}\cdot\frac{e^{t}}{(D^2-4D+13)}=\frac{1}{2}\cdot\frac{e^{t}}{1-4+13}=\frac{e^{t}}{20}$$

$$p_3=\frac{2^t}{(D^2-4D+13)}=\frac{e^{\log 2\cdot t}}{(D^2-4D+13)}=\frac{e^{\log 2\cdot t}}{(\log 2)^2-4(\log 2)+13}$$

$$y=y_c+y_p \quad \text{i.e.,} \quad y=y_c+p_1+p_2+p_3$$

or $$y=e^{2t}(c_1\cos 3t+c_2\sin 3t)+\frac{e^{5t}}{36}+\frac{e^{t}}{20}+\frac{2^t}{(\log 2)^2-4(\log 2)+13}$$

EXERCISES

Solve the following equations.

1. $y'' - 3y' + 2y = e^{2x}$ $\left[Ans.\quad xe^{2x}\right]$

2. $y'' - \frac{1}{9}y = e^{-x}$ $\left[Ans.\quad c_1 e^{\frac{x}{3}} + c_2 e^{\frac{-x}{3}} + \frac{9}{8}e^{-x}\right]$

3. $y'' - 10y' + 16y = 2\cosh(3x)$

$$\left[Ans.\quad y = c_1 e^{8x} + c_2 e^{2x} - \left(\frac{1}{5}\right)e^{3x} + \left(\frac{1}{55}\right)e^{-3x}\right]$$

4. $y'' - 7y' + 10y = \left(1 + e^x\right)^2$

$$\left[Ans.\quad y = c_1 e^{x} + c_2 e^{5x} + \left(\frac{1}{10}\right) + \left(\frac{1}{2}\right)e^{x} - \left(\frac{x^2}{3}\right)e^{2x}\right]$$

5. $y'' - y' + y = \sinh x$

$$\left[Ans.\quad y = e^{\frac{x}{2}}\left(c_1 \cos\frac{\sqrt{3}}{2}x + c_2 \sin\frac{\sqrt{3}}{2}x\right) + \frac{1}{6}\left(3e^{x} - e^{-x}\right)\right]$$

6. $(D^2 - p^2)y = \sinh px$ $\left[Ans.\quad y = c_1 e^{px} + c_2 e^{-px} + \frac{x}{2p}\cosh px\right]$

7. $(D^2 + 4D + 5)y = -2\cosh x$, with $y(0) = 0,\ y'(0) = 1$

$$\left[Ans.\quad y = \frac{3}{5}e^{-2x}(\cos x + 3\sin x) - \frac{e^{x}}{10} - \frac{e^{-x}}{2}\right]$$

8. $(D^2 - 2aD + a^2)y = e^{ax}$ $\left[Ans.\quad y = (c_1 + c_2 x)e^{ax} + \frac{x^2}{2}e^{ax}\right]$

9. $(D^3 + 6D^2 + 9D)y = e^{-3x}$ $\left[Ans.\quad y = c_1 + (c_2 + xc_3)e^{-3x} - \frac{x^2 e^{-3x}}{6}\right]$

10. $(D^3 - 5D^2 + 8D - 4)y = e^{2x} + 2e^x + 3e^{-x}$

$$\left[\textit{Ans.}\quad y = c_1e^x + c_2e^{2x} + c_3xe^{2x} + \frac{1}{2}x^2e^{2x} + 2xe^x - \frac{e^{-x}}{6}\right]$$

11. $(D+2)(D-1)^2 y = e^{-2x} + 2\sinh x$

$$\left[\textit{Ans.}\quad y = c_1e^{-2x} + (c_2 + c_3x)e^x + xe^{-2x} + \frac{x^2e^x}{6} + \frac{e^{-x}}{4}\right]$$

12. $(D^3 - 12D + 16)y = (e^x + e^{-2x})^2$

$$\left[\textit{Ans.}\quad y = (c_1 + c_2x)e^{2x} + c_3e^{-4x} + \frac{x^2e^{2x}}{12} + \frac{2e^{-x}}{27} + \frac{xe^{-4x}}{36}\right]$$

II. Particular integral when $F(x) = \sin(ax+b)$ or $\cos(ax+b)$

Let $f(D^2)$ be a rational function of D^2

$$D\sin(ax+b) = a\cos(ax+b)$$
$$D^2\sin(ax+b) = -a^2\sin(ax+b)$$
$$D^3\sin(ax+b) = -a^3\cos(ax+b)$$
$$D^4\sin(ax+b) = a^4\sin(ax+b) = (-a^2)^2\sin(ax+b)$$

Thus, we find that $f(D^2)\sin(ax+b) = f(-a^2)\sin(ax+b)$

where $f(-a^2)$ is obtained by putting $D^2 = -a^2$ in $f(D^2)$. Operating with $\frac{1}{f(D^2)}$ on both sides,

$$\frac{1}{f(D^2)}\cdot f(D^2)\sin(ax+b) = \frac{1}{f(D^2)}f(-a^2)\sin(ax+b)$$

$$\Rightarrow\quad \sin(ax+b) = f(-a^2)\cdot\frac{1}{f(D^2)}\sin(ax+b)\quad \text{(since } f(-a^2) \text{ is a constant.)}$$

Thus, $\frac{1}{f(D^2)}\sin(ax+b) = \frac{1}{f(-a^2)}\sin(ax+b)$ provided $f(-a^2) \neq 0$

Similarly, $\frac{1}{f(D^2)}\cos(ax+b)=\frac{1}{f(-a^2)}\cos(ax+b)$

Note: When $f(D)$ contains terms of odd powers $D, D^3 \ldots$ etc., even after application of the above result, multiply by D so as to get even powers of $D^2, D^4 \ldots$ etc., interpreting D as differentiation and $\frac{1}{D}$ as integration.

Corollary: If $\phi(-a^2)=0$ then $\phi(D^2)=(D^2+a^2)\psi(D^2)$

with $\psi(-a^2)\neq 0$.

$$\text{Thus, } \phi\frac{1}{(D^2)}\sin(ax+b)=\frac{1}{(D^2+a^2)\psi(D^2)}\sin(ax+b)$$

$$=\frac{1}{\psi(-a^2)}\cdot\frac{1}{D^2+a^2}\sin(ax+b)$$

$$=\frac{1}{\psi(-a^2)}\cdot\left(\frac{-x\cos(ax+b)}{2a}\right)$$

$$\text{Similarly, } \phi\frac{1}{(D^2)}\cos(ax+b)=\frac{1}{(D^2+a^2)}\cdot\frac{1}{\psi(D^2)}\cos(ax+b)$$

$$=\frac{1}{\psi(-a^2)}\frac{1}{(D^2+a^2)}\cos(ax+b)$$

$$=\frac{1}{\psi(-a^2)}\left(\frac{x\sin(ax+b)}{2a}\right).$$

(F) EXAMPLES

Solve the following equations.

1. $y''+4y=\cos 3x$

Solution: The auxiliary equation is $(m^2+4)=0$

$\Rightarrow \quad m=\pm 2i$

$\therefore \quad y_c=(c_1\cos 2x+c_2\sin 2x)$

Particular integral: $(D^2+4)y=\cos 3x$;

$$\Rightarrow \quad y_p = \frac{1}{(D^2+4)}\cos 3x = \frac{1}{(-3^2)+4}\cos 3x$$

$$= \frac{\cos 3x}{-5}$$

$$\therefore \quad y = c_1 \cos 2x + c_2 \sin 2x - \left(\frac{1}{5}\right)\cos 3x.$$

2. $y'' - 6y' + 13y = 2\cos^2 x$

Solution: Given $(D^2 - 6D + 13)y = 2\cos^2 x$

The auxiliary equation is $(m^2 - 6m + 13) = 0 \Rightarrow m = 3 \pm 2i$

$\therefore$ Complementary function $= y_c = e^{3x}(c_1 \cos 2x + c_2 \sin 2x)$

particular integral: $y_p = \dfrac{1}{(D^2-6D+13)}1 + \dfrac{1}{(D^2-6D+13)}\cos 2x$

(a) $\dfrac{1}{(D^2-6D+13)}1 = \dfrac{1}{13};$

(b) $\dfrac{1}{(D^2-6D+13)}\cos 2x = \dfrac{1}{(D^2+13)-6D}\cos 2x$

$$y_p = \frac{1}{\left[(D^2+13)-6D\right]}\cos 2x = \frac{\left[(D^2+13)+6D\right](\cos 2x)}{\left[(D^2+13)^2 - 36D^2\right]}\cos 2x$$

$$= \left\{(D^2+13)+6D\right\}\frac{\cos 2x}{\left[(-2^2+13)^2 - 36(-2^2)\right]}$$

$$= (D^2+13+6D)\left[\frac{(\cos 2x)}{225}\right]$$

$$= \left(\frac{1}{225}\right)\left[D^2(\cos 2x) + 13(\cos 2x) + 6D(\cos 2x)\right]$$

$$= \left(\frac{1}{225}\right)[-4\cos 2x + 13\cos 2x + 6(-2\sin 2x)]$$

$$= \left(\frac{1}{225}\right)[9\cos 2x - 12\sin 2x]$$

$$\therefore \quad y = y_c + y_p$$

$$= e^{3x}\left[c_1 \cos 2x + c_2 \sin 2x\right] + \left(\frac{1}{13}\right) - \left(\frac{1}{225}\right)[9\cos 2x - 12\sin 2x]$$

3. $y'' = 4y' + 4y = 2\sin^2 x$

Solution: The auxiliary equation is $(m^2 - 4m + 4) = 0$

$\Rightarrow \quad (m-2)^2 = 0 \quad \Rightarrow \quad m = 2, 2$

$\therefore$ Complementary function $= y_c = (c_1 + c_2 x)e^{2x}$

particular integral $y_p = \dfrac{1}{(D^2 - 4D + 4)}(2\sin^2 x) = \dfrac{1}{(D^2 - 4D + 4)}(1 - \cos 2x)$

$$= \frac{1}{(D^2 - 4D + 4)}(1) - \frac{1}{(D^2 - 4D + 4)}(\cos 2x)$$

$$= \frac{1}{4} - \frac{1}{-(2)^2 - 4D + 4)}(\cos 2x)$$

$$= \frac{1}{4} - \frac{1}{-4 - 4D + 4)}(\cos 2x) = -\frac{1}{4D}(\cos 2x)$$

$$= \left(-\frac{1}{4}\right)\int \cos 2x\,dx = \left(-\frac{1}{4}\right)\left(\frac{1}{2}\right)\sin 2x = \left(-\frac{1}{8}\right)\sin 2x$$

$\therefore$ The solution is $y = y_c + y_p = (c_1 + c_2 x)e^{2x} + \left(\frac{1}{4}\right) - \left(\frac{1}{8}\right)\sin 2x$

4. $(D^2 - 4D - 5)y = e^{2x} + 3\cos(4x + 3)$

Solution: Complementary function: The auxiliary equation is $m^2 - 4m - 5 = 0$. The roots are $m = -1, 5$ so that the complementary function is

$$y_c = c_1 e^{-x} + c_2 e^{5x}.$$

particular integral $y_p = \dfrac{1}{D^2 - 4D - 5}\left[e^{2x} + 3\cos(4x + 3)\right]$

$$= I_1 + I_2$$

$$I_1 = \frac{1}{2^2 - 4\cdot 2 - 5}e^{2x} = -\frac{1}{9}e^{2x}$$

$$I_2 = \frac{3}{-4^2 - 4D - 5}\cos(4x+3)$$

$$= -\frac{3}{4D+21}\cos(4x+3)$$

Since D^2 terms are not present, we rewrite

$$I_2 = \frac{3(4D-21)}{(4D+21)(4D-21)}\cdot\cos(4x+3)$$

$$I_2 = \frac{-12D+63}{16D^2 - 21^2}\cdot\cos(4x+3)$$

Replace D^2 by -4^2. Then

$$I_2 = \frac{-12D+63}{16(-4)^2 - 21^2}\cdot\cos(4x+3)$$

$$= \frac{1}{697}[-12D\cos(4x+3) + 63\cos(4x+3)]$$

$$= \frac{-1}{697}[48\sin(4x+3) + 63\cos(4x+3)].$$

5. Find particular integral of $y'' + 9y = \cos 3x$

Solution:

Particular integral $\left(\frac{1}{(D^2+9)}\right)\cos 3x$ = (real part or r. p. of) $\left(\frac{1}{D^2+9}\right)e^{i(3x)}$

$\left(\because\ e^{ix} = \cos x + i\sin x \Rightarrow \cos x = \text{(r. p. of) } e^{ix}\right)$

$$= \text{(r. p. of)}\left(\frac{1}{(D+3i)(D-3i)}\right)e^{i(3x)}$$

$$= \text{(r. p. of)}\left(\frac{1}{6i(D-3i)}\right)e^{i(3x)}$$

$$= \text{(r. p. of)}\left(\frac{xe^{i(3x)}}{6i}\right) = \text{(r. p. of)}\left[-\frac{1}{6}ix(\cos 3x + i\sin 3x)\right]$$

$$= \frac{1}{6} x \sin x \text{ is the required particular integral.}$$

6. $(D^2 + 4)y = \sin 3x + \cos 2x$

Solution: Complementary function: Here auxiliary equation is $m^2 + 4 = 0$ having complex roots $m = \pm 2i$. Thus complementary function is

$$y_c = c_1 \cos 2x + c_2 \sin 2x$$

particular integral $y_p = \frac{1}{D^2 + 4}(\sin 3x + \cos 2x)$

$$= \frac{1}{D^2 + 4} \sin 3x + \frac{1}{D^2 + 4} \cos 2x$$

$$= I_1 + I_2$$

Consider $I_1 = \frac{1}{D^2 + 4} \sin 3x$

Replace D^2 by -3^2 since $f(-a^2) = f(-3^2) \neq 0$

$$I_1 = \frac{1}{-3^2 + 4} \sin 3x = -\frac{1}{5} \sin 3x$$

Consider $I_2 = \frac{1}{D^2 + 4} \cdot \cos 2x$

Since $f(-a^2) = f(-2^2) = 0$ we apply case $(D^2 = -a^2)$

$$I_2 = \frac{x \sin 2x}{4}$$

$$y_p = I_1 + I_2 = -\frac{1}{5} \sin 3x + \frac{x}{4} \sin 2x$$

General solution: $y = y_c + y_p = c_1 \cos 2x + c_2 \sin 2x - \frac{1}{5} \sin 3x + \frac{x}{4} \sin 2x.$

7. $y'' + a^2 y = \sin ax$

Solution: Here, $y_c = (c_1 \cos ax + c_2 \sin ax)$

particular integral: $y = \frac{1}{(D^2 + a^2)} \sin ax$

$$= \text{Im}\left\{\frac{1}{D^2 + a^2} e^{iax}\right\}$$

$$= \text{Im}\left\{\frac{1}{2D} xe^{iax}\right\} \quad = \text{Im}\left\{\frac{1}{2ia} xe^{iax}\right\}$$

$$= \text{Im}\left\{-\frac{i}{2a} x(\cos ax + i\sin ax)\right\}$$

$$= \text{Im}\left\{\frac{x}{2a}(\sin ax - i\cos ax)\right\}$$

$$= -\frac{x}{2a}\cos ax.$$

$\therefore \quad y$ = complementary function + particular integral

$$= c_1 \cos ax + c_2 \sin ax - \frac{x}{2a}\cos ax$$

8. $(D^2 + 5D - 6)y = \sin 4x \cdot \sin x.$

***Solution*:** Complementary function: Here auxiliary equation is $m^2 + 5m - 6 = 0$. It has roots $m_1 = 1, -6$. So complementary function is

$$y_c = c_1 e^x + c_2 e^{-6x}$$

particular integral: $y_p = \dfrac{1}{D^2 + 5D - 6} \sin 4x \cdot \sin x$

$$= \frac{1}{D^2 + 5D - 6} \cdot \frac{1}{2}[\cos 3x - \cos 5x]$$

$$= \frac{1}{2}[I_1 - I_2]$$

Consider $I_1 = \dfrac{1}{D^2 + 5D - 6} \cdot \cos 3x$

Replacing D^2 by -3^2, we get

$$I_1 = \frac{1}{-3^2 + 5D - 6}\cos 3x = \frac{1}{5(d-3)}\cos 3x$$

To get D^2 terms in denominator, rewrite as

$$I_1 = \frac{1}{5}\frac{(D+3)}{(D-3)(D+3)}\cdot\cos 3x = \frac{1}{5}\frac{(D+3)}{D^2-3^2}\cos 3x$$

Replacing $D^2 = -3^2$, we get

$$I_1 = \frac{1}{5}\frac{(D+3)}{-3^2-3^2}\cos 3x = -\frac{1}{90}(D+3)(\cos 3x)$$

$$= -\frac{1}{90}[-3\sin 3x + 3\cos 3x] = \frac{1}{30}[\sin 3x - \cos 3x]$$

Consider $I_2 = \dfrac{1}{D^2+5D-6}\cdot\cos 5x$

Replacing D^2 by -5^2, we get

$$I_2 = \frac{1}{-5^2+5D-6}\cos 5x = \frac{1}{5D-31}\cos 5x$$

In order to get D^2 terms, we rewrite

$$I_2 = \frac{5D+31}{(5D-31)(5D+31)}\cdot\cos 5x$$

$$= \frac{(5D+31)}{25D^2-31^2}\cdot\cos 5x$$

Replacing D^2 by -5^2, we get

$$= \frac{(5D+31)}{25(-5)^2-31^2}\cdot\cos 5x$$

$$= \frac{1}{1586}(5D+31)(\cos 5x)$$

$$= \frac{1}{1586}[-25\sin 5x + 31\cos 5x)]$$

Thus $y_p = \dfrac{1}{2}[I_1 - I_2] = \dfrac{1}{2}\left[\dfrac{\sin 3x - \cos 3x}{30} + \dfrac{31\cos 5x - 25\sin 3x}{1586}\right]$

Hence general solution: $y = y_c + y_p$

9. $y'' + a^2 y = \cos ax$

Solution: Here, $y_c = (c_1\cos ax + c_2\sin ax)$ [1]

particular integral: $\dfrac{1}{D^2+a^2}\cos ax = \dfrac{1}{-a^2+a^2}\cos ax = \dfrac{1}{0}\cos ax$ [2]

(a case of failure)

We recall that $e^{iax} = \cos ax + i\sin ax$

So that $\cos ax = (\text{r.p. of})e^{iax}$

$\therefore$ [2] becomes

$$\text{particular integral} = (\text{r.p. of})\left\{\frac{1}{D^2+a^2}e^{iax}\right\}$$

$$= (\text{r.p. of})\left\{\frac{1}{2D}xe^{iax}\right\} = (\text{r.p. of})\left\{\frac{1}{2ia}xe^{iax}\right\}$$

$$= (\text{r.p. of})\left\{-\frac{i}{2a}x(\cos ax + i\sin ax)\right\} = (\text{r.p. of})\left\{-\frac{x}{2a}(\sin ax - i\cos ax)\right\}$$

$$= \frac{x}{2a}\sin ax$$

$$\therefore \quad y = y_c + y_p = c_1\cos ax + c_2\sin ax + \frac{x}{2a}\sin ax$$

10. $y'' - 4y' + y = \cos x.\cos 4x + \sin^2 x$

Solution: Complementary function: Here auxiliary equation is $m^2 - 4m + 1 = 0$ with real distinct roots $m_1 = 2+\sqrt{3}$ and $m_2 = 2-\sqrt{3}$ so that the complementary function is

$$y_c = c_1 e^{(2+\sqrt{3})x} + c_2 e^{(2-\sqrt{3})x}$$

We know that $\cos x\cdot\cos 4x = \frac{1}{2}\cos 3x + \cos x$ and $\sin^2 x = \frac{1}{2}[1-\cos 2x]$

so that $\cos x\cdot\cos 4x + \sin^2 x = \frac{1}{2}[\cos 3x + \cos x + 1 - \cos 2x]$

$$y_p = \frac{1}{D^2-4D+1}[\cos x\cdot\cos 4x + \sin^2 x]$$

$$= \frac{1}{2}\frac{1}{D^2-4D+1}[\cos 3x + \cos x + 1 - \cos 2x]$$

$$= \frac{1}{2}[I_1 + I_2 + I_3 - I_4]$$

Here $I_1 = \dfrac{1}{D^2 - 4D + 1}\cos 3x$

Replacing D^2 by -3^2

$$I_1 = \frac{1}{-3^2 - 4D + 1}\cos 3x = \frac{-1}{4}\frac{1}{D+2}\cos 3x$$

To introduce D^2 terms we rewrite the above as

$$I_1 = \frac{1}{4}\frac{D-2}{(D+2)(D-2)}\cdot\cos 3x$$

$$= -\frac{1}{4}\frac{D-2}{(D^2 - 2^2)}\cdot\cos 3x$$

Replacing D^2 by -3^2, we get

$$I_1 = \frac{1}{4}\frac{D-2}{-3^2 - 2^2}\cdot\cos 3x = \frac{1}{52}(D-2)\cos 3x$$

$$= \frac{1}{52}[-3\sin 3x - 2\cos 3x]$$

Similarly, $I_2 = \dfrac{1}{D^2 - 4D + 1}\cdot\cos x$

$$= \frac{1}{-1^2 - 4D + 1}\cos x, \qquad (\because D^2 = -1^2)$$

$$= \frac{1}{4}\frac{1}{D}\cos x = -\frac{1}{4}\int \cos x dx = -\frac{1}{4}\sin x$$

Also $I_3 = \dfrac{1}{D^2 - 4D + 1}\cdot 1 = \dfrac{1}{D^2 - 4D + 1}e^{0\cdot x} \qquad (\because e^{0x} = e^0 = 1)$

$$= \frac{1}{0 - 4\cdot D + 1} = \frac{1}{1} = 1$$

where D is replaced by $a = 0$

Finally $I_4 = \dfrac{1}{D^2 - 4D + 1}\cdot\cos 2x = \dfrac{1}{-2^2 - 4D + 1}\cos 2x$

where D^2 is replaced by -2^2

$$I_4 = \frac{1}{(4D+3)}\cos 2x$$

Rewriting this to get D^2 terms we have

$$I_4 = -\frac{(4D-3)}{(4D+3)(4D-3)} \cdot \cos 2x$$

$$= -\frac{(4D-3)}{16D^2 - 3^2} \cos 2x$$

Now replacing D^2 by -2^2, we get

$$I_4 = -\frac{(4D-3)}{16(-2^2)-3^2} \cos 2x$$

$$= \frac{(4D-3)}{73} \cos 2x$$

$$= \frac{1}{73}(4(-2)\sin 2x - 3\cos 2x)$$

$$I_4 = \frac{1}{73}(8\sin 2x + 3\cos 2x)$$

Thus the particular integral $y_p = \frac{1}{2}[I_1 + I_2 + I_3 + I_4]$

$$y_p = \frac{1}{2}\left[\frac{1}{52}(3\sin 3x + 2\cos 3x) - \frac{1}{4}\sin x + 1 - \frac{1}{73}(8\sin 2x + 3\cos 2x)\right]$$

General solution: $y = y_c + y_p$

$$y = c_1 e^{(2+\sqrt{3})x} + c_2 e^{(2-\sqrt{3})x} + 1 - \frac{1}{4}\sin x - \frac{1}{73}(3\cos 2x + 8\sin 2x)$$
$$- \frac{1}{52}(3\sin 3x + 3\cos 3x).$$

EXERCISES

Solve the following equations.

1. $y'' + y' + y = \sin 2x$

$$\left[\textit{Ans. } y = e^{\frac{-x}{2}}\left(c_1 \cos\left(\frac{\sqrt{3}}{2}x\right) + c_2 \sin\left(\frac{\sqrt{3}}{2}x\right)\right) - \frac{2\cos 2x}{13} - \frac{3\sin 2x}{13}\right]$$

2. $y'' + 9y = 4\sin 3x$ $\left[\textit{Ans.}\ \ y = \text{complementary function} + \left(-\dfrac{2x\cos 3x}{3}\right)\right]$

3. $y'' - 8y' + 9y = 8\sin 5x$ $\left[\textit{Ans.}\ \ y = e^{4x}(c_1 e^{\sqrt{7}x} + c_2 e^{-\sqrt{7}x}) + \dfrac{1}{29}(5\cos 5x - 2\sin 5x)\right]$

4. $y'' + 5y' + 6y = e^{-2x} + \sin x$

$$\left[\textit{Ans.}\ \ y = y_c + y_p = c_1 e^{-3x} + c_2 e^{-2x} + xe^{-2x} - \frac{1}{10}(\cos x - \sin x)\right]$$

5. $y'' + 4y = \sin^2 x$ $\left[\textit{Ans.}\ \ y = c_1\cos 2x + c_2\sin 2x + \dfrac{1}{8}(1 - x\sin 2x)\right]$

6. $y'' - y' - 2y = 2^{x+3} + \cos^2(x+3)$

$$\left[\textit{Ans.}\ \ y = c_1 e^{-x} + c_2 e^{2x} + \frac{2^{x+3}}{(\log 2)^2 - \log 2 - 2} - \left(1 + \frac{1}{10}\{\sin 2(x+3) + 3\cos 2(x+3)\}\right)\right]$$

7. $(D^4 + 10D^2 + 9)y = \cos(2x+3)$

$$\left[\textit{Ans.}\ \ y = c_1\cos x + c_2\sin x + c_3\cos 3x + c_4\sin 3x - \frac{1}{15}\cos(2x+3)\right]$$

8. $(D^2 + 2D + 5)y = 6\sin 2x + 7\cos 2x$

$$\left[\textit{Ans.}\ \ y = e^{-x}(c_1\sin 2x + c_2\cos 2x) + 2\sin 2x - \cos 2x\right]$$

9. $(D^3 + D^2 + D + 1)y = \sin 2x + \cos 3x$

$$\left[\textit{Ans.}\ \ y = c_1 e^{-x} + c_2\cos 2x + c_3\sin x + \frac{1}{15}\cos 2x - \sin 2x) - \frac{1}{80}(3\sin 3x + \cos 3x)\right]$$

10. $(D^2 + 2D + 1)y = e^{2x} - \cos^2 x$ $\left[\textit{Ans.}\ \ y = (c_1 + c_2 x)e^{-x} + \dfrac{1}{2} + \dfrac{1}{5}(2\sin 2x + \cos 2x)\right]$

11. $(D^2 + 1)y = \cos x$ $\left[\textit{Ans.}\ \ y = (c_1\cos x + c_2\sin x + \sin x\ln\sin x - x\cos x\right]$

12: $(D^2 - 4D + 13)y = 8\sin x,\ y(0) = 1,\ y(0) = 2$

$$\left[\textit{Ans.}\ \ y = \frac{1}{5}\left(e^{2x}(\sin 3x + 2\cos 3x) + \sin 3x - 3\cos 3x\right)\right]$$

III. Particular integral when $F(x) = x^m$, m being a positive integer consider $f(D)y = x^m$ so that

Particular integral $= \dfrac{1}{f(D)} \cdot x^m = [f(D)]^{-1} \cdot x^m$

From $f(D)$ take out the lowest degree term (LDT) and the remaining factor shall be of the $[1 \pm \phi(D)]$. We shall take this factor in the number with a negative index and expand it binomially, the expansion shall be made only upto the terms where the index of D is m. Because D^{m+1} when operated on x^m, i.e., $D^{m+1}(x^m)$ will be zero. Also, we use the expansions:

(a) $(1-D)^{-1} = 1 + D + D^2 + D^3 + \ldots\infty$

(b) $(1-D)^{-2} = 1 + 2D + 3D^2 + \ldots + (r+1)D^r + \ldots\infty$

(c) $(1-D)^{-3} = 1 + 3D + 6D^2 + 10D^3 + \ldots + \dfrac{(r+1)(r+2)}{2} D^2 + \ldots\infty$

Examples: (i) For $(D+1)y = x^3$

Particular integral $= \dfrac{1}{(D+1)} \cdot x^3$

$= (1+D)^{-1} \cdot x^3$

$= (1 - D + D^2 - D^3 + \ldots)x^3$

$x^3 - 3x^2 + 6x - 6.$

(ii) For $D(D^2+1)y = x^2 + 2$

Particular integral: $\dfrac{1}{D(D^2+1)} \cdot (x^2+2)$

$= \dfrac{1}{D} \cdot (1+D^2)^{-1} \cdot (x^2+2)$

$= \dfrac{1}{D} \cdot (1 - D^2 + \ldots)(x^2+2) = \dfrac{1}{D}(x^2 + 2 - 2) = \dfrac{1}{D} \cdot x^2 = \int x^2 dx = \dfrac{x^3}{3}.$

IV. Particular integral: When $F(x) = e^{ax}V_{(x)}$.

(Exponential Shift)

To prove that $\frac{1}{f(D)} e^{ax} V(x) = e^{ax} \frac{1}{f(D+a)} V(x)$, i.e., e^{ax} which is on the right side of $\frac{1}{f(D)}$ may be taken out to the left side provided D is replaced by $(D+a)$ in $f(D)$.

Proof: Let u be a function of x.

$$\therefore \quad D(e^{ax}u) = e^{ax} \cdot Du + ae^{ax} \cdot u$$

$$= e^{ax}(D+a)u$$

$$\text{Similarly,} \quad D^2(e^{ax} \cdot u) = e^{ax} \cdot (D+a)^2 \cdot u$$

$$\vdots \qquad \vdots$$

$$D^n(e^{ax} \cdot u) = e^{ax} \cdot (D+a)^n \cdot u$$

$$\therefore \quad f(D)(e^{ax} \cdot u) = e^{ax} \cdot f(D+a)u.$$

Operating on both sides by $\frac{1}{f(D)}$

$$\frac{1}{f(D)} \cdot f(D)(e^{ax} \cdot u) = \frac{1}{f(D)} \cdot \left(e^{ax} \cdot f(D+a)u\right)$$

$$\Rightarrow \quad e^{ax} \cdot u = \frac{1}{f(D)} \cdot \left(e^{ax} \cdot f(D+a)u\right)$$

$\because \quad f(D+a)u$ is a function of x,

let $f(D+a)u = V$ and $u = \frac{1}{f(D+a)} \cdot V$.

Hence, from above, we obtain

$$e^{ax} \frac{1}{f(D+a)} \cdot V = \frac{1}{f(D)} \cdot (e^{ax} \cdot u)$$

$\therefore$ particular integral $= e^{ax} \frac{1}{f(D+a)} \cdot V$

Example: For $(D-1)^2 = xe^x$

particular integral: $\frac{1}{(D^2 - 2D + 1)} \cdot (xe^x)$

$$= e^x \cdot \frac{1}{(D+1)^2 - 2(D+1) + 1} \cdot x$$

$$= e^x \cdot \frac{1}{D^2} \cdot x = e^x \cdot \frac{1}{D} \cdot \frac{x^2}{2} = e^x \cdot \frac{x^3}{3}.$$

V. Particular integral when $F(x) = x \cdot V(x) = xV$

To prove that

$$\frac{1}{f(D)}\{xV\} = \left[x - \frac{1}{f(D)} \cdot f'(D)\right] \cdot \frac{1}{f(D)} \cdot V$$

where V is a function of x and $f'(D)$ is the derivative of $f(D)$ w.r.t. D.

Proof: Let V_1 be a function of x, then, we have

$$D[xV_1] = x \cdot DV_1 + V_1 = xDV_1 + (D)' \cdot V_1$$

$$D^2[xV_1] = x \cdot D^2 V_1 + 2DV_1 = xD^2 V_1 + (D^2)' \cdot V_1$$

$$\cdots \quad \cdots \quad \cdots \quad \cdots$$

$$\cdots \quad \cdots \quad \cdots \quad \cdots$$

$$D^n[xV_1] = x \cdot D^n \cdot V_1 + (D^r)' \cdot V_1$$

$$\therefore \quad f(D)xV_1 = x f(D) \cdot V_1 + f'(D) \cdot V_1 \qquad [1]$$

Let us take V_1 such that

$$f(D)V_1 = V \Rightarrow V_1 = \frac{1}{f(D)} \cdot V \qquad [2]$$

From [1] and [2] we obtain:

$$f(D) \cdot x \frac{1}{f(D)} \cdot V = x \cdot V + f'(D) \cdot \frac{1}{f(D)} \cdot V$$

Operating both sides by $\frac{1}{f(D)}$, we obtain:

$$x \cdot \frac{1}{f(D)} \cdot V = \frac{1}{f(D)}(xV) + \frac{1}{f(D)} \cdot f'(D) \cdot \frac{1}{f(D)} \cdot V$$

$$\Rightarrow \quad \frac{1}{f(D)}(xV) = x \cdot \frac{1}{f(D)} \cdot V - \frac{1}{f(D)} \cdot f'(D) \cdot \frac{1}{f(D)} \cdot V$$

$\Rightarrow$ Particular integral: $= \left[x - \frac{1}{f(D)} \cdot f'(D)\right] \cdot \frac{1}{f(D)} \cdot V$

Table 1.3

Form of $F(x)$	Particular Integral: y_p	Subject to the condition that	Remarks/ observation
Any function of x	$y_p = \frac{1}{D-a} F(x) = e^{ax} \int F(x) e^{-ax} dx$	$c=0$	Substitute $F(x)$ and integrate w.r.t. x
$F(x) = b = \text{constant}$	$y_p = \frac{1}{D-a} b = -\frac{b}{a}$	$a \neq 0$	
$F(x) = e^{ax}$	$y_p = \frac{1}{D-a} e^{ax} = x e^{ax}$		
$F(x) = e^{ax}$	$y_p = \frac{1}{(D-a)^n} e^{ax} = \frac{x^n}{n!} e^{ax}$		Repeated factor n times
Any function of x	If $f(D) = (D-a_1)(D-a_2)\ldots(D-a_n)$ Resolving into partial fractions		Most general case
	$y_p = \frac{1}{f(D)} F(x) = \left(\frac{A_1}{D-a_1} + \frac{A_2}{D-a_2} + \ldots + \frac{A_n}{D-a_n} \right) F(x)$		Resolving $f(D)$ into factors

	Applying result I_a $y_p = A_1 e^{a_1 x}\int F(x)e^{-a_1 x}dx + \ldots$ $+ A_n e^{a_n x}\int F(x)e^{-a_1 x}dx$		Resolve $\frac{1}{f(D)}$ into partial fractions Applying result I_a to each term
e^{ax+b} constant $= b = be^{0x}$ with $a = 0$ e^{ax+b}	$y_p = \frac{1}{f(D)}e^{ax+b} = \frac{1}{f(a)}e^{ax+b}$ $y_p = \frac{1}{f(D)}b = \frac{b}{f(0)}$ $y_p = \frac{1}{f(D)}e^{ax+b} = \frac{1}{(D-a)^n g(D)}e^{ax+b}$ $= \frac{1}{g(a)}\frac{1}{(D-a)^n}e^{ax+b}$ $y_p = \frac{1}{g(a)}\frac{x^n}{n!}e^{ax+b}$	$f(a) \neq 0$ $f(0) \neq 0$ $g(a) \neq 0$	Replace D by 'a' in $f(D)$ Replace D by zero in $f(D)$ Replace D by 'a' in $g(D)$ and applying result of I_d
$\sin(ax+b)$	$y_p = \frac{1}{f(D^2)}\sin(ax+b)$ $= \frac{1}{f(-a^2)}\sin(ax+b)$	$f(-a^2) \neq 0$	Replace D^2 by $-a^2$ in $f(D)$ when terms of $D, D^3 \cdots$ are present,

$\cos(ax+b)$	$y_p = \frac{1}{f(D^2)}\cos(ax+b)$ $= \frac{1}{f(-a^2)}\cos(ax+b)$	$f(-a^2) \neq 0$	Multiply by D in order to get $D^2, D^4 \cdots$ etc., terms. Then replace D^2 by $-a^2$, etc.
$\sin(ax+b)$	$y_p = \frac{1}{D^2+a^2}\sin(ax+b)$ $= -\frac{x\cos(ax+b)}{(2a)}$	$a \neq 0$	sin is replaced by cos and multiplied by $-\frac{x}{2a}$
$\sin(ax.+b)$	$y_p = \frac{1}{(D^2+a^2)^n}\sin(ax+b)$ $= \frac{x^n}{(2a)^n n!}\sin\left(ax - n\frac{\pi}{2}\right)$	$a \neq 0$	
$\cos(ax+b)$	$y_p = \frac{1}{D^2+a^2}\cos(ax+b)$ $= \frac{x\sin(ax+b)}{(2a)}$	$a \neq 0$	cos is replaced by sin and multiplied by $\frac{x}{2a}$
$\cos(ax+b)$	$y_p = \frac{1}{(D^2+a^2)^n}\cos(ax+b)$ $= \frac{x^n}{(2a)^n}\frac{1}{n!}\cos\left(ax - \frac{n\pi}{2}\right)$	$a \neq 0$	

x^m	$y_p = \frac{1}{f(D)} x^m = [f(D)]^{-1} x^m$ $y_p = [a_0 + a_1 D + a_2 D^2 + \ldots + a_m D^m] x^m$	$a_0 \neq 0$	Expand $[f(D)]^{-1}$ is ascending powers of D and delete all terms beyond D^m (since $D^n x^m = 0$ when $n > m$)
$e^{ax} V_{(x)}$	$y_p = \frac{1}{f(D)} e^{ax} V(x) = e^{ax} \frac{1}{f(D+a)} V_{(x)}$ (Exponential shift)	$f(D+a) \neq 0$	Replace D by $D+a$ in $f(D)$. Then evaluate $\frac{1}{f(D+a)} V_{(x)}$ by the above methods.
$xV_{(x)}$	$y_p = \frac{1}{f(D)} xV(x)$ $= x \frac{1}{f(D)} V(x) - \frac{f'(D)}{[f(D)]^2} V_{(x)}$		

Note: $e^{iax} = \cos ax + i \sin ax$

Thus particular integral for e^{iax} = particular integral for $(\cos ax + i \sin ax)$

$\therefore$ real part of particular integral for e^{iax} = particular integral for $\cos ax$
imaginary part of particular integral for e^{iax} = particular integral for $\sin ax$

$\therefore \quad \dfrac{\cos ax}{D^2+a^2}$ and $\dfrac{\sin ax}{D^2+a^2}$ are respectively real and imaginary parts of

$$\frac{e^{iax}}{D^2+a^2} \equiv \frac{e^{iax}}{(D+ai)(D-ai)}$$

$$\equiv \frac{e^{aix}}{(ai+ai)(D-ai)} \equiv \frac{e^{aix}}{2ai}\left[\frac{1}{D+ai-ai}\right]\cdot 1$$

$$\equiv \frac{e^{aix}}{2ai}\cdot\frac{1}{D}\cdot 1 \equiv \frac{x}{2ai}(e^{iax})$$

$$\equiv -\frac{ix(\cos ax + i\sin ax)}{2a}, \qquad \left(\because \frac{1}{i} = \frac{1}{i^2}i = -i\right)$$

$$\equiv -\frac{ix}{2a}\cos ax + \frac{x}{2a}\sin ax$$

Real part $\dfrac{x}{2a}\sin ax \equiv \dfrac{\cos ax}{D^2+a^2}$;

Imaginary part $-\dfrac{x}{2a}\cos ax \equiv \dfrac{\sin ax}{D^2+a^2}$

$$\frac{\cos ax}{D^2+a^2} \equiv \frac{x}{2}\int \cos ax\, dx \equiv \frac{x}{2}\cdot\frac{1}{a}\sin ax$$

$$\frac{\sin ax}{D^2+a^2} \equiv \frac{x}{2}\int \sin ax\, dx \equiv -\frac{x}{2}\cdot\frac{1}{a}\cos ax$$

$$\therefore \frac{\sin 3x}{D^2+9} \equiv \frac{x}{2}\int \sin 3x\, dx \equiv \frac{x}{2}\cdot\frac{-1}{3}\cos 3x$$

$$\frac{\cos 3x}{D^2+9} \equiv \frac{x}{2}\int \cos 3x\, dx \equiv \frac{x}{2}\cdot\frac{1}{3}\sin 3x.$$

(G) EXAMPLES

Solve the following equations.

1. $y'' + y = x$

Solution: The auxiliary equation is $m^2 + 1 = 0$

$\therefore \quad m = \pm i = 0 \pm i = \alpha \pm i\beta$, where $\alpha = 0, \beta = 1$

$\therefore \quad y_c$ = complementary function $= c_1 \cos x + c_2 \sin x$

$\therefore$ particular integral: $\left(\dfrac{1}{D^2 + 1}\right)x$

$$= (1 + D^2)^{-1}x$$

$$= (1 - D^2)x \qquad \text{(using binomial series)}$$

$$= x$$

$\therefore$ The general solution is $= y_c + y_p$

or $\quad y = c_1 \cos x + c_2 \sin x + x.$

2. $y'' + 3y' + 2y = x^2$

Solution: The auxiliary equation is $m^2 + 3m + 2 = 0$

$\Rightarrow \quad (m+2)(m+1) = 0$

$\Rightarrow \quad m = -2, -1$

$\therefore \quad y_c$ = complementary function $= c_1 e^{-2x} + c_2 e^{-x}$

y_p = particular integral: $\left(\dfrac{1}{D^2 + 3D + 2}\right)x^2$

$$= \frac{1}{2}\left(1 + \frac{3D + D^2}{2}\right)^{-1} \cdot x^2$$

$$= \frac{1}{2}\left(1 - \frac{3D + D^2}{2} + \left(\frac{3D + D^2}{2}\right)^2\right)x^2$$

$$= \frac{1}{2}\left(1 - \frac{3D}{2} + \frac{7D^2}{4}\right)x^2$$

$$= \frac{1}{2}\left(x^2 - 3x + \frac{7}{2}\right)$$

$\therefore$ General solution y = Complementary function + Particular integral

$= y_c + y_p$

$$y = c_1 e^{-2x} + c_2 e^{-x} + \frac{1}{2}\left(x^2 - 3x + \frac{7}{2}\right)$$

3. $(D^6 - D^4)y = x^2$

Solution: Complementary function: Here auxiliary equation is $m^6 - m^4 = m^4(m^2 - 1) = 0$.

$\Rightarrow$ $m = 0$ is root repeated 4 times and $m \pm 1$ are the other roots. So the complementary function is

$$y_c = (c_1 - c_2 x + c_3 x^2 + c_4 x^3) + c_5 e^x + c_6 e^{-x}$$

particular integral: Rewriting $(D^6 - D^4)y = D^4(D^2 - 1)y = x^2$ so that

$$y_p = \frac{1}{D^4(D^2 - 1)} x^2 = \frac{1}{D^4}\left[\frac{1}{(D^2 - 1)} x^2\right]$$

Consider $y_p = \frac{1}{(D^2 - 1)} x^2 = -\frac{1}{(1 - D^2)} x^2$

Expanding in Binomial series

$$= -(1 - D^2)^{-1} x^2$$

$$= -[1 + D^2 + D^4 + D^6 + \cdots] x^2$$

$$= -x^2 - D^2 x^2 + 0 + 0 \cdots = -x^2 - 2$$

Thus $y_p = \frac{1}{D^4}\left[\frac{1}{(D^2 - 1)} x^2\right] = \frac{-1}{D^4}[x^2 + 2]$

$$\frac{1}{D^4} x^2 = \frac{1}{D^3}\int x^2 dx = \frac{1}{D^3}\frac{x^3}{3} = \frac{1}{D^2}\frac{1}{D}\frac{x^3}{3}$$

$$= \frac{1}{D^2}\int \frac{x^3}{3} dx$$

$$= \frac{1}{D^2}\frac{x^4}{12} = \frac{1}{D}\frac{1}{D}\frac{x^4}{12} = \frac{1}{D}\int \frac{x^4}{12} dx = \frac{1}{D}\frac{x^5}{60}$$

$$= \int \frac{x^5}{60} dx = \frac{x^6}{360}$$

Similarly, $\frac{1}{D^4}[2] = \frac{1}{D^3} 2x = \frac{1}{D^2} x^2 = \frac{1}{D}\frac{x^3}{3} = \frac{x^4}{12}$

$$y_p = -\left[\frac{x^6}{360} + \frac{x^4}{12}\right].$$

Hence the general solution is $y = y_c + y_p$

$$y_c = (c_1 + c_2 x + c_3 x^2 + c_4 x^3 + c_5 e^x + c_6 e^{-x}) - \left[\frac{x^6}{360} + \frac{x^4}{12}\right]$$

4. $y'' - 2y' + y = e^x \sin 2x + x^2$

Solution: The auxiliary equation is $m^2 - 2m + 1 = 0$

$\Rightarrow \quad (m-1)^2 = 0$

$\Rightarrow \quad m = 1, 1$

$\therefore \quad y_c = \text{complementary function} = (c_1 + c_2 x)e^x$

particular integral (P.I.): $\frac{1}{D^2 - 2D + 1}(e^x \sin 2x + x^2)$

$$= \frac{1}{D^2 - 2D + 1} \cdot e^x \sin 2x + \frac{1}{D^2 - 2D + 1} \cdot x^2$$

$$= \text{P.I.}_1 + \text{P.I.}_2$$

$$\text{P.I.}_1 = \frac{1}{D^2 - 2D + 1} \cdot e^x \sin 2x$$

$$= \frac{1}{(D-1)^2} \cdot e^x \sin 2x$$

$$= e^x \cdot \frac{1}{(D+1-1)^2} \cdot \sin 2x; \qquad \left(\because \frac{1}{f(D)} e^{ax} V = e^{ax} \cdot \frac{1}{f(D+a)} V\right)$$

$$= e^x \cdot \frac{1}{D^2} \cdot \sin 2x \quad = -\frac{e^x \sin 2x}{4}$$

$$\text{P.I.}_2 = \frac{1}{D^2 - 2D + 1} \cdot x^2$$

$$= \frac{1}{(D-1)^2} \cdot x^2 = (1-D)^{-2} \cdot x^2$$

$$= (1 + 2D + 3D^2 + \cdots)x^2$$

$$= x^2 + 4x + 6$$

$\therefore$ General solution is $y = (c_1 + c_2 x)e^x - \dfrac{e^x \sin 2x}{4} + (x^2 + 4x + 6)$

5. $(D^3 - 3D^2 + 3D - 1)y = x^2 e^x$

or

$y''' - 3y'' + 3y' - y = x^2 e^x$

***Solution*:** The auxiliary equation is $(m^3 - 3m^2 + 3m - 1) = 0$

$\Rightarrow \quad (m-1)^3 = 0$

$\therefore \quad m = -1$ (thrice)

$\therefore \quad y_c =$ Complementary function $= e^{-x}(c_1 + c_2 x + c_3 x^2)$

$$y_p = \text{Particular integral} = \left(\frac{1}{D^3 - 3D^2 + 3D - 1}\right) x^2 e^x$$

$$= \left(\frac{e^x}{(D+1)^3 - 3(D+1)^2 + 3(D+1) - 1}\right) x^2$$

$$= e^x \left(\frac{1}{D^3}\right) x^2 = \left(\frac{e^x x^5}{60}\right)$$

$\therefore$ The general solution is $y = y_c + y_p$

$$y = e^{-x}(c_1 + c_2 x + c_3 x^2) + \frac{x^5 e^x}{60}.$$

6. $(D^2 - 1)y = 2x^4 - 3x + 1.$

***Solution*:** Complementary function: Here auxiliary equation is $m^2 - 1 = 0$ so that $m = \pm 1$ are the two roots. Thus the complementary function is

$$y_c = c_1 e^x + c_2 e^{-x}$$

Particular integral: $y_p = \dfrac{1}{(D^2-1)}(2x^4 - 3x + 1)$

$$= \frac{1}{1-D^2}(2x^4 - 3x + 1)$$

$$= -\left[1 + D^2 + D^4 + D^6 + \cdots\right]\left[2x^4 - 3x + 1\right]$$

$$= -\left[1 + D^2 + D^4\right]\left[2x^4 - 3x + 1\right]$$

$$= -\left[(2x^4 - 3x + 1) + (24x^2) + 48\right].$$

General solution is $y = y_c + y_p = c_1 e^x + c_2 e^{-x} - [2x^4 + 24x^2 - 3x + 49]$

7. $(D^4 - a^4)y = x^4 + \sin bx$

Solution: Complementary function is: $c_1 e^{ax} + c_2 e^{-ax} + (c_3 \cos ax + c_4 \sin ax)$

particular integral of $x^4 = \dfrac{x^4}{D^4 - a^4} = \dfrac{-1}{a^4}\left[1 - \dfrac{D^4}{a^4}\right](x^4)$

$$= \frac{-1}{a^4}\left[1 + \frac{D^4}{a^4}\right](x^4) = -\frac{1}{a^4}\left[x^4 + \frac{1}{a^4}(4!)\right]$$

$$= -\frac{1}{a^4}\left[x^4 + \frac{24}{a^4}\right]$$

particular integral of $\sin bx = \dfrac{\sin bx}{D^4 - a^4}$. Put $D^2 = -b^2 \quad \therefore D^4 = b^4$

$$= \frac{\sin bx}{b^4 - a^4}$$

$\therefore$ General solution is $y = c_1 e^{ax} + c_2 e^{-ax} + (c_3 \cos ax + c_4 \sin ax) + \dfrac{\sin bx}{b^4 - a^4}$.

8. $(D^3 - 1)y = x^5 + 3x^4 - 2x^3$.

Solution: Complementary function: Here auxiliary equation is $m^3 - 1 = 0$ having 3 roots $m = 1, \dfrac{-1 \pm \sqrt{3}i}{2}$. Thus the complementary function y_c is

$$y_c = c_1 e^x + e^{-\frac{x}{2}}\left[c_2 \cos\frac{\sqrt{3}}{2}x + c_3 \sin\frac{\sqrt{3}}{2}x\right]$$

particular integral: $y_p = \dfrac{1}{(D^3-1)}(x^5+3x^4-2x^3)$

$$= \frac{-1}{(1-D^3)}(x^5+3x^4-2x^3)$$

$$= -\left[1+D^3+D^6+D^9+\cdots\right]\left[x^5+3x^4-2x^3\right]$$

$$= -\left[(x^5+3x^4-2x^3)+D^3(x^5+3x^4-2x^3)+0+0+\cdots\right]$$

$$= -\left[(x^5+3x^4-2x^3)+(60x^2+72x+12)\right]$$

9. $y''+4y = x\sin x$

***Solution*:** Complementary function is: $c_1\cos 2x + c_2 \sin 2x$

particular integral for $x\sin x = \dfrac{x\sin x}{D^2+4}$

$= \text{imaginary part (I.P. of)} \dfrac{xe^{ix}}{D^2+4}$ $\quad \left(\begin{array}{l}\because e^{ix} = \cos x + i\sin x \\ \therefore \sin x = (\text{I.P. of}) e^{ix}\end{array}\right)$

$= (\text{I.P. of}) e^{ix}\left(\dfrac{1}{(D+i)^2+4}\right)x$ $\quad \left(\because \dfrac{1}{f(D)}(e^{ax}V) = e^{ax}\cdot\dfrac{1}{f(D+a)}\cdot V\right)$

$= (\text{I.P. of}) e^{ix}\left(\dfrac{1}{D^2+2Di+3}\right)x$ $\quad (\because i^2 = -1)$

$= \left(\text{I.P. of } \dfrac{e^{ix}}{3}\right)\left(1+\dfrac{2Di}{3}\cdots\right)^{-1} x$ (neglecting D^2)

$$= (\text{I.P. of}) \frac{e^{ix}}{3}\left(1-\frac{2D}{3}i\cdots\right)x$$

$$= (\text{I.P. of}) \frac{e^{ix}}{3}\left(1-\frac{2}{3}i\right)$$

$$= (\text{I.P. of}) \frac{(\cos x + i\sin x)(3x-2)}{9}$$

$$= \frac{3x\sin x - 2\cos x}{9}$$

$\therefore$ General solution $y = c_1 \cos 2x + c_2 \sin 2x + \dfrac{3x\sin x - 2\cos x}{9}$

10. $(D^2 - 2D + 2)y = e^x \sin x$

Solution: The auxiliary equation is $m^2 - 2m + 2 = 0$

Hence, $m = 1 \pm \mathrm{i}$; $y_c =$ Complementary function $= e^x c_1 \sin x - 2\cos x$

particular integral $= \left[\dfrac{1}{D^2 - 2D + 2}\right](e^x \sin x)$

$$= \left[\frac{1}{(D+1)^2 - 2(D+1) + 2}\right] \sin x = \left[\frac{e^x}{D^2 + 1}\right] \sin x$$

$$= \left[\frac{e^x}{(D+\mathrm{i})(D-\mathrm{i})}\right] \sin x$$

$$= e^x (\text{I.P. of}) \left[\frac{1}{(D+\mathrm{i})(D-\mathrm{i})}\right] e^{\mathrm{i}x} = e^x (\text{I.P. of}) \left[\frac{1}{2\mathrm{i}} x e^{\mathrm{i}x}\right]$$

$$= e^x (\text{I.P. of}) \left[-\frac{1}{2} \mathrm{i}x(\cos x + \mathrm{i}\sin x)\right] = -\frac{1}{2} x e^x \cos x$$

$\therefore$ General solution: $y = e^x (c_1 \cos x + c_2 \sin x) - \dfrac{1}{2} x e^x \cos x$

11. $(D^2 - 2D + 1)y = xe^x \sin x$

Solution: The auxiliary equation is $m^2 - 2m + 1 = 0 \Rightarrow (m-1)^2 = 0$

$\Rightarrow m = 1, 1$ (a double root)

so $y_c =$ Complementary function $= (c_1 + c_2 x)e^x$

particular integral $y_p = \dfrac{1}{(D^2 - 2D + 1)} x(e^x \sin x)$

$$= \frac{1}{(D-1)^2} e^x (x \sin x)$$

$$= e^x \frac{1}{[(D-1)-1]^2}(x\sin x)$$

$$= \frac{e^x}{D^2}(x\sin x)$$

and $\frac{1}{D^2}x(\sin x) = x\cdot\frac{1}{D^2}\sin x - \frac{2D}{D^4}\sin x$

$$= x(-\sin x) - 2\cos x$$

$\therefore \quad y_p = e^x[-x\sin x - 2\cos x]$

$\therefore$ General solution $y = y_c + y_p$

$$= (c_1 + c_2 x)e^x - e^x(x\sin x + 2\cos x).$$

12. $(D^2+2)y = x^2e^{3x} + e^x\cos 2x$

Solution: The auxiliary equation is $m^2+2=0$ with roots $m=\pm\sqrt{2}\mathrm{i}$ so that the complementary function is $y_c = (c_1\cos\sqrt{2}x + c_2\sin\sqrt{2}x)$

particular integral: $y_p = \frac{1}{D^2+2}[x^2e^{3x} + e^x\cdot\cos 2x]$

$$= \frac{1}{D^2+2}[e^{3x}\cdot x^2] + \frac{1}{D^2+2}[e^x\cdot\cos 2x]$$

$$= I_1 + I_2$$

where $\quad I_1 = \frac{1}{D^2+2}[e^{3x}\cdot x^2] = \frac{e^{3x}}{(D+3)^2+2}[x^2]$

$$= \frac{e^{3x}}{(D^2+6D+11)}x^2 \qquad = \frac{e^{3x}}{11\left[1+\frac{D^2+6D}{11}\right]}\cdot x^2$$

Expanding in powers of $\frac{D^2+6D}{11}$, we get

$$I_1 = \frac{e^{3x}}{11}\left[1 - \left(\frac{D^2+6D}{11}\right) + \left(\frac{D^2+6D}{11}\right)^2 - \left(\frac{D^2+6D}{11}\right)^3 + \cdots\right]x^2$$

Since x^2 is the highest power, discard terms of D^3 and higher orders. Then

$$I_1 = \frac{e^{3x}}{11}\left[1 + \left(-\frac{D^2}{11} - \frac{6D}{11}\right) + \left(\frac{36}{11^2}D^2\right)\right]x^2$$

$$= \frac{e^{3x}}{11}\left[x^2 - \frac{12x}{11} + \frac{50}{121}\right]$$

Now consider, $I_2 = \dfrac{1}{D^2+2}e^x \cdot \cos 2x$

$$I_2 = \frac{e^x}{(D+1)^2+2} \cdot \cos 2x = \frac{e^x}{D^2+2D+3}\cos 2x$$

Replace D^2 by -2^2, we get

$$I_2 = \frac{e^x}{-2^2+2D+3}\cos 2x = \frac{e^x}{2}\frac{1}{2D-1}\cos 2x$$

To have D^2 terms, rewrite the above as

$$I_2 = e^x \cdot \frac{2D+1}{(2D-1)(2D+1)}\cos x \quad = \frac{e^x(2D+1)}{4D^2-1^2}\cos 2x$$

Replace D^2 by -2^2, we get

$$I_2 = \frac{e^x(2D+1)}{4\cdot(-2^2)-1^2}\cos 2x = \frac{e^x(2D+1)}{-17}\cos 2x$$

$$I_2 = \frac{e^x}{17}[-4\sin 2x + \cos 2x].$$

Thus $y_p = I_1 + I_2 = \dfrac{e^{3x}}{11}\left[x^2 - \dfrac{12x}{11} + \dfrac{50}{121}\right] + \dfrac{e^x}{17}[4\sin 2x - \cos 2x]$

Hence general solution: $y = y_c + y_p$

$$y = [c_1 \cos\sqrt{2}x + c2\sin\sqrt{2}x] + \frac{e^{3x}}{11} \times \left[x^2 - \frac{12x}{11} + \frac{50}{121}\right] + \frac{e^x}{17}[4\sin 2x - \cos 2x].$$

13. $(D-1)^2(D^2+1)^2 y = \sin\dfrac{2x}{2} + e^x + x$

***Solution*:** The auxiliary equation is $(m-1)^2(m^2+1)^2 = 0$

$\Rightarrow \quad m = 1,1, \;\; m^2 = -1, \;\; m = \pm i, \;\; m^2 = -1, \;\; m = \pm i$

$$y_c = (c_1 + c_2x)e^x + \left[(c_3 + c_4x)\cos x + (c_5 + c_6x)\sin x\right]$$

$$\text{P.I. of } e^x = \frac{e^x}{(D-1)^2(D^2+1)^2} = \frac{x^2}{2!}\cdot\frac{e^x}{(1+1)^2} = \frac{x^2}{8}\cdot e^x.$$

$$\text{P.I. of } x = \frac{x}{(D-1)^2(D^2+1)^2}$$

$$= \frac{x}{(1-2D+D^2)(1+2D^2+D^4)}$$

$$= \frac{x}{(1-2D+\cdots)} = (1-2D)^{-1}x$$

$$= (1-2D+\cdots)x = x+2$$

$$\text{P.I. of } \sin^2\frac{x}{2} = \text{P.I. of } \frac{1}{2}(1-\cos x)$$

$$= \text{P.I. of } \left(\frac{1}{2} - \frac{1}{2}\cos x\right)$$

$$\text{P.I. of } \frac{1}{2} = \frac{-\frac{1}{2}e^{0x}}{(D-1)^2(D^2+1)^2} = \frac{1}{2}$$

$$\text{P.I. of } -\frac{1}{2}\cos x = \frac{-\frac{1}{2}\cos x}{(D^2-2D+1)(D^2+1)^2} \qquad \text{Put } D^2 = -1$$

$$= \frac{-\frac{1}{2}\cos x}{-2D(D^2+1)^2} = \frac{1}{4}\cdot\frac{D\cos x}{D^2(D^2+1)^2} \qquad \text{Put } D^2 = -1$$

$$= -\frac{1}{4}\cdot\frac{-\sin x}{(D^2+1)^2} = \frac{1}{4}\frac{\sin x}{(D^2+1)^2}$$

$$= \frac{1}{4}\cdot \text{I} \quad \text{where} \quad \text{I} = \frac{\sin x}{(D^2+1)^2}$$

$$\text{I} = (\text{I.P. of})\ \frac{e^{ix}}{(D^2+1)^2} = (\text{I.P. of})\ \frac{e^{ix}}{(D^2+\text{i}^2)^2}$$

$$= (\text{I.P. of})\ \frac{e^{ix}}{(D+\text{i})^2(D+\text{i})^2}$$

$$= \text{(I.P. of)}\ \frac{x^2}{2!}\cdot\frac{e^{ix}}{(\mathrm{i}+\mathrm{i})^2}$$

$$= \text{(I.P. of)}\ -\frac{x^2}{8}e^{ix}$$

$$= \text{(I.P. of)}\ -\frac{x^2}{8}(\cos x + \mathrm{i}\sin x) - \frac{x^2}{8}\sin x$$

$$\therefore \quad \text{(I.P. of)}\ \frac{1}{2}\cos x = \frac{1}{4}\cdot \mathrm{I} = -\frac{x^2}{32}\sin x$$

$\therefore$ General solution is $y = (c_1 + c_2 x)e^x + \left[(c_3 + c_4 x)\cos x + (c_5 + c_6 x)\sin x\right]$

$$+\frac{x^2}{8}e^x + x + 2 + \frac{1}{2} - \frac{x^2}{32}\sin x.$$

14. $(D^2 - 1)y = x\sin x + x^2 e^x$

Solution: Complementary function: Here auxiliary equation is $m^2 - 1 = 0$ with $m = \pm 1$ as roots so that the complementary function $y_c = c_1 e^x + c_2 e^{-x}$.

Particular integral $y_p = \dfrac{1}{D^2 - 1}x\sin x + \dfrac{1}{D^2 - 1}x^2 e^x = I_1 + I_2$

Applying the case $D^2 = -a^2$ for I_1, we get

$$I_1 = \frac{1}{D^2 - 1}x\sin x$$

$$= \frac{x}{D^2 - 1}\sin x - \frac{2D}{(D^2 - 1)^2}\sin x$$

Replacing D^2 by -1^2 in the denominator

$$I_1 = \frac{x}{-1^2 - 1}\sin x - \frac{2}{(-1^2 - 1)^2}.D\sin x$$

$$= -\frac{x\sin x}{2} - \frac{\cos x}{2}$$

$$= -\frac{1}{2}(x\sin x + \cos x)$$

$$I_2 = \frac{1}{D^2 - 1}x^2 e^x$$

$$= x\cdot\frac{1}{D^2-1}(xe^x)-\frac{2D}{(D^2-1)^2}(xe^x)$$

$$I_2 = x\left[x\frac{1}{D^2-1}e^x-\frac{2D}{(D^2-1)}e^x\right]-2D\left[x\frac{1}{(D^2-1)^2}e^x-\frac{2(D^2-1)2D}{(D^2-1)^4}e^x\right]$$

$$\because \frac{1}{f(D)}\{xV\}=\left[x\frac{1}{f(D)}-\frac{f'(D)}{\{f(D)\}^2}\right]V$$

$$=\left[x^2\frac{1}{(D+1)(D-1)}e^x-\frac{2xD}{(D+1)^2(D-1)^2}e^x\right]-2D\left[x\frac{1}{(D+1)^2(D-1)^2}e^x-\frac{4D}{(D+1)^3(D-1)^3}e^x\right]$$

Replacing D by 1, we get

$$=\left[x^2\frac{1}{2(D-1)}e^x-\frac{2xD}{4(D-1)^2}e^x\right]-2D\left[\frac{x}{4}\frac{1}{(D-1)^2}e^x-\frac{4}{8}\frac{D}{(D-1)^3}e^x\right].$$

$$I_2=\frac{x^2}{2}xe^x-\frac{x}{e}D.\left(\frac{x^2e^x}{2!}\right)\left[\left(\because\frac{1}{(D-a)^1}e^{ax}=xe^{ax}\right)\text{ and }\frac{1}{(D-a)^r}e^{ax}=\frac{x^2e^{ax}}{r!}\right]$$

$$-2D\left[\frac{x}{4}\cdot\frac{x^2e^x}{2!}-\frac{1}{2}D\frac{x^3e^x}{3!}\right]$$

$$=\frac{x^3e^x}{2}-\frac{x}{4}\left[x^2e^x+2xe^x\right]-\frac{1}{4}\left[x^3e^x+3x^2e^x\right]+\frac{1}{6}\left[x^3e^x+6x^2e^x+6xe^x\right]$$

$$=xe^x\left[\frac{x^2}{2}-\frac{x^2}{4}-\frac{x}{2}-\frac{x^2}{4}-\frac{3}{4}x+\frac{1}{6}x^2+x+1\right]$$

$$I_2=xe^x\left[\frac{x^2}{6}-\frac{x}{4}+1\right]=\frac{xe^x}{12}[2x^2-3x+12]$$

$$y_p=I_1+I_2=-\frac{1}{2}(x\sin x+\cos x)+\frac{xe^x}{12}(2x^2-3x+12).$$

Thus general solution is $y=y_c+y_p$

$$y=c_1e^x+c_2e^{-x}-\frac{1}{2}(x\sin x+\cos x)+\frac{xe^x}{12}(2x^2-3x+12).$$

15. $\dfrac{d^2y}{dx^2}+n^2y=k\sin\mu x;\quad(\mu\neq n)$

***Solution*:** $(D^2+n^2)y=k\sin\mu x$ [1]

Complementary function: $m^2+n^2=0;\ \ m\pm(in)$

$\therefore\quad y=c_1\cos nx+c_2\sin nx$ [2]

Particular integral: $y=\left[\dfrac{1}{D^2+n^2}\right](k\sin\mu x)$

$$=\frac{k\cdot\sin\mu x}{-\mu^2+n^2}$$ [3]

$$\therefore\quad y=(c_1\cos nx+c_2\sin nx+\frac{k\sin\mu x}{(n^2-\mu^2)}.$$

16. $y''+y=\text{cosec}\,x$

***Solution*:** Equation: $(D^2+1)y=\text{cosec}\,x$ [1]

Complementary function: $(m^2+1)=0;\ \therefore m=\pm\text{i}$

$\therefore\quad y=c_1\cos x+c_2\sin x$ [2]

Particular integral: $y=\left[\dfrac{1}{(D^2+1)}\right](\text{cosec}\,x)$

The function (cosec x) is not in the usual familiar form, and so the particular integral is to be found by the general method of partial fractions.

$$\frac{1}{(D^2+1)}=\frac{1}{(D+\text{i})(D-\text{i})}=\frac{1}{2\text{i}}\left[\frac{1}{(D-\text{i})}-\frac{1}{(D+\text{i})}\right]$$

$$\frac{1}{(D-\text{i})}\text{cosec}\,x=e^{\text{i}x}\int e^{\text{i}x}\text{cosec}\,x\,dx$$

$$=e^{\text{i}x}\int(\cos x-\text{i}\sin x)\text{cosec}\,x\,dx$$

$$=e^{\text{i}x}\int(\cot x-\text{i})\,dx$$

$$=e^{\text{i}x}\left[\log\sin-\text{i}x\right]$$

Similarly, $\dfrac{1}{(D-\text{i})}\text{cosec}\,x=e^{-\text{i}x}\left[\log x+\text{i}x\right]$

$$\therefore \quad \frac{1}{(D+\mathrm{i})}(\operatorname{cosec} x) = \frac{1}{2\mathrm{i}}\left[\frac{1}{(D-\mathrm{i})} - \frac{1}{(D-\mathrm{i})}\right]\operatorname{cosec} x$$

$$= \left[e^{\mathrm{i}x}(\log\sin x - \mathrm{i}x) - e^{-\mathrm{i}x}(\log\sin x + \mathrm{i}x)\right]\left(\frac{1}{2\mathrm{i}}\right)$$

$$= (\log\sin x)\frac{e^{ix} - e^{-\mathrm{i}x}}{2\mathrm{i}} - \frac{x}{2}(e^{ix} + e^{-\mathrm{i}x})$$

$$= (\log\sin x)\cdot(\sin x) - x\cos x$$

$\therefore$ General solution is $y = (c_1\cos x + c_2\sin x) + [\sin x(\log\sin x) - x\cos x]$

17. Solve $y'' + a^2 y = \sec ax$

Solution: The auxiliary equation is $(D^2 + a^2) = 0$

or $\quad (m^2 + a^2) = 0$

$\Rightarrow \quad m = \pm a\mathrm{i}$

y_c = complementary function = $(c_1\cos ax + c_2\sin ax)$

particular integral $= \dfrac{\sec ax}{D^2 + a^2} = \dfrac{1}{(D+a\mathrm{i})(D-a\mathrm{i})}\cdot\sec ax$

$$= \frac{1}{2a\mathrm{i}}\left[\frac{1}{(D-a\mathrm{i})} - \frac{1}{(D+a\mathrm{i})}\right]\sec ax \qquad [1]$$

Now, $\dfrac{1}{(D-a\mathrm{i})}\sec ax = \dfrac{1}{(D-a\mathrm{i})}e^{\mathrm{i}ax}\cdot\dfrac{e^{-\mathrm{i}ax}}{\cos ax}$

Using $\dfrac{e^{ax}V}{f(D)} = e^{ax}\left[\dfrac{1}{f(D+a)}\right]V$, we get

$$= e^{\mathrm{i}ax}\int\frac{\cos ax - \mathrm{i}\sin ax}{\cos ax}dx$$

$$= e^{\mathrm{i}ax}\int(1 - \mathrm{i}\tan ax)\,dx = e^{\mathrm{i}ax}\left[x + \frac{\mathrm{i}}{a}\log\cos ax\right]$$

Similarly, $\dfrac{1}{(D+a\mathrm{i})}\cdot\sec ax = e^{-\mathrm{i}ax}\left[x - \dfrac{\mathrm{i}}{a}\log\cos ax\right]$

Putting in [1], we get the required particular integral as

$$\frac{1}{2ai}\left[e^{iax}\left\{x+\frac{i}{a}\log\cos ax\right\}-e^{-iax}\left\{x-\frac{i}{a}\log\cos ax\right\}\right]$$

$$=\frac{x}{a}\cdot\frac{e^{iax}-e^{-iax}}{2i}+\frac{1}{a^2}\log(\cos ax)\cdot\frac{e^{iax}-e^{-iax}}{2}$$

$$=\frac{x}{a}\sin ax+\frac{1}{a^2}\log(\cos ax)\cdot\cos ax$$

$\therefore$ The general solution is :

$$y=c_1\cos ax+c_2\sin ax+\frac{x}{a}\sin ax+\frac{1}{a^2}\log(\cos ax)\cos ax.$$

18. $(D^2-1)y=x^2\cos x$

***Solution*:** Complementary function: Here auxiliary equation is $m^2-1=0$ with $m=\pm 1$ as the roots so that the complementary function is

$$y_c=c_1e^x+c_2e^{-x}$$

particular integral: $y_p=\dfrac{1}{(D^2-1)}\cdot x^2\cos x=\dfrac{1}{D^2-1}x\cdot(x\cos x)$

Apply, case $D^2=-a^2$

$$y_p=\frac{x}{D^2-1}(x\cos x)-\frac{2D}{(D^2-1)^2}\cdot x\cos x=I_1-2I_2$$

Consider $I_1=\dfrac{x}{D^2-1}(x\cos x)$

Applying case $D^2=-a^2$ once again

$$I_1=x\cdot\left[x\cdot\frac{1}{D^2-1}\cos x-\frac{2D}{(D^2-1)^2}\cos x\right]$$

Replacing D^2 by -1^2 in the denominator

$$I_1=x^2\cdot\frac{1}{-1^2-1}\cos x-\frac{x2D}{(-1^2-1)^2}\cos x$$

$$=-\frac{x^2\cos x}{2}+\frac{x\sin x}{2}$$

Consider

$$I_2=\frac{D}{(D^2-1)^2}x\cos x,\text{ by case } D^2=-a^2$$

$$= D\left[\frac{x}{(D^2-1)^2}\cdot\cos x - \frac{2(D^2-1)\cdot 2D}{(D^2-1)^4}\cos x\right]$$

$$I_2 = D\left[\frac{x}{(-1^2-1)^2}\cdot\cos x - \frac{4D}{(-1^2-1)^3}\cos x\right]$$

$$= \frac{1}{4}D[x\cos x] + \frac{1}{2}D^2(\cos x)$$

$$= \frac{1}{4}[\cos x - x\sin x] - \frac{1}{2}\cos x.$$

Thus $y_p = I_1 - 2I_2 = \frac{1}{2}\left[-x^2\cos x + x\sin x\right] - \frac{1}{2}[\cos x - x\sin x] + \cos x$

$$= \left(\frac{1-x^2}{2}\right)\cos x + x\sin x$$

$$y = c_1e^x + c_2e^{-x} + x\sin x + \frac{(1-x^2)\cos x}{2}.$$

19. $y'' + 4y = 4\tan 2x$

Solution: The auxiliary equation is $m^2 + 4 = 0$

$\Rightarrow \quad m^2 = -4, \Rightarrow m = \pm 2\mathrm{i}$

$\therefore \quad y_c =$ complementary function $= c_1\cos 2x + c_2\sin 2x$

particular integral: $y_p = \frac{4\tan 2x}{D^2+4} = \frac{1}{\mathrm{i}}\left[\frac{1}{D-2\mathrm{i}} - \frac{1}{D+2\mathrm{i}}\right]\tan 2x$ [1]

Now, let $\frac{1}{D-2\mathrm{i}}\tan 2x = v$

$\therefore \quad (D2\mathrm{i})v = \tan 2x$

or, $\quad \frac{dv}{dx} - 2\mathrm{i}v = \tan 2x$

integral factor $= e^{-2\mathrm{i}x}$

Multiplying both sides by integral factor and integrating, we get

$$v\cdot e^{-2\mathrm{i}x} = \int e^{-2\mathrm{i}x}\cdot\tan 2x\,dx$$

$$= \int(\cos 2x - \mathrm{i}\sin 2x)\frac{\sin 2x}{\cos 2x}dx$$

$$= \int \sin 2x\, dx - \mathrm{i}\int \frac{1-\cos^2 2x}{\cos 2x} dx$$

$$= -\frac{\cos 2x}{2} - \mathrm{i}\int (\sec 2x - \cos 2x)\, dx$$

$$= -\frac{\cos 2x}{2} - \mathrm{i}\cdot\frac{1}{2}\cdot \log\tan\left(\frac{\pi}{4}+x\right) + \frac{\mathrm{i}}{2}\sin 2x.$$

$$= -\frac{1}{2}(\cos 2x - \mathrm{i}\sin 2x) - \frac{\mathrm{i}}{2}\log\tan\left(\frac{\pi}{4}+x\right)$$

or $\quad v\cdot e^{-2\mathrm{i}x} = -\frac{1}{2}e^{-2\mathrm{i}x} - \frac{\mathrm{i}}{2}\log\tan\left(\frac{\pi}{4}+x\right)$

$\therefore \quad v = -\frac{1}{2} - \frac{\mathrm{i}}{2}\log\tan\left(\frac{\pi}{4}+x\right)e^{2\mathrm{i}x}$

Similarly if we calculate $\frac{1}{D+2\mathrm{i}}\tan 2x = u,$

we will get $u = -\frac{1}{2} + \frac{\mathrm{i}}{2}\log\tan\left(\frac{\pi}{4}+x\right)e^{-2\mathrm{i}x}$

$\therefore$ particular integral $\frac{1}{i}(v-u),$ using [1]

$$= \frac{1}{i}\left[-\frac{\mathrm{i}}{2}\log\tan\left(\frac{\pi}{4}+x\right)\left(e^{2\mathrm{i}x}+e^{-2\mathrm{i}x}\right)\right]$$

$$= -\frac{1}{2}\log\tan\left(\frac{\pi}{4}+x\right)\cdot 2\cos 2x$$

$$= -\log\tan\left(\frac{\pi}{4}+x\right)\cos 2x$$

$\therefore$ General solution is $y = c_1\cos 2x + c_2\sin 2x - \log\tan\left(\frac{\pi}{4}+x\right)\cos 2x$

EXERCISES

Solve the following equations.

1. $(D^2+D+1)y = x\cos x$ $\left[\textit{Ans.}\quad y = (c_1+c_2x)e^{-x} + \frac{x}{2}\sin x + \frac{1}{2}(\cos x - \sin x)\right]$

2. $(D^2-4D+4)y = x^2+e^x+\cos 2x$

$$\left[Ans.\quad y=(c_1+c_2x)e^{2x}+e^x-\frac{1}{8}\sin 2x+\frac{1}{8}(2x^2+4x+3)\right]$$

3. $(D^2-4)y = e^x+\sin 2x+\cos^2 x$

$$\left[Ans.\quad y=c_1e^{2x}+c_2e^{-2x}-\frac{e^x}{8}-\frac{\sin 2x}{8}-\frac{1}{16}\cos 2x-\frac{1}{8}\right]$$

4. $(D^2-2D+1)y = x^2e^{3x}$ $\left[Ans.\quad y=(c_1+c_2x)e^x+\frac{e^x}{8}(2x^2-4x+3)\right]$

5. $(D^2-3D+2)y = e^x$ when $y=3,\ y'=3$ at $x=0$ $\left[Ans.\quad y=2e^x+e^{2x}-xe^x\right]$

6. $(D^2+2D+1)y = \dfrac{e^{-x}}{(x+2)}$

$$\left[Ans.\quad y=\text{Complementary function}+e^{-x}[x\log(x+2)-x+2\log(x+2)]\right]$$

7. $(D^2+4)y = x\sin^2 x$ $\left[Ans.\quad y=c_1\cos 2x+c_2\sin 2x+\frac{x}{8}-\frac{x\cos 2x+2x^2}{32}\right]$

8. $(D^2+1)y = x^2e^{2x}+x\cos x$

$$\left[Ans.\quad y=(c_1\cos x+c_2\sin x)+\frac{e^{2x}}{5}\left(x^2-\frac{8}{5}x+\frac{22}{25}\right)+\left(\frac{x^2}{4}-\frac{1}{8}\right)\sin x+\frac{1}{4}\cos x\right]$$

9. $(D^4-2D^3-3D^2+4D+4)y = x^2e^x$

$$\left[Ans.\quad y_1=(c_1+c_2x)e^{-x}+(c_3+c_4x)e^{2x}+\frac{e^x}{4}\left(x^2-2x-\frac{3}{2}\right)\right]$$

10. $(D^2-1)y = x^2\sin 3x$ $\left[Ans.\quad y=c_1e^x+c_2e^{-x}\frac{-x^2}{10}\cdot\sin 3x-\frac{3x}{25}\cos 3x+\frac{13}{250}\sin 3x\right]$

11. $(D^3-3D^2-6D+8)y = xe^{-3x}$ $\left[Ans.\quad y=c_1e^x+c_2e^{4x}+c_3e^{-2x}-\frac{e^{-3x}}{784}(28x+39)\right]$

12. $(D^2-4)y = x\sinh x$ $\left[Ans.\quad y=c_1e^{2x}+c_2e^{-2x}-\frac{x}{2}\sinh x-\frac{2}{9}\cosh x\right]$

13. $(D^3 - 2D + 4)y = e^x \sin x$

$$\left[Ans. \quad y = c_1 e^{-2x} + (c_2 \cos x + c_3 \sin x)e^x + (3\sin x + \cos x)\frac{xe^x}{20} \right]$$

14. $(D^2 - 2D + 1)y = x^2 e^{3x}$ $\qquad \left[Ans. \quad y = (c_1 + c_2 x)e^x + (2x^2 - 4x + 3)\frac{e^{3x}}{8} \right]$

15. $(D^2 + 4D + 3)y = e^{-x} \sin x + x$

$$\left[Ans. \quad y = c_1 e^x + c_2 e^{-3x} + \frac{1}{16} e^x (2x^2 - x) - \frac{1}{27}(9x^3 + 18x^2 + 42x + 40) \right]$$

16. $(D^4 - 1)y = \cos x \cdot \cosh x$

$$\left[Ans. \quad y = c_1 e^x + c_2 e^{-x} + c_3 \cos x + c_4 \cos x - \frac{1}{5}\cos x \cdot \cosh x \right]$$

17. $(D^2 - 4D + 4)y = e^{2x} \cos^2 x$ $\qquad \left[Ans. \quad y = (c_1 + c_2 x)e^{2x} + e^{2x}\left(\frac{x^2}{4} - \frac{1}{8}\cos 2x \right) \right]$

18. $(D^2 + 4D + 5)y = e^{-2x}(1 + \cos x)$ $\qquad \left[Ans. \quad y = e^{-2x}\left(c_1 \cos x + c_2 \sin x + 1 + \frac{x}{2}\sin x \right) \right]$

19. $(D^2 - 6D + 13)y = 8e^{3x} \sin 2x$ $\qquad \left[Ans. \quad y = e^{3x}(c_1 \cos 2x + c_2 \sin 2x - 2x\cos 2x) \right]$

20. $(D^2 - 4D + 3)y = 2xe^{3x} + 3e^x \cos 2x$

$$\left[Ans. \quad y = c_1 e^x + c_2 e^{3x} + \frac{e^{3x}}{2}(x^2 - x) - \frac{3e^x}{8}(\cos 2x + \sin 2x) \right]$$

21. $(D^2 + 5D + 6)y = e^{-2x}(\sec^2 x)(1 + 2\tan x)$ $\qquad \left[Ans. \quad y = c_1 e^{-2x} + c_2 e^{-3x} + e^{-2x} \tan x \right]$

22. $(D^2 + 4)y = e^x \sin^2 x$

$$\left[Ans. \quad y = c_1 \cos 2x + c_2 \sin 2x + \frac{e^x}{2}\left(\frac{1}{5} - \frac{1}{17}(4\sin 2x + \cos 2x) \right) \right]$$

23. $\dfrac{d^2x}{dt^2} + 2n\cos\alpha \dfrac{dx}{dt} + n^2 x = a\cos nt,\ x - 0$ and $\dfrac{dx}{dt} = 0$

when $t = 0$

$$\left[Ans. \quad x = e(-n\cos\alpha)t + \left[-\frac{a}{n^2 \sin 2\alpha x} \cdot \sin(n\sin\alpha)t \right] + \frac{a}{2n^2 \cos\alpha}\sin nt \right]$$

24. $(D^4 + D^2 + 1)y = e^{-\frac{x}{2}} \cos \frac{\sqrt{3}}{2} x$

$$\left[\textit{Ans.} \quad y = \text{Complementary function} + \frac{1}{4\sqrt{3}} x e^{-\frac{x}{2}} \left(\sin \frac{\sqrt{3}}{2} x + \sqrt{3} \cos \frac{\sqrt{3}}{2} x \right) \right]$$

Application of linear second order differential equations to electrical and mechanical systems

(I) Electric Circuits

In this section we consider linear differential equations that govern the flow of electricity in an electric circuit which consists of

(i) a source of electromotive force (e.m.f) E.
E(volts) $\equiv$ Voltage, (perhaps a battery or generator) which drives electric charge and produces a current I.

(ii) a resistance, R
R(ohms) $\equiv$ Resistance, which opposes current by producing a drop in e.m.f of magnitude $E_1 \equiv RI$, this equation is called ohm's law.

(iii) an inductor of inductance L
L(henrys) $\equiv$ Inductance, which opposes any change in the current by producing a drop in e.m.f of magnitude $E_2 \equiv L\frac{dI}{dt}$.

(iv) a condenser of capacitance, C
C(farads) $\equiv$ Capacitance, which stores a charge Q,
Q(coloumbs) $\equiv$ charge, quantity of electricity. The charge accumulated by the capacitor resists the inflow of additional charge and the drop in e.m.f arising in this way is $E_3 \equiv \frac{1}{C}Q$. (Such a circuit is called an *LCR* circuit).

Note: The elements of a circuit, the units and the names are given below.

t (seconds)	$\equiv$	Time
Q (coloumbs)	$\equiv$	Quantity of electricity, charge
I (amperes)	$\equiv$	Current, rate of change of Q
E (volts)	$\equiv$	Voltage, (e.m.f.)
R (ohms)	$\equiv$	Resistance

L	(henrys)	$\equiv$	Inductance
C	(farads)	$\equiv$	Capacitance

Also Note: $E_1 \equiv RI; \quad E_2 \equiv L\dfrac{dI}{dt}; \quad E_3 \equiv \dfrac{Q}{C}$

Statements of Kirchoff's laws:

(1) The algebraic sum of all the currents flowing into a junction is equal to zero.

(2) The algebraic sum of all the potential drops in a closed electric circuit is equal to zero.

Now, from the second law, we see that the sum of the voltage drops in a circuit is equal to the total e.m.f.

We also know that

(i) Voltage drop across a resistance of R ohms $\Big\}$ $\equiv RI \equiv (E_2)$

(ii) Voltage drop across a coil of inductance L henrys $\Big\}$ $\equiv L\dfrac{dI}{dt} \equiv (E_1)$

(iii) Voltage drop across a condenser of capacitance C farads $\Big\}$ $\equiv L\dfrac{Q}{C} \equiv (E_3)$

Since the current is the rate of flow of charge, the rate at which charge builds up on the capacitor is given by $I = \dfrac{dQ}{dt}$ [1]

By Kirchoff's law the algebraic sum of electromotive forces around a closed circuit is zero.

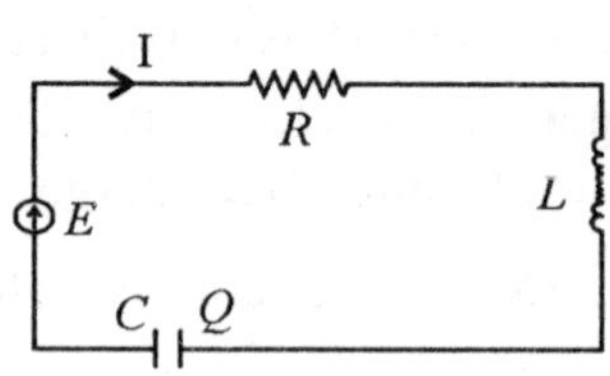

Hence, it follows that

$E \equiv E(t) \equiv \text{e.m.f} \equiv \text{Sum of voltage drops} \equiv E_1 + E_2 + E_3$

$$\Rightarrow \quad E - E_1 - E_2 - E_3 \equiv 0$$

$$\therefore \quad E - RI - L\frac{dI}{dt} - \frac{1}{C}Q \equiv 0$$

$$\therefore \quad L\frac{dI}{dt}+RI+\frac{Q}{C}\equiv E, \text{ or } E(t). \tag{2}$$

or, since $I=\left(\frac{dQ}{dt}\right),\left(\frac{dI}{dt}\right)=\left(\frac{d^2Q}{dt^2}\right)$

$$\therefore \quad L\left(\frac{d^2Q}{dt^2}\right)+R\left(\frac{dQ}{dt}\right)+\left(\frac{Q}{C}\right)=E, \text{ or } E(t). \tag{3}$$

Again, differentiating [2] w.r.t '*t*'

We get, $L\left(\frac{d^2I}{dt^2}\right)+R\left(\frac{dI}{dt}\right)+\left(\frac{1}{C}\right)I=\left(\frac{dE}{dt}\right).$ [4]

Oscillatory Electrical Circuit

1. *LC* circuit

In *LC* circuit there contains an inductance *L* and a capacitance *C* (Figure 1). since there is no e.m.f. in the circuit, therefore, by Kirchhoff's law of voltage.

we have $L\frac{d^2I}{dt^2}+\frac{Q}{C}=0 \Rightarrow \frac{d^2Q}{dt^2}+\frac{Q}{LC}=0$ [1]

which represent free electrical oscillations of the current having period $=2\pi\sqrt{LC}$.

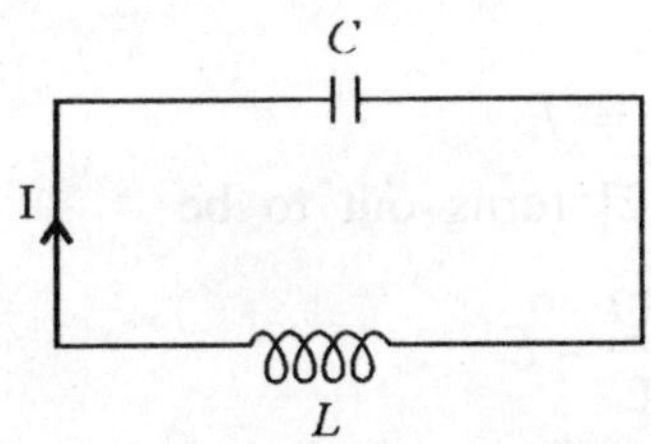

Figure (1) LC-circuit

2. *LCR* circuit

Consider the *LCR* circuit as given in Figure 2

When the switch is close we have by Kirchoff's law of voltage,

$$L\frac{d^2Q}{dt^2}+\frac{dQ}{dt}+\frac{Q}{C}=0 \tag{2}$$

The initial conditions of the problem are

$$Q=Q_0,\ I=\frac{dQ}{dt}=0 \text{ at } t=0$$

Equation [2] is similar to the equation of damped spring mass free oscillations.

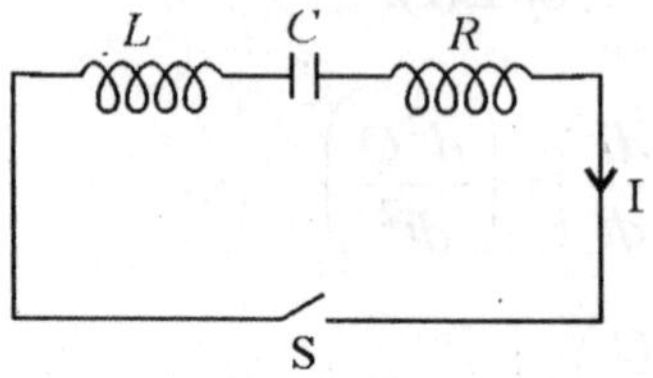

Figure (2)

3. *LC* circuit with e.m.f. *E*

In this case, Equation [1] turns out to be (Figure 3)

$$\frac{d^2Q}{dt^2}+\frac{Q}{LC}=\frac{E}{L} \qquad [3]$$

Equation [3] is similar to the equation of an undamped spring mass forced oscillations.

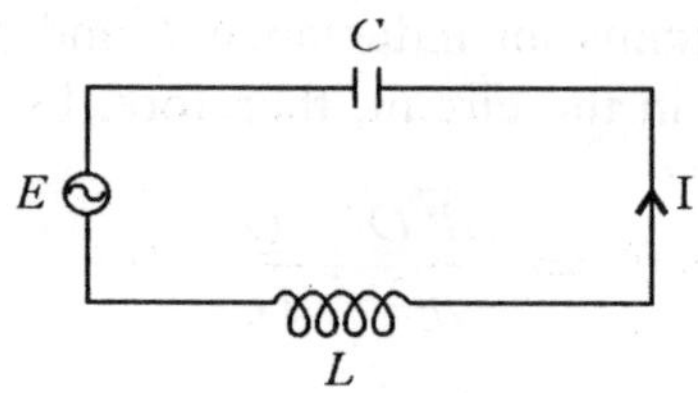

Figure (3)

4. *LCR* circuit with e.m.f. = *E*

In this case, Equation [2] turns out to be

$$L\frac{d^2Q}{dt^2}+R\frac{dQ}{dt}+\frac{Q}{C}=E \qquad [4]$$

The initial conditions are now $Q=0, \dfrac{dQ}{dt}=0$ at $t=0$.

Complementary function of [4] represent the free oscillations of the circuit whereas the particular integral represent the forced oscillations.

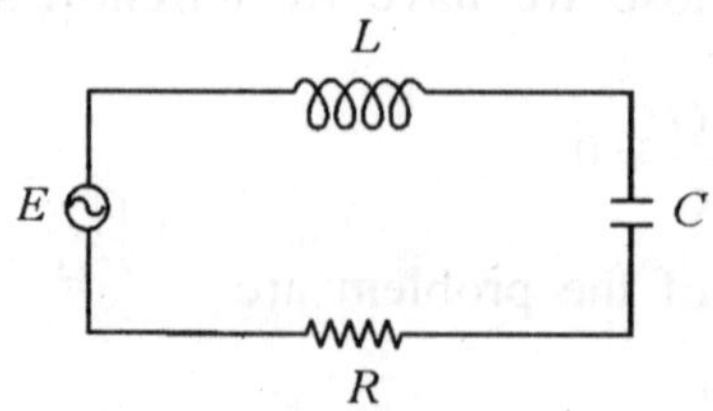

Figure (4)

Some important results.

Since $E \equiv E(t)$, we have

a) $L\dfrac{dI}{dt} + RI + \dfrac{Q}{C} = E$

b) $L\dfrac{d^2Q}{dt^2} + \dfrac{Q}{C} = E$

c) $L\dfrac{d^2I}{dt^2} + R\dfrac{dI}{dt} + \dfrac{1}{C}I = \dfrac{dE}{dt}$

If the current is charging the condenser then $I = (dQ/dt)$, where Q is the charge on the condenser. If the current is caused by the condenser discharging, then $I = -(dQ/dt)$, we know the standard results,

(i) $Lt\, e^t = \infty$, $Lt\, e^{-t} = 0$; *as* $t \to \infty$

(ii) $Lt\, e^{-at}(\sin bt) = 0$, and $Lt(e^{-at}\cos bt) = 0$, *as* $t \to \infty$. In the solution to certain problems we get results $Q = e^{-kt}[A\cos nt + B\sin nt] + f(t)$.

Here, as $t \to \infty$, the first part tends to be zero by 8.7.

So, very quickly, it becomes negligible. This part is called the *transient part*. After a while, the first part can be ignored, only the second part is significant, i.e., $Q = j(b)$. This is called the 'steady state' part of the solution.

(H) EXAMPLES

1. Solve $L\dfrac{dI}{dt} + RI = E_0$ with the following initial conditions. I_0 is the current flow initially. E_0 is the constant e.m.f. impressed on the circuit at $t = 0$.

Solution: $L\dfrac{dI}{dt} + RI = E_0$

$$\therefore \quad \frac{dI}{E_0 - RI} = \frac{dt}{L}$$

$$\therefore \quad -\frac{1}{R}\log(E_0 - RI) = \frac{t}{L} + C$$

$$\therefore \quad \log(E_0 - RI) = \frac{Rt}{L} + CR.$$

Initially, (i.e., when $t=0$), $I=I_0$.

$$\therefore \quad \log(E_0 - RI_0) = CR$$

$$\therefore \quad \log(E_0 - RI) = \frac{Rt}{L} + \log(E_0 - RI_0)$$

$$\therefore \quad E_0 - RI = (E_0 - RI_0)e^{-\frac{Rt}{L}}$$

$$\therefore \quad I = \frac{E_0}{R} + \left(I_0 - \frac{E_0}{R}\right)e^{-\frac{Rt}{L}}.$$

2. An electromotive force $E\cos\omega t$ is applied at $t=0$ to a circuit consisting of capacitance C and inductance L in series. The initial charge and current are zero. Find the current at time t.

Solution: We obtain the differential equation of the figure by using Kirchoff's law of voltage as:

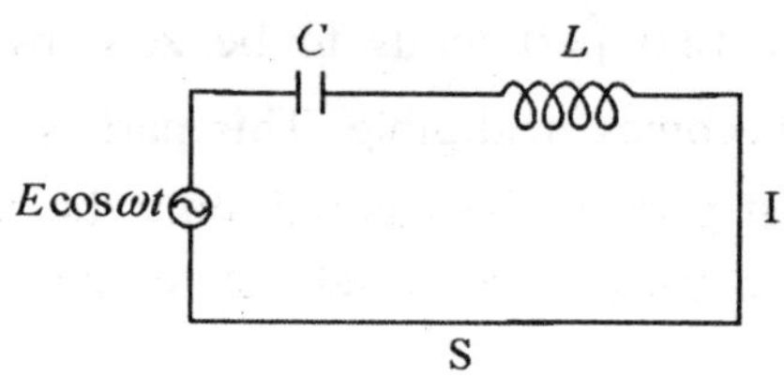

Figure Ex. 2

$$\frac{d^2Q}{dt^2} + \frac{Q}{LC} = \frac{E\cos\omega t}{L}$$

$$\Rightarrow \quad \frac{d^2Q}{dt^2} + a\cdot Q = b\cos\omega t \qquad [1]$$

$$\left(\text{where } a = \frac{1}{LC};\quad b = \frac{E}{L}\right)$$

Initial charges are: $Q=0$, $\frac{dQ}{dt}=0$ at $t=0$

The auxiliary equation of [1] is

$$(m^2 + a) = 0$$

$$\Rightarrow \quad m = \pm\sqrt{a}\,i$$

$\Rightarrow$ $\boxed{\text{Complementary function} = c_1 \cos\sqrt{a}t + c_2 \sin\sqrt{a}t}$

To find particular integral P.I: $\dfrac{1}{D^2+a}(b\cos\omega t)$

$$= \frac{b}{-\omega^2+a}(\cos\omega t) \qquad \left(D = \frac{d}{dt}\right)$$

$\therefore$ General solution of [1] is:

$$Q = \left[c_1 \cos\left(\sqrt{a}\right)t + c_2 \sin\left(\sqrt{a}\right)t\right] + \frac{b}{-\omega^2+a}\cos\omega t$$

by the given conditions, we obtain:

$$0 = c_1 + \frac{b}{-\omega^2+a} \Rightarrow c_1 = \frac{-b}{-\omega^2+a}$$

Again, $\dfrac{dQ}{dt} = -c_1\sqrt{a}\sin\sqrt{a}t + c_2\cos\sqrt{a}t - \dfrac{b\omega}{-\omega^2+a}\sin\omega t$

$$0 = 0 + c_2\sqrt{a} + 0 \Rightarrow c_2 = 0$$

$\therefore$ $$Q = \frac{b}{-\omega^2+a}\left(-\cos\left(\sqrt{a}\right)t + \cos\omega t\right)$$

$\therefore$ current at time 't' is

$$I = \frac{dQ}{dt} = \frac{b}{-\omega^2+a}\left(\sqrt{a}\sin\left(\sqrt{a}\right)t - \omega\cdot\sin\omega t\right)$$

where $a = \dfrac{1}{LC}$ and $b = \dfrac{E}{L}$.

3. For a certain circuit $L = 1, R = 40, C = \left(\dfrac{1}{20{,}000}\right)$. **Find the steady state current,** when $E = e\cos(150t)$.

Solution: The equation for the current deduced by Kirchoff's laws.

$$L\frac{d^2I}{dt^2} + R\frac{dI}{dt} + F\frac{1}{C}I = \frac{dE}{dt} \qquad [1]$$

$$\left.\begin{aligned} &\text{Here, } L = 1, R = 40, \left(\frac{1}{C}\right) = 20{,}000 \\ &E = 120\cos 150t; \left(\frac{dE}{dt}\right) = -18000\sin 150t \end{aligned}\right\} \qquad [2]$$

$$\therefore \quad \frac{d^2I}{dt^2}+40\frac{dI}{dt}+20{,}000I=-18000\sin 150t. \qquad [3]$$

Solving this equation by methods of linear differential equation (LDE) with constant coefficients,

$$I=e^{-20t}(A\cos 140t+B\sin 140t)-\left(\frac{86}{13}\right)\sin(150t+\alpha);\ \tan\alpha=\left(\frac{12}{5}\right).$$

The complementary function is transient, and so the steady state current is given by

$$I=-\left(\frac{86}{13}\right)[\sin(150t+\alpha)].$$

4. An inductor of 2 henrys, resistor of 16 ohms and capacitor of 0.02 microfarads are connected in series with a battery of e.m.f. $E=100\sin 3t$. At $t=0$, the charge on the capacitor and current in the circuit are zero. Find the charge and current at $t>0$.

Solution: We know that the differential equation of LCR circuit is given by

$$L\frac{d^2Q}{dt^2}+R\frac{dQ}{dt}+\frac{Q}{C}=E. \qquad [1]$$

$$\therefore \quad 2\frac{d^2Q}{dt^2}+16\frac{dQ}{dt}+\frac{Q}{0.02}=100\sin 3t$$

$$\Rightarrow \quad \frac{d^2Q}{dt^2}+8\frac{dQ}{dt}+25Q=50\sin 3t.$$

The auxiliary equation is: $(m^2+8m+25)=0$.

$$\Rightarrow \quad m=\frac{-8\pm\sqrt{64-100}}{2}=-4\pm 3i.$$

Hence, complementary function $=e^{-4t}(c_1\cos 3t+c_2\sin 3t)$.

To find particular integral P.I: $\left(\frac{1}{D^2+8D+25}\right)50\sin 3t$

$$=\left(\frac{50}{-9+8D+25}\right)\sin 3t=\left(\frac{50}{8D+16}\right)\sin 3t$$

$$=\frac{25}{4}\left(\frac{D-2}{D^2-4}\right)\sin 3t=-\frac{25}{52}(3\cos 3t-2\sin 3t)$$

$$=\frac{25}{52}(2\sin 3t-3\cos 3t).$$

$\therefore$ General solution: $Q = e^{-4t}(c_1 \cos 3t + c_2 \sin 3t) + \frac{25}{52}(2\sin 3t - 3\cos 3t)$.

$$\therefore \quad \frac{dQ}{dt} = -4e^{-4t}(c_1 \cos 3t + c_2 \sin 3t) + e^{-4t}(-3c_1 \sin 3t - 3c_2 \cos 3t) + \frac{25}{52}(6\sin 3t + 9\cos 3t)$$

Initially when $t = 0$, $Q = 0$ and $\frac{dQ}{dt} = 0$.

$\therefore$ we have $c_1 - \frac{75}{52} = 0$

and $3c_2 - 4c_1 + \frac{150}{52} = 0$.

On solving we get $c_1 = \frac{75}{52}$ and $c_2 = \frac{50}{52}$.

$$\therefore \quad Q = \frac{25}{52}e^{-4t}(3\cos 3t + 2\sin 3t) + \frac{25}{52}(2\sin 3t - 3\cos 3t)$$

and $I = \frac{dQ}{dt} = \frac{75}{52}(2\cos 3t + 3\sin 3t) - \frac{25}{52}e^{-4t}(17\sin 3t - 6\cos 3t)$.

5. A circuit consists of an inductance of 0.05 henrys, a resistance of 20 ohms and a condenser of capacitance 100 microfarads, and an e.m.f. of 100 volts. If initially $Q = 0$ and $I = 0$, when $t = 0$, find Q and I at any time.

Solution: $L\frac{d^2Q}{dt^2} + R\frac{dQ}{dt} + \frac{Q}{C} = E$; $\quad \frac{1}{20}\frac{d^2Q}{dt^2} + 20\frac{dQ}{dt} + 10{,}000Q = 100$

or $\left(\frac{d^2Q}{dt^2}\right) + 400\frac{dQ}{dt} + 200{,}000Q = 2000 \qquad [1]$

Solving the equation by methods of LDE with constant coefficients we get

$$Q = e^{-200t}[A\cos(400t) + B\sin s(400t)] + 0.01 \qquad [2]$$

$$I = \left(\frac{dQ}{dt}\right) = e^{-200t}[(2B - A)\cos 400t - (2A + B)\sin 400t]200 \qquad [3]$$

Using initial conditions, $I = 0$, $Q = 0$, $t = 0$, in [2] and [3],

$$\left.\begin{aligned} 0 &= 1.(A + B.\,0) + 0.01; \quad A = -0.01 \\ 0 &= 200[(2B - A) + 0] \quad B = -0.005 \end{aligned}\right\}$$

$$Q = e^{-200t}[0.01\cos(400t) + 0.005\sin(400t)] + 0.01$$

$$I = e^{-200t}[0-(-0.025)\sin 400t]200 = 5e^{-200t}\sin(400t)$$

I is transient (and $\to 0$), becomes negligible quickly.

Q is practically, equal to (0.01), the first part is negligible.

6. A circuit consists of an inductance L, a condenser of capacitance C in series. An alternating e.m.f. $\equiv e \sin nt$ is applied to the circuit at time $t = 0$. Find initial charge as the condenser and the current are both equal to zero, when $t = 0$. Find the current at time t, given $C \cdot L,\ \omega^2 = 1$.

***Solution*:** The standard equation by Kirchoff's law is,

$$L\frac{d^2Q}{dt^2} + R\frac{dQ}{dt} + \frac{Q}{C} = E \qquad [1]$$

$$\left.\begin{array}{l}\text{Here, } R = 0, \text{(since there is no resistance)}, e \sin nt \\ \text{and } CL\omega^2 = 1. \quad \therefore \left(\dfrac{1}{CL}\right) = \omega^2\end{array}\right\} \qquad [2]$$

$$\therefore \quad L\frac{d^2Q}{dt^2} + 0 + \frac{Q}{C} = e \sin nt; \quad \frac{d^2Q}{dt^2} + \frac{1}{CL}Q = \frac{e}{L}\sin nt$$

i.e. $$\frac{d^2Q}{dt^2} + \omega^2 Q = \frac{e}{L}\sin nt \qquad [3]$$

Complementary function: $m^2 + \omega^2 = 0; \quad m = \pm i\omega$

$$\therefore \quad Q = (A\cos\omega t + B\sin s\,\omega t) \qquad [4]$$

Particular integral: $Q = \dfrac{1}{(D^2+\omega^2}\left(\dfrac{e}{L}\sin nt\right) = \dfrac{e}{C}\dfrac{\sin nt}{\omega^2 - n^2}; (\omega \neq n)$

$$\therefore \quad Q = (A\cos\omega t + B\sin s\,\omega t) + \left[\frac{e}{L}(\omega^2 - n^2)\right]\sin nt \qquad [5]$$

when $t = 0,\ Q = 0;\ \therefore\ 0 = A + 0 + 0;\ A = 0$

$$\therefore \quad I = \frac{dQ}{dt} = B\cdot\omega\cos\omega t + \frac{e}{L}\frac{n}{\omega^2 - n^2} : \cos nt$$

$$(t = 0, I = 0); \quad 0 = B\omega + \left[\frac{en}{L}(\omega^2 - n^2)\right]$$

$$\therefore \quad B\omega = \left[\frac{-en}{L(\omega^2 - n^2)}\right]; \quad \therefore\ I = \frac{en}{(\omega^2 - n^2)L}(\cos\omega t - \cos nt)$$

7. A capacitance of $4\cdot 10^{-3}$ farads is in series with a 25 ohm resistance and an e.m.f. of 20 cos $6t$. Determine the charge and current at any time, given that the charge is zero at $t=0$.

Solution: The equation obtained by Kirchoff's law is

$$L\frac{d^2Q}{dt^2}+R\frac{dQ}{dt}+\frac{1}{C}Q=E \qquad [1]$$

Here, $L=0,\ R=25,\ E=20\cos 6t;\ C=4\cdot 10^{-3}$

$$\therefore \quad 25\frac{dQ}{dt}+\frac{1000}{4}Q=20\cos 6t;\ 25\frac{dQ}{dt}+250Q=20\cos 6t$$

$$\frac{dQ}{dt}+10Q=\frac{4}{5}\cos 6t$$

$$\therefore \quad \text{solution: } Qe^{10t}=A+\int\left(\frac{4}{5}\right)e^{10t}\cos 6t\,dt$$

$$=A+\frac{4}{5}e^{10t}\,\frac{(10\cos 6t+6\sin 6t)}{(10^2+6^2)}$$

$$\therefore \quad Qe^{10t}=A+\left(\frac{1}{170}\right)e^{10t}(10\cos 6t+6\sin 6t)$$

or $\quad Q=Ae^{-10t}+\left(\frac{1}{85}\right)[5\cos 6t+3\sin 6t];$

When $t=0$, we have $Q=0$: $\therefore \quad 0=A+\left(\frac{1}{85}\right)(5+0)$: $A=\left(-\frac{1}{17}\right)$

$$\therefore \quad Q=\left(-\frac{1}{17}\right)e^{-10t}+\left(\frac{1}{85}\right)[5\cos 6t+3\sin 6t]$$

$$\therefore \quad I=\left(\frac{dQ}{dt}\right)=\left(-\frac{1}{17}\right)(-10)e^{-10t}+\left(\frac{1}{85}\right)[5t(-6\sin 6t)+3\cdot 6\cos 6t]$$

$$=\frac{10}{17}e^{-10t}+\frac{1}{85}(-30\sin 6t+18\cos 6t).$$

8. In an *LCR* circuit, the charge Q on a plate of a condenser is given by

$$L\frac{d^2Q}{dt^2}+R\frac{dQ}{dt}+\frac{Q}{C}=E\sin pt.$$

The circuit is turned to resonance so that $p^2 = \frac{1}{LC}$. If initially the current I and the charge Q are zero, show that for small values of $\frac{R}{L}$, the current in the circuit at time t is given by $\frac{Et}{2L}\sin pt$.

Solution: Given: $\left(LD^2 + RD + \frac{1}{C}\right)Q = E\sin pt.$ [1]

The auxiliary equation is $\left(Lm^2 + Rm + \frac{1}{C}\right) = 0.$

$$\Rightarrow \quad m = \frac{-R \pm \sqrt{R^2 - \frac{4L}{C}}}{2L} = \frac{-R}{2L} \pm \sqrt{\frac{R^2}{4L^2} - \frac{1}{LC}}.$$

Since $\frac{R}{L}$ is small, the term $\frac{R^2}{L^2}$ may be omitted.

Hence $m = -\frac{R}{2L} \pm \sqrt{-\frac{1}{LC}}$

$$= -\frac{R}{2L} \pm i\sqrt{\frac{1}{LC}} \quad = -\frac{R}{2L} \pm ip\left(\because\ p^2 = \frac{1}{LC} > 0\right).$$

$\therefore$ Complementary function $= e^{-\frac{Rt}{2L}}(c_1 \cos pt + c_2 \sin pt)$

$$= \left(1 - \frac{Rt}{2L} + \frac{R^2t^2}{8L^2} + \cdots\right)(c_1 \cos pt + c_2 \sin pt)$$

$$= \left(1 - \frac{Rt}{2L}\right)(c_1 \cos pt + c_2 \sin pt) \qquad \left(\text{omitting higher power of } \frac{R}{L}\right)$$

Particular integral $= \left(\frac{1}{LD^2 + RD + \frac{1}{C}}\right)(E\sin pt)$

$$= E\left[\frac{1}{L(-p^2) + RD + \frac{1}{C}}\sin pt\right]$$

$$= \frac{E}{RD}(\sin pt) \qquad \left(\because P^2 = \frac{1}{LC}\right)$$

$$= \frac{E}{R}\int \sin pt\, dt \quad = -\frac{E}{Rp}\cos pt.$$

$\therefore$ the complete solution is given by

$$Q = \left(1 - \frac{Rt}{2L}\right)(c_1 \cos pt + c_2 \sin pt) - \frac{E}{Rp}\cos pt \qquad [2]$$

$$\therefore \quad I = \frac{dQ}{dt} = \left(1 - \frac{Rt}{2L}\right)(-c_1 \sin pt + c_2 \cos pt)p$$

$$-\frac{Rt}{2L}(c_1 \cos pt + c_2 \sin pt) + \frac{E}{R}\sin pt$$

Initially when $t = 0$, $Q = 0$, $I = 0$.

From [1] we get $c_1 = \frac{E}{Rp}$

and from [2] we get $c_2 p - \frac{Rc_1}{2L} = 0$.

$$\therefore \quad c_2 = \frac{E}{2Lp^2}.$$

Substituting c_1 and c_2 in [2] we get,

$$I = \left(1 - \frac{Rt}{2L}\right)\left(-\frac{E}{Rp}\sin pt + \frac{E}{2Lp^2}\cos pt\right)p$$

$$-\frac{R}{2L}\left(-\frac{E}{Rp}\cos pt + \frac{E}{2Lp^2}\sin pt\right) + \frac{E}{R}\sin pt$$

$$= \frac{Et}{2L}\sin pt \qquad \left(\because \frac{R}{L}\text{ is small}\right).$$

9. An electric circuit consists of an inductance 0.1 henry, a resistance of 20 ohms, and a condenser of capacitance 25 microfarads. Find the charge Q and the current I at time t, given that the initial conditions Q = 0.05 coloumbs, and I = 0 when t = 0. (1 farad = 10^6 microfarads)

Solution: The given governing differential equation is

$$L\frac{d^2Q}{dt^2} + R\frac{dQ}{dt} + \frac{Q}{C} = E(t). \qquad [1]$$

Here, $R = 20,\ L = 0\cdot 1,\ C = 25\cdot 10^{-6};\ E(t) = 0$ [2]

Also, initial conditions $\equiv Q = 0.05;\ I = 0,\ t = 0$ [3]

$$\therefore \quad \frac{1}{10}\frac{d^2Q}{dt^2} + 20\frac{dQ}{dt} + \frac{Q}{(25)10^{-6}} = 0$$

or $$\frac{d^2Q}{dt^2} + 200\frac{dQ}{dt} + 400{,}000\,Q \equiv 0 \quad [4]$$

To solve this differential equation

$$m^2 + 200m + 400{,}000 = 0; \quad m = (-100 \pm i100\sqrt{39})$$

$$\therefore \quad Q = e^{-100t}[A\cos(100\sqrt{39})t + B\sin(100\sqrt{39})t] \quad [5]$$

From [3], $Q = 0.05$, when $t = 0$.

$$\therefore \quad 0.05 = e^0[A\cos 0 + B\sin 0] = A;\ \therefore A = 0.05 \quad [6]$$

Also differentiating [5], we get $I = \dfrac{dQ}{dt}$

$$= 100e^{-100t}[(B\sqrt{39} - A)\cos(100\sqrt{39})t + B\sin(100\sqrt{39})t] \quad [7]$$

[5] and [7] give the charge and current at any time.

Again $I = 0$, when $t = 0$, by [3]. Substitute in [7].

$$0 = 100[(B\sqrt{39} - A) - 0];\ B = \left(\frac{A}{\sqrt{39}}\right) = \left(\frac{.05}{\sqrt{39}}\right) \quad [8]$$

So Q and I are given by [5] and [7], with [6] and [8].

It is observed that $e^{-x} \to 0$ as $x \to \infty$; hence both Q and I are transients; each becomes negligibly small very quickly.

EXERCISES

1. An e.m.f. $E\sin pt$ is applied at $t = 0$ to a circuit containing a capacitance C and inductance L. The current I satisfies the equation $L\dfrac{dI}{dt} + \dfrac{1}{C}Idt = E\sin pt$. If $p^2 = \dfrac{1}{LC}$ and initially there is no charge and no current in the circuit, show that the current at time t is $\left(\dfrac{Et}{2L}\right)\sin pt$ where $I = \dfrac{dQ}{dt}$.

2. The differential equation for a circuit in which self inductance and capacitance neutralize each other is $L\dfrac{d^2I}{dt^2} + \dfrac{1}{C} = 0$. Find the current I as a function of t given that I_0 is the maximum current and $I = 0$ when $t = 0$.

$$\left[\textit{Ans.}\quad I = I_0\sin\left(\frac{t}{\sqrt{LC}}\right)\right]$$

3. A constant e.m.f. E at $t = 0$ is applied to a circuit consisting of inductance L, resistance R and capacitance C in series. The initial values of the current and the charge being zero, find the current at any time t, if $CR^2 < 4L$.

$$\left[\textit{Ans.}\quad I = 2Eke^{-\frac{RT}{2L}}\sin\left(\frac{kt}{2L}\right), \text{ where } k = \sqrt{\frac{4L - CR^2}{C}}\right]$$

4. A circuit consists of an inductance of 0.05 henry, a resistance of 20 ohms, a condenser of capacitance 100 microfarads and e.m.f. of E = 100 volts. Find I and Q, given the initial conditions $Q = 0$, $I = 0$ when $t = 0$.

$$\left[\textit{Ans.}\quad Q = e^{-200t}(-0.01\cos 400t - 0.005\sin 400t) + 0.01 \text{ and } I = 5e^{-200t}\sin 400t\right]$$

5. A condenser of capacity C discharged through an inductance L and resistance R in series and the charge Q at time t satisfies the equation $L\dfrac{d^2Q}{dt^2} + R\dfrac{dQ}{dt} + \dfrac{Q}{C} = 0$. Given that L = 0.25 henry, R = 250 ohms, $C = 2\times 10^{-6}$ farads, and that when $t = 0$, charge Q is 0.002 coulombs and the current $\dfrac{dQ}{dt} = 0$ obtain the value of Q in terms of t.

$$\left[\textit{Ans.}\quad Q = e^{-500t}(0.002\cos 1323t + 0.0008\cdot\sin 1323t)\right]$$

6. Given that $L = 1H$, $R = 1200\,\Omega$, $C = 10^{-6}f$, $i(0) = q(0) = 0$ *and* $E = 100\sin 600tV$, determine the transient current and the steady-state current.

$$\left[\begin{aligned} \textit{Ans.}\quad I_{\text{transient}} &= \frac{e^{-600t}}{2320}(-153\sin 800t - 96\cos 800t);\\ I_{\text{steady state}} &= \frac{1}{580}(24\cos 600t + 27\sin 600t)\end{aligned}\right]$$

7. Suppose in a $R\,L\,C$ circuit, $L = 1$ henry (H), $R = 100$ ohms (Ω), $C = 10^{-4}$ farad (f) and $E = 100$ volts (V) are taken and there is no charge present and no current is flowing at time $t = 0$. Determine an expression for charge and an expression for current.

$$\left[\begin{aligned} \textit{Ans.}\quad Q &= \frac{1}{10} - \frac{1}{10\sqrt{3}}e^{-50t}(\sin 50\sqrt{3}\,t + \sqrt{3}\cos 50\sqrt{3}\,t)\\ I &= \frac{20}{\sqrt{3}}e^{-50t}\sin 50\sqrt{3}\,t\end{aligned}\right]$$

II. Mechanical systems—motion of a particle in resisting medium

In this section, we will discuss the motion of a particle moving in a straight line or in a plane in resisting medium.

(i) Velocity and acceleration of a moving point.

Consider the case of a particle moving in a straight line. If the distance of the particle from a fixed point O in the straight line at time 't' be equal to S, we have,

(a) Velocity = v = rate of displacement $=\left(\frac{dS}{dt}\right)$

(b) Acceleration = rate of change of velocity $=\left(\frac{dv}{dt}\right)$

$$=\frac{d}{dt}\left(\frac{dS}{dt}\right)=\frac{d^2S}{dt^2}$$

$$=\frac{dv}{dt}=\frac{dv}{dS}\cdot\frac{dS}{dt}=\frac{dv}{dS}\cdot v=v\frac{dv}{ds}$$

(ii) Deflection (bending) of beams

The bending of beam is another problem which can be formulated using differential equations.

Let $y=\text{deflection}=f(x)$

$\theta=\frac{dy}{dx}=\text{slope}$

E = Young's Modulus

I = Moment of Inertia about the axis of bending

M_x = Bending moment

V_x = Shear force

W_x = Intensity of loading

R = Radius of curvature at (x, y)

For small deflection, $\frac{d^2y}{dx^2}=\frac{I}{R}$

Also, $M_x=EI\frac{d^2y}{dx^2}$ [1]

$$V_x=\frac{dM_x}{dx}=EI\cdot\frac{d^3y}{dx^3} \quad [2]$$

$$W_x = \frac{dV_x}{dx} = EI \cdot \frac{d^4 y}{dx^4} \qquad [3]$$

The sum of the moments about a section NN' due to external forces on the left of the section, if anticlockwise, is taken as positive and, if clockwise, is taken as negative. (see figure below).

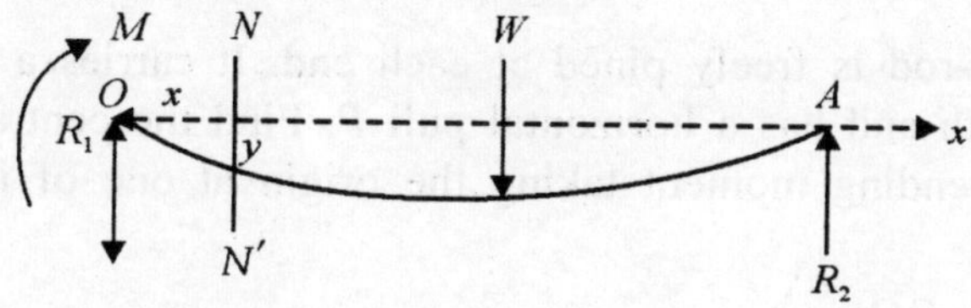

For the differential Equation [1], the arbitrary constants are found by using the end conditions.

For example, (a) There is no deflection and no bending moment at a freely supported end $\left(\text{i.e., } y = 0, \text{ and } \frac{d^2 y}{dx^2} = 0\right)$

(b) There is no deflection and the slope of the beam at a horizontal fixed end $\left(\text{i.e., } y = 0, \text{ and } \frac{dy}{dx} = 0\right)$.

(iii) Spring-Mass System: Free Motion

Consider the simple spring mass system shown in the figure below.

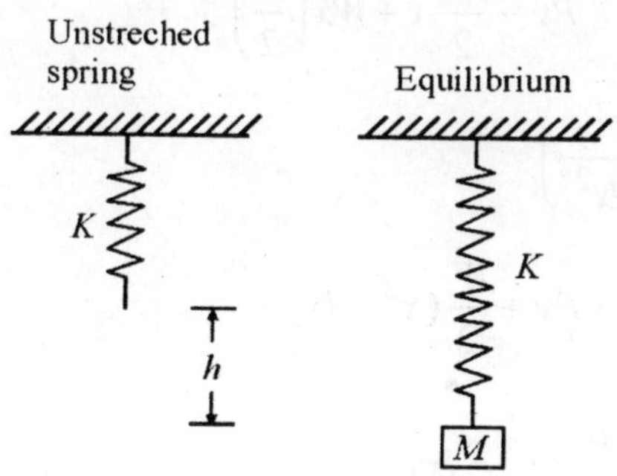

Spring-Mass System

We assume:

1. The mass m in kg is constrained to move in the vertical direction only.
2. The viscous damping C, with units of kg/s, is proportional to the velocity $\frac{dy}{dt}$.

3. The force in the spring is Kd where K = spring modulus (N/m) and d = distance in metres from the unstretched position.
4. The mass of the spring is negligible as compared to m.
5. No external forces act on the system.

(I) EXAMPLES

1. A horizontal tie-rod is freely pined at each end. It carries a uniform load W lb per unit length and has a horizontal pull P, Find the central deflection and the maximum bending moment taking the origin at one of its ends.

Solution: Let $OQ = x$.

The forces acting on the segment OQ are the horizontal pull P, the reaction $R = \dfrac{Wl}{2}$ and the weight Wx at the midpoint of OQ.

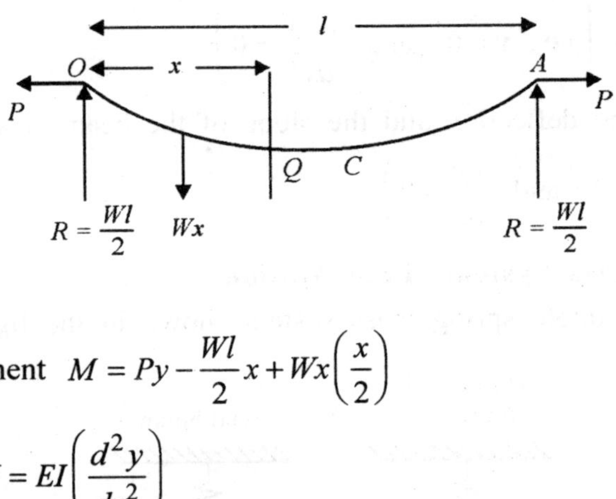

$\therefore$ the moment $M = Py - \dfrac{Wl}{2}x + Wx\left(\dfrac{x}{2}\right)$

We have $M = EI\left(\dfrac{d^2y}{dx^2}\right)$

Hence $\quad EI\left(\dfrac{d^2y}{dx^2}\right) = Py + \dfrac{W}{2}(x^2 - lx)$

i.e., $\quad EI\dfrac{d^2y}{dx^2} - Py = \dfrac{W}{2}(x^2 - lx)$

$$\boxed{\frac{d^2y}{dx^2} - a^2y = \frac{W}{2EI}(x^2 - lx)} \qquad [1]$$

where $\quad a^2 = \dfrac{P}{EI}$.

$\therefore$ The auxiliary equation is $m^2 - a^2 = 0$.

$\therefore$ $m = \pm a$.

$\therefore$ Complementary function $= \lambda e^{ax} + \mu e^{-ax}$

$$= c_1 \cosh ax + c_2 \sinh ax$$

$$\left(\because \sinh ax = \frac{e^{ax} - e^{-ax}}{2} \text{ and } \cosh ax = \frac{e^{ax} + e^{-ax}}{2}\right)$$

Particular integral $= \dfrac{1}{D^2 - a^2}\left[\dfrac{W}{2EI}(x^2 - lx)\right]$

$$= \frac{W}{2EIa^2}\left(1 - \frac{D^2}{a^2}\right)^{-1} (x^2 - lx)$$

$$= \frac{W}{2P}\left(1 + \frac{D^2}{a^2}\right)(x^2 - lx) \text{ (omitting higher powers of D)}$$

$$= \frac{W}{2P}\left(x^2 - lx + \frac{2}{a^2}\right).$$

$\therefore$ the complete solution of [1] is

$$y = c_1 \cosh ax + c_2 \sinh ax - \frac{W}{2P}\left(x^2 - lx + \frac{2}{a^2}\right). \qquad [2]$$

At the end O, $y = 0$ when $x = 0$. (we do not need the other condition $\dfrac{dy}{dx} = 0$.)

$\therefore$ [2] gives $0 = c_1 - \dfrac{W}{Pa^2}$.

$$\therefore \quad c_1 = \frac{W}{Pa^2}. \qquad [3]$$

At the end A, $y = 0$ when $x = l$.

$\therefore$ [2] gives, $0 = c_1 \cosh al + c_2 \sinh al - \dfrac{W}{Pa^2}$

$$\therefore \quad c_2 \sinh al = \frac{W}{Pa^2}(1 - \cosh al)$$

$$\therefore \quad c_2 = -\frac{W}{Pa^2}\tanh\left(\frac{al}{2}\right). \qquad [4]$$

Substituting c_1 and c_2 in [2] we get,

$$y = \frac{W}{Pa^2}\left(\cosh ax - \tanh\left(\frac{al}{2}\right)\sinh ax\right) - \frac{W}{Pa}\left(x^2 - lx + \frac{2}{a^2}\right)$$

which gives the deflection of the beam.

Thus the central deflection $= y\left(\text{at } x = \frac{l}{2}\right)$

$$= \frac{W}{Pa^2}\left(\cosh ax\left(\frac{al}{2}\right) - \tanh\left(\frac{al}{2}\right)\sinh\left(\frac{al}{2}\right) - 1\right) + \frac{Wl^2}{8P}.$$

$$= \frac{W}{Pa^2}\left(\frac{\cosh^2\left(\frac{al}{2}\right) - \sinh^2\left(\frac{al}{2}\right) - 1}{\cosh\left(\frac{al}{2}\right)}\right) + \frac{Wl^2}{8P}.$$

$$= \frac{W}{Pa^2}\left(\operatorname{sech}\frac{al}{2} - 1\right) + \frac{Wl^2}{8P}. \qquad [5]$$

Also the bending moment is maximum at the point of maximum deflection $\left(x = \frac{l}{2}\right)$.

$$(M_x)_{\max} = EI\frac{d^2y}{dx^2} \text{ at } x = \frac{l}{2}$$

$$= Py + \frac{W}{2}(x^2 - lx) \text{ at } x = \frac{l}{2}$$

$$= \frac{W}{a^2}\left(\operatorname{sech}\frac{al}{2} - 1\right) \qquad \text{from [5]}$$

2. A horizontal tie rod of length $2l$ with concentrated load W at the centre and ends freely hinged, satisfies the differential equation: $EI\frac{d^2y}{dx^2} = Py - \frac{W}{2}x$. With conditions $x = 0,\ y = 0$ and $x = l,\ \frac{dy}{dx} = 0$, prove that the deflection δ and the bending moment M at the centre ($x = l$) are given by $\delta = \frac{W}{2Pn}(nl - \tanh nl)$ and $m = -\frac{W}{2n}\tanh nl$, where $n^2EI = P$.

Solution: The given equation is rewritten as

$$\frac{d^2y}{dx^2} - \frac{Py}{EI} = -\frac{W}{2EI}\cdot x$$

$$\frac{d^2y}{dx^2} - a^2y = bx \text{ where } a^2 = \frac{P}{El} \text{ and } b = -\frac{W}{2EI} \tag{1}$$

The general solution of [1] is

$$y = Ae^{ax} + Be^{-ax} - \frac{b}{a^2}\cdot x \tag{2}$$

Given $x = 0,\ y = 0 \Rightarrow 0 = A + B \Rightarrow B = -A$

$\therefore$ Equation [2] is reduced to $y = 2A\left(\dfrac{e^{ax} - e^{-ax}}{2}\right) - \dfrac{b}{a^2}\cdot x = 2A \sinh ax - \dfrac{b}{a^2}\cdot x$

Now, at $x = l,\ \dfrac{dy}{dx} = 0,\ \Rightarrow 2A\cdot \cosh al\cdot a - \dfrac{b}{a^2} = 0$

$$\Rightarrow A = \frac{b\cdot \operatorname{sech} al}{2a^3}$$

$\therefore$ the particular solution is $y = \dfrac{b}{a^3}\cdot\dfrac{\sinh ax}{\cosh ax} - \dfrac{b}{a^2}\cdot x$

Now, the deflection at $x = l$ is

$$y(l) = \delta = \frac{b}{a^2}\left\{\frac{1}{a}\tanh al - l\right\} \tag{3}$$

$$= \frac{b}{a^3}\{\tanh al - al\} \tag{4}$$

$$\therefore \quad a = n = \sqrt{\frac{P}{EI}},\ b = -\frac{W}{2EI},\ El = \frac{P}{n^2} = -\frac{W}{2b},\ \therefore \frac{b}{a^3} = -\frac{W}{2Pn}$$

$$\therefore \text{ From (iv), } \delta = \frac{W}{2Pn}(nl - \tanh nl) \tag{5}$$

Now, bending moment at $x = l$ is

$$M = EI\left[\frac{d^2y}{dx^2}\right]_{x=l} = EI\frac{b}{a}\tanh al$$

$$= \frac{EIWn}{2P}\tanh nl \qquad \left(\frac{P}{n^2} = -\frac{W}{2b},\ n = a,\ l\frac{b}{n} = -\frac{Wn}{2P}\right)$$

$$= -\frac{P}{n^2}\cdot\frac{Wn}{2P}\tanh nl \qquad \left(\text{and } EI\frac{P}{n^2} = -\frac{W}{2b}\right)$$

$$= -\frac{W}{2n}\tanh nl.$$

EXERCISES

1. A uniform horizontal rod of length l freely supported at both ends, carries a uniformly distributed load W per unit length. If the thrust at each end is P, prove that the maximum deflection is

$$\frac{W}{Pa^2}\left(\sec\frac{al}{2}-1\right)-\frac{Wl^2}{8P} \text{ where } \frac{P}{El}=a^2.$$

Prove also that the maximum bending moment is of the magnitude $\frac{W}{a^2}\left(\sec\frac{al}{2}-1\right)$.

2. Find the equation of the elastic curve and its maximum deflection for the following horizontal beams.

(i) Cantilever beam of length l, having uniform load of W unit/length.

(ii) Built in or encastre beam (having both ends fixed) carrying a uniform load of W unit/length.

3. A horizontal beam of length $2l$ ft, carrying a uniform load of W 1bs, per foot of length, is freely supported at both ends, satisfying the differential equation.

$$EI\cdot\frac{d^2y}{dx^2}=\frac{1}{2}Wx^2-\frac{W}{x},$$

y being the deflection at a distance x from one end. If $y = 0$ at $x = 0$ and $y_1 = 0$ at $x = l$, find the deflection at any point and also find the maximum deflection.

$$\left[\textit{Ans.}\quad y=\frac{W}{24EI}(x^4-4lx^3-8l^3x),\ y_{\max}=\frac{5Wl^4}{24EI}\right]$$

4. A uniform horizontal rod of length l freely supported at both ends, carries a uniformly distributed load W per unit length. If the thrust at each end is P, prove that the maximum deflection is

$$\frac{W}{Pa^2}\left(\sec\frac{al}{2}-1\right)-\frac{Wl^2}{8P}, \text{ where } \frac{P}{El}=a^2.$$

1.3 *Z*-TRANSFORMS

A transform of sampled signal or sequence was defined in 1947 by W. Hurewicz. It was later denoted in 1952 as a 'Z-transform' by a sampled data control group at Colombia university.

The *Z*-transform is useful in solving difference equations which represent a discrete system. *Z*-transform is the discrete analogue of Laplace transform. *Z*-transform operates on a sequence u_n where n takes the discrete values 0,1,2,.... To every operational rule of Laplace transforms, there exists an analogue rule for *Z*-transform.

Definition: The *Z*-transform of a sequence u_n defined for discrete values $n = 0,1,2,\cdots$. (and $u_n = 0$ for $n < 0$) is denoted by $Z(u_n)$ and is defined as

$$Z(u_n) = \bar{u}(z) = \sum_{n=0}^{\infty} u_n z^{-n}, \qquad [1]$$

where $\bar{u}$ is a function of *Z*.

Z-transform exists only when the infinite series in [1] is convergent.

Inverse *Z*-transform: It is denoted by $z^{-1}(\bar{u}(z)) = u_n$, determines the sequence u_n which generates the given *Z*-transform.

(J) EXAMPLES

1. $u_n = \{a^n\} = \{1, a, a^2, a^3, \cdots a^n \cdots\}$

By definition,

$$Z(u_n) = Z(a^n) = \sum_{n=0}^{\infty} u_n z^{-n} = \sum_{n=0}^{\infty} a^n z^{-n}$$

$$= 1 + \frac{a}{z} + \frac{a^2}{z^2} + \frac{a^3}{z^3} + \ldots + \frac{a^n}{z^n} + \cdots = \frac{1}{1 - \left(\frac{a}{z}\right)}$$

$$\therefore \quad Z(a^n) = \frac{Z}{Z - a}$$

> **Note:** (i) Put $a = 1$, $z(1) = \left(\frac{z}{z-1}\right)$
>
> (ii) $a = -1$, $z(-1) = \left(\frac{z}{z+1}\right)$

2. Find the *Z*-transform of the following.

(i) $\frac{a^n}{n!}e^{-a}$ (ii) $\frac{1}{(n+1)!}$

***Solution*:** (i) $Z\left(\frac{a^n}{n!}\right)=\sum_{n=0}^{\infty}\left(\frac{a^n}{n!}e^{-a}\right)z^{-n}=e^{-a}\sum_{n=0}^{\infty}\frac{\left(\frac{a}{z}\right)^n}{n!}$

$$=e^{-a}\left[1+\frac{\left(\frac{a}{z}\right)}{1!}+\frac{\left(\frac{a}{z}\right)^2}{2!}+\frac{\left(\frac{a}{z}\right)^3}{3!}+\cdots\right]$$

$$=e^{-a}\cdot e^{\frac{a}{z}}$$

$$=e^{a}(z^{-1}-1)$$

(ii) $Z\left(\frac{1}{(n+1)!}\right)=\sum_{n=0}^{\infty}\frac{1}{(n+1)!}z^{-n}$

$$=\frac{1}{1!}+\frac{z^{-1}}{2!}+\frac{\left(z^{-1}\right)^2}{3!}+\frac{\left(z^{-1}\right)^3}{4!}+\cdots$$

$$=\frac{1}{z^{-1}}\left[\frac{z^{-1}}{1!}+\frac{\left(z^{-1}\right)^2}{2!}+\frac{\left(z^{-1}\right)^3}{3!}+\cdots\right]$$

$$=z\left[\left(1+\frac{z^{-1}}{1!}+\frac{\left(z^{-1}\right)^2}{2!}+\frac{\left(z^{-1}\right)^3}{3!}+\cdots\right)-1\right]$$

$$=z(e^{-z}-1).$$

3. Find the *Z*-transform of a^{n+3}

***Solution*:** $Z(a^{n+3})=Z(a^3\cdot a^n)$

$$=a^3Z(a^n)$$

$$=\frac{a^3z}{z-a}$$

4. Find the Z-transform of $\cos n\theta$ and $\sin n\theta$

***Solution*:** $Z(a^{n+3}) = Z(a^3 \cdot a^n)$

$$= \frac{z}{z - e^{-i\theta}}$$

$$= \frac{z(z - e^{i\theta})}{(z - e^{-i\theta})(z - e^{i\theta})}$$

$$= \frac{z^2 - ze^{i\theta}}{z^2 - z(e^{i\theta} + e^{-i\theta}) + 1}$$

$$= \frac{z^2 - z(\cos\theta + i\sin\theta)}{z^2 - 2z\cos\theta + 1}$$

$$Z(\cos n\theta - i\sin n\theta) = \frac{z(z - \cos\theta) - iz\sin\theta}{z^2 - 2z\cos\theta + 1}$$

Equating real and imaginary parts we get,

$$Z(\cos n\theta) = \frac{z(z - \cos\theta)}{z^2 - 2z\cos\theta + 1}$$

$$Z(\sin n\theta) = \frac{z\sin\theta}{z^2 - 2z\cos\theta + 1}.$$

Properties

P_1: *Z*-transform is linear

i.e., $Z(au_n + bv_n) = aZ(u_n) + bZ(v_n)$

Proof: $Z(au_n + bv_n) = \sum_{n=0}^{\infty} (au_n + bv_n) Z^{-n}$

$$= \sum_{n=0}^{\infty} au_n z^{-n} \sum_{n=0}^{\infty} bv_n z^{-n}$$

$$= a\sum_{n=0}^{\infty} u_n z^{-n} + b\sum_{n=0}^{\infty} v_n z^{-n}$$

$$= aZ(u_n) + bZ(v_n).$$

$\therefore \quad Z(au_n + bv_n) = aZ(u_n) + bZ(v_n)$

P$_2$: (Damping rule) If $Z(u_n) = \bar{u}(z)$, then

$$Z(a^{-n}u_n) = \bar{u}(az) \text{ and } Z(a^n u_n) = \bar{u}\left(\frac{z}{a}\right).$$

Proof: $Z(a^{-n}u_n) = \sum_{n=0}^{\infty} a^{-n}u_n z^{-n}$

$$= \sum_{n=0}^{\infty} u_n (az)^{-n}$$

$$= \bar{u}(az).$$

Similarly $Z(a^n u_n) = \bar{u}\left(\frac{z}{a}\right)$.

P$_3$: (Shifting theorem). Let $Z(u_n) = \bar{u}(z)$ and $k > 0$. Then

(i) $Z(u_{n-k}) = z^{-k}\bar{u}(z)$

(ii) $Z(u_{n+k}) = z^k\left[\bar{u}(z) - u_0 - u_1 z^{-1} - u_2 z^{-2} - \cdots - u_{k-1}z^{-(k-1)}\right]$

Proof: (i) $Z(u_{n-k}) = \sum_{n=0}^{\infty} u_{n-k} z^{-n}$

$$= u_0 z^{-k} - u_1 z^{-(k+1)} + \cdots$$

$$= z^{-k}\left[u_0 - u_1 z^{-1} - u_2 z^{-2} + \cdots\right]$$

$$= z^{-k}\sum_{n=0}^{\infty} u_n z^{-n}$$

$$= z^{-k}\bar{u}(z).$$

(ii) $Z(u_{n+k}) = \sum_{n=0}^{\infty} u_{n+k} z^{-n}$

$$= z^k \sum_{n=0}^{\infty} u_{n+k} z^{-(n+k)}$$

$$= z^k\left[\sum_{n=0}^{\infty} u_n z^{-n} - \sum_{n=0}^{k-1} u_n z^{-n}\right]$$

$$= z^k\left[\bar{u}(z) - u_0 - u_1 z^{-1} - u_2 z^{-2} - \cdots - u_{k-1}z^{-(k-1)}\right].$$

Remark. By putting $k = 1,2,3$ in (ii) of the above theorem, we get

$$Z(u_{n+1}) = z[\overline{u}(z) - u_0]$$

$$Z(u_{n+2}) = z^2[\overline{u}(z) - u_0 - u_1 z^{-1}]$$

and $$Z(u_{n+3}) = z^3[\overline{u}(z) - u_0 - u_1 z^{-1} - u_2 z^{-2}].$$

Taking limit as $z \to 1$ we get,

$$\lim_{z\to 1}\left[(z-1)\overline{u}(z)\right] - u_0 = \lim_{z\to 1}\sum_{n=0}^{\infty}(u_{n+1} - u_n)z^{-n}$$

$$= \sum_{n=0}^{\infty}(u_{n+1} - u_n)$$

$$= \lim_{n\to\infty}\left[(u_1 - u_0) + (u_2 - u_1) + \cdots + (u_{n+1} - u_n)\right]$$

$$= \lim_{n\to\infty} u_{n+1} - u_0$$

$$\therefore \quad \lim_{z\to 1}(z-1)\overline{u}(z) = \lim_{n\to\infty} u_{n+1}$$

$$= \lim_{n\to\infty} u_n.$$

Thus $\lim_{n\to\infty} u_n = \lim_{z\to 1}(z-1)\overline{u}(z).$

P$_4$: (Convolution Theorem) If $Z^{-1}[\overline{u}(z)] = u_n$ and $Z^{-1}[\overline{v}(z)] = v_n$ then

$$Z^{-1}[\overline{u}(z)\cdot\overline{v}(z)] = \sum_{m=0}^{n} u_m \cdot u_{n-m} = u_n * v_n$$

where * denotes the convolution operation.

Proof: By definition $\overline{u}(z) = \sum_{n=0}^{\infty} u_n z^{-n}$ and

$$\overline{v}(z) = \sum_{n=0}^{\infty} v_n z^{-n}$$

$$\therefore \quad \overline{u}(z)\cdot\overline{v}(z) = (u_0 + u_1 z^{-1} + u_2 z^{-2} + \cdots + u_n z^{-n} + \cdots)$$
$$\times (v_0 + v_1 z^{-1} + v_2 z^{-2} + \cdots + v_n z^{-n} + \cdots)$$

$$= \sum_{n=0}^{\infty}(u_0 v_n + u_1 v_{n-1} + u_2 v_{n-2} + \cdots + + u_n v_0)z^{-n}$$

$$= Z(u_0 v_n + u_1 v_{n-1} + \cdots + u_n v_0)$$

$$\therefore \quad Z^{-1}[\bar{u}(z)\cdot\bar{v}(z)] = u_0 v_n + u_1 v_{n-1} + \cdots + u_n v_0$$

$$= \sum_{m=0}^{n} u_m \cdot u_{n-m}$$

$$= u_n * v_n$$

P$_5$: (Initial Value theorem). Let $Z(u_n) = \bar{u}(z)$. Then $u_0 = \lim_{z\to\infty} \bar{u}(z)$.

Proof: $\bar{u}(z) = Z(u_n) = u_0 + u_1 z^{-1} + u_2 z^{-2} + \cdots$

$$\therefore \quad \lim_{z\to\infty} \bar{u}(z) = u_0.$$

Remark. Similarly the values of u_1, u_2 etc., can be found as given below.

$$u_1 = \lim_{z\to\infty} \left[z(\bar{u}(z) - u_0) \right]$$

$$u_2 = \lim_{z\to\infty} \left[z^2 (\bar{u}(z) - u_0 - u_1 z^{-1}) \right]$$

and so on.

P$_6$: (Final Value theorem). If $Z(u_n) = \bar{u}(z)$ then $\lim_{n\to\infty} u_n = \lim_{z\to 1} (z-1)\bar{u}(z)$

Proof: By definition we have

$$Z(u_{n+1} - u_n) = \sum_{n=0}^{\infty} (u_{n+1} - u_n) z^{-n}$$

i.e., $$Z(u_{n+1}) - Z(u_n) = \sum_{n=0}^{\infty} (u_{n+1} - u_n) z^{-n}$$

i.e., $$z[\bar{u}(z) - u_0] - \bar{u}(z) = \sum_{n=0}^{\infty} (u_{n+1} - u_n) z^{-n} \qquad \text{(By shifting theorem)}$$

$$\therefore \quad \bar{u}(z)(z-1) - u_0 z = \sum_{n=0}^{\infty} (u_{n+1} - u_n) z^{-n}.$$

Proceeding to the limit as $z \to 1$, we get

$$\lim_{z\to 1} [(z-1)\bar{u}(z)] - u_0 = \lim_{z\to 1} \sum_{n=0}^{\infty} (u_{n+1} - u_n) z^{-n}$$

$$= \sum_{n=0}^{\infty} (u_{n+1} - u_{n.})$$

$$= \lim_{n\to\infty} \left[(u_1 - u_0) + (u_2 - u_1) + \cdots + (u_{n+1} - u_n)\right]$$

$$= \lim_{n\to\infty} u_{n+1} - u_n$$

$\therefore \quad \lim_{z\to 1}(z-1)\bar{u}(z) = \lim_{n\to\infty} u_{n+1} = \lim_{n\to\infty} u_n.$

Thus $\lim_{n\to\infty} u_n = \lim_{z\to 1}(z-1)\bar{u}(z).$

5. Find $Z(e^{-an} \cdot \cos n\theta)$

Solution: We know that

$$Z(\cos n\theta) = \frac{z(z - \cos n\theta)}{z^2 - 2z\cos\theta + 1}$$

$\therefore$ by damping rule,

$$Z\left[(e^a)^{-n} \cos n\theta\right] = \frac{ze^a(ze^a - \cos\theta)}{z^2 e^{2a} - 2ze^a \cos\theta + 1}$$

6. Evaluate $Z\left(\frac{1}{(n+1)!}\right)$ by using shift theorem and hence evaluate $Z\left(\frac{1}{(n+2)!}\right)$.

Solution: We have, $Z\left(\frac{1}{n!}\right) = \sum_{n=0}^{\infty} \frac{1}{n!} z^{-n}$

$$= 1 + \frac{z^{-1}}{1!} + \frac{z^{-2}}{2!} + \frac{z^{-3}}{3!} + \cdots$$

$$= e^{\frac{1}{z}} = \bar{u}(z) \qquad \text{(say)}$$

By shift theorem, $Z(u_{n+1}) = z\left[\bar{u}(z) - u_0\right]$

and $Z(u_{n+2}) = z^2\left[\bar{u}(z) - u_0 - u_1 z^{-1}\right]$

Here $u_n = \frac{1}{n!}$ and $\bar{u}(z) = e^{\frac{1}{z}}$

$$\therefore \quad Z\left(\frac{1}{(n+1)!}\right) = z\left(e^{\frac{1}{z}} - 1\right)$$

$$Z\left(\frac{1}{(n+2)!}\right) = z^2\left(e^{\frac{1}{z}} - 1 - z^{-1}\right)$$

7. Find the Z-transform of $\sin(n+1)\theta$ and $\cos(n+1)\theta$

Solution: $Z(\sin n\theta) = \dfrac{z\sin\theta}{z^2 - 2z\cos\theta + 1}$ and

$$Z(\cos n\theta) = \frac{z(z-\cos\theta)}{z^2 - 2z\cos\theta + 1}$$

$$\therefore \quad Z[(\sin(n+1)\theta] = z\left(\frac{z\sin\theta}{z^2 - 2z\cos\theta + 1}\right) \qquad \text{(by shifting theorem)}$$

$$= \frac{z^2\sin\theta}{z^2 - 2z\cos\theta + 1}.$$

Similarly $Z[(\cos(n+1)\theta] = z\left[\dfrac{z(z-\cos\theta)}{z^2 - 2z\cos\theta + 1} - 1\right]$

$$= \frac{z^2 - \cos\theta - z}{z^2 - 2z\cos\theta + 1}$$

8. If $u_n = 1$ find the Z-transform of u_{n-1} and $u_n - u_{n-1}$.

Solution: We know that $Z(u_n) = Z(1) = \dfrac{z}{z-1} = \overline{u}(z)$ (say)

$$\therefore \quad Z(u_{n-1}) = z^{-1}\overline{u}(z) \quad \text{(by shifting theorem)}$$

$$= \frac{z^{-1}z}{z-1} \qquad = \frac{1}{z-1}.$$

Now, $Z(u_n - u_n - 1) = Z(u_n) - Z(u_{n-1})$

$$= \frac{z}{z-1} - \frac{1}{z-1}$$

$$= 1.$$

9. Find the Z-transform of $5\sin\left(\frac{n\pi}{4}\right)$.

Solution: We know that $Z(\sin n\theta) = \dfrac{z\sin\theta}{z^2 - 2z\cos\theta + 1}$

$$Z\left[5\sin\left(\frac{n\pi}{4}\right)\right] = 5z\left[\sin\left(\frac{n\pi}{4}\right)\right] = 5\left[\frac{z\sin\left(\frac{\pi}{4}\right)}{z^2 - 2z\cos\left(\frac{\pi}{4}\right) + 1}\right]$$

$$= \frac{5z}{\sqrt{2}}\left(\frac{1}{z^2 - \frac{2z}{\sqrt{2}} + 1}\right) = \frac{5z}{\sqrt{2}\left(z^2 + \sqrt{2}z + 1\right)}$$

10. Let $u_n = \{n^p\} = \{0, 1^p, 2^p, 3^p, \cdots n^p, \cdots\}$ where p is a positive integer. Then, prove that $Z\{n^p\} = -z\frac{d}{dz}\{Z(n^{p-1})\}$ where p is any positive integer. Deduce that $Z(n) = \dfrac{z}{(z-1)^2}$ and $Z(n^2) = \dfrac{z^2+2}{(z-1)^3}$.

Solution: $Z(n^p) = \sum\limits_{n=0}^{\infty} n^p z^{-n} = z\sum\limits_{n=0}^{\infty} n^{p-1} \cdot n \cdot z^{-(n+1)}$ [1]

Changing p to $p - 1$ in [1] we get, $Z(n^{p-1}) = \sum\limits_{n=0}^{\infty} n^{p-1} z^{-n}$.

Differentiating this w.r.t z we get,

$$\frac{d}{dz}\left[Z(n^{p-1})\right] = z\sum_{n=0}^{\infty} n^{p-1}(-n)z^{-(n+1)}. \qquad [2]$$

Substituting [2] in [1] we get,

$$Z(n^p) = -z\frac{d}{dz}\left[Z(n^{p-1})\right] \qquad [3]$$

This can be taken as a recurrence formula.

Putting, $p = 1$ in [3] we get,

$$Z(n) = -z\frac{d}{dz}\left[Z(n^0)\right]$$

$$= -z\frac{d}{dz}Z(1) \qquad = -z\frac{d}{dz}\left(\frac{z}{z-1}\right)$$

$$= -z\left(\frac{(z-1)-z}{(z-1)^2}\right) = \frac{z}{(z-1)^2}.$$

$$Z(n^2) = -z\frac{d}{dz}[Z(n)]$$

$$= -z\frac{d}{dz}\left(\frac{z}{(z-1)^2}\right) \qquad = -z\left[\frac{(z-1)^2 - 2z(z-1)}{(z-1)^4}\right]$$

$$= -z\left[\frac{z-1-2z}{(z-1)^3}\right]$$

$$= \frac{z^2+z}{(z-1)^3}.$$

Remark. Putting $p = 3,4$ in [2] we get

$$Z(n^3) = \frac{z^3+4z^2+2}{(z-1)^4} \quad \text{and}$$

$$Z(n^4) = \frac{z^4+11z^3+11z^2+z}{(z-1)^5}.$$

11. $Z(k) = \dfrac{kz}{z-1}$ (where k is a constant)

Proof: $Z(k) = \sum\limits_{n=0}^{\infty} kz^{-n}$

$$= k\left[1+\frac{1}{z}+\frac{1}{z^2}+\cdots+\frac{1}{z^n}+\cdots\right]$$

$$= k\left[\frac{1}{\left(1-\frac{1}{z}\right)}\right] = \frac{kz}{z-1}.$$

12. $Z\left(\frac{1}{n!}\right)=\sum_{n=0}^{\infty} z^{-n}$

Proof: $Z\left(\frac{1}{n!}\right)=\sum\left(\frac{1}{z}\right)^{n}\frac{1}{n!} \quad = e^{\frac{1}{z}}.$

13. $Z(n)=\sum_{n=0}^{\infty} nz^{-n}$

Proof: $$Z(n)=\frac{1}{z}+\frac{2}{z^2}+\frac{3}{z^3}+\cdots$$

$$=\frac{1}{z}\left(1+\frac{2}{z}+\frac{3}{z^2}+\cdots\right)$$

$$=\frac{1}{z}\left(1-\frac{1}{z}\right)^{-1} \quad =\frac{z}{(z-1)^2}$$

14. Find $Z(na^n)$

Solution: We know that $Z(n)=\dfrac{z}{(z-1)^2}=u(z)$ (say)

$\therefore \quad Z(na^n)=\bar{u}\left(\frac{z}{a}\right)$ (By damping rule)

$$=\frac{\frac{z}{a}}{\left[\left(\frac{z}{a}\right)-1\right]^2} \quad =\frac{az}{(z-a)^2}.$$

15. Find the Z-transform of $a^n \sin n\theta$.

Solution: We know that $Z(\sin n\theta)=\dfrac{z\sin\theta}{z^2-2z\cos\theta+1}$

$\therefore$ By Damping rule,

$$Z(a^n \sin n\theta) = \frac{\left(\frac{z}{a}\right)\sin\theta}{\left(\frac{z}{a}\right)^2 - 2\left(\frac{z}{a}\right)\cos\theta + 1}$$

$$= \frac{az\sin\theta}{z^2 - 2az\cos\theta + a^2}.$$

16. Show that $Z(\cosh n\theta) = \dfrac{z(z\cosh\theta)}{z^2 - 2z\cosh\theta + 1}$

***Solution*:** $z(\cosh n\theta) = z(\cos in\theta)$.

We know that $Z(\cos n\theta) = \dfrac{z(z-\cos\theta)}{z^2 - 2z\cos\theta + 1}$

$$\therefore \quad Z(\cos in\theta) = \frac{z(z-\cos i\theta)}{z^2 - 2z\cos i\theta + 1}$$

$$= \frac{z(z\cosh\theta)}{z^2 - 2z\cosh\theta + 1}$$

Hence $z(\cos \text{h}\, n\theta) = \dfrac{z(z\cosh\theta)}{z^2 - 2z\cosh\theta + 1}$.

17. Find the Z-transform of $2n + 5\sin\dfrac{n\pi}{4} - 3a^4$

***Solution*:** By linearity property $Z\left\{2n + 5\sin\dfrac{n\pi}{4} - 3a^4\right\}$

$$= 2Z(n) + 5Z\left(\sin\frac{n\pi}{4}\right) - 3a^4 Z(1)$$

$$= \frac{2\cdot z}{(z-1)^2} + \frac{5\cdot z\cdot\sin\frac{\pi}{4}}{z^2 - 2z\cdot\cos\frac{\pi}{4} + 1} - 3a^4\frac{z}{z-1}$$

$$= \frac{2z}{(z-1)^2} + \frac{5\left(\frac{z}{\sqrt{2}}\right)}{z^2 - \sqrt{2}z + 1} - 3a^4\frac{z}{z-1}$$

18. Find $Z\{(\cos\theta + i\sin\theta)^n\}$

Solution: We know that $(\cos\theta + i\sin\theta) = e^{i\theta}$

$$Z\{(\cos\theta + i\sin\theta)^n\} = Z\left[\left(e^{i\theta}\right)^n\right] = \frac{z}{z - e^{i\theta}}$$

$$Z(a^n) = \frac{z}{z-a}$$

19. Find $Z\left\{(n-1)^2\right\}$

Solution: $Z\left\{(n-1)^2\right\} = Z(n^2 - 2n + 1)$

$$= Z(n^2) - Z(2n) + Z(1)$$

$$= Z(n^2) - 2Z(n) + Z(1)$$

$$= \frac{z^2+z}{(z-1)^3} - \frac{2z}{(z-1)^2} + \frac{z}{z-1}$$

$$= \frac{z^3 - 3z^2 + 4z}{(z-1)^3}.$$

20. Find $Z\left\{(n+1)^2\right\}$

Solution: $Z\left\{(n+1)^2\right\} = Z(n^2 + 2n + 1)$

$$= Z(n^2) - 2Z(n) + Z(1)$$

$$= \frac{z^2+z}{(z-1)^3} - \frac{2\cdot z}{(z-1)^2} + \frac{z}{z-1}$$

$$= \frac{(z^2+z) + 2z(z-1) + z(z-1)^2}{(z-1)^3}.$$

$$= \frac{z^3+z^2}{(z-1)}.$$

21. Find $Z(e^{-an} \cdot \sin n\theta)$

Solution: We know that $Z(\sin n\theta) = \dfrac{z \sin\theta}{z^2 - 2z\cos\theta + 1}$

Now $Z(e^{-an} \cdot \sin n\theta) = Z\left[(e^a)^{-n} \sin n\theta\right]$

By damping rule (p_2), $Z[a^{-n} \cdot u_n] = \bar{u}(az)$

Replace z by az i.e., $e^a z$ then

$$Z(e^{-an} \sin n\theta) = \left.\frac{z\sin\theta}{z^2 - 2z\cos\theta + 1}\right]_{z = e^a z}$$

$$= \frac{e^a z \cdot \sin\theta}{e^{2a} z^2 - 2e^a z\cos\theta + 1}$$

22. If $\bar{u}(z) = \dfrac{2z^2 + 3z + 12}{(z-1)^4}$ find the values of u_2 and u_3

Solution: We use initial value theorem to find u_2 and u_3. First we find u_0 and u_1

$$u_0 = \lim_{z\to\infty} \bar{u}(z)$$

$$= \lim_{z\to\infty} \frac{2z^2 + 3z + 12}{(z-1)^4}$$

$$= \lim_{z\to\infty} \frac{z^2(2 + 3z^{-1} + 12z^{-2})}{z^4(1 - z^{-1})^4}$$

$$= \lim_{z\to\infty} \frac{2 + 3z^{-1} + 12z^{-2}}{z^2(1 - z^{-1})^4} = 0.$$

$$u_1 = \lim_{z\to\infty} \left[z(\bar{u}(z) - u_0)\right]$$

$$= \lim_{z\to\infty} \left[z\bar{u}(z)\right] \qquad (\because u_0 = 0)$$

$$= \lim_{z\to\infty} \frac{2 + 3z^{-1} + 12z^{-2}}{z(1 - z^{-1})^4} = 0.$$

$$u_2 = \lim_{z\to\infty} \left[z^2\left(\bar{u}(z) - u_0 - u_1 z^{-1}\right)\right]$$

$$= \lim_{z\to\infty}\left[z^2\left(\frac{2z^2+3z+12}{(1-z^{-1})^4}\right)\right] \qquad (\because u_0 = u_1 = 0)$$

$$= \lim_{z\to\infty}\left[\frac{2+3z^{-1}+12z^{-2})}{(1-z^{-1})^4}\right] = 2.$$

$$u_3 = \lim_{z\to\infty}\left[z^3\left(\bar{u}(z)-u_0-u_1z^{-1}-u_2z^{-2}\right)\right]$$

$$= \lim_{z\to\infty}\left[z^3\left(\frac{2z^2+3z+12}{(z-1)^4}-\frac{2}{z^2}\right)\right]$$

$$= \lim_{z\to\infty} z^3\left[\frac{2z^4+3z^3+12z^2-2(z-1)^4}{z^2(z-1)^4}\right]$$

$$= \lim_{z\to\infty} z^3\left[\frac{2z^4+3z^3+12z^2-2(z^4-4z^3-6z^2-4z-1)}{z^2(z-1)^4}\right]$$

$$= \lim_{z\to\infty}\frac{z^3}{z^6}\left(\frac{11z^3+8z-2}{C1-z^{-1})^4}\right) = 11.$$

23. Find $Z\{u(n)\}$, where $u(n) = \begin{cases} 0, & \text{for } n<0 \\ 1, & \text{for } n\ge 0 \end{cases}$

Solution: $Z\left(u(n)\right) = \sum_{n=0}^{\infty} u(n)z^{-n} = \sum_{n=0}^{\infty} 1\cdot z^{-n}$

$$= 1+\frac{1}{z}+\frac{1}{z^2}+\cdots+\frac{1}{z^n}+\cdots.$$

$$Z = \left[u(n)\right] = \frac{1}{1-\dfrac{1}{z}} = \frac{z}{z-1}. \qquad \text{(Unit step sequence)}$$

24. Find $Z\left[\delta(n)\right]$ (Unit impulse sequence)

where $\delta(n) = \begin{cases} 1, & \text{for } n=0 \\ 0, & \text{for } n\ne 0 \end{cases}$

Solution: $Z[\delta(n)] = \sum_{n=0}^{\infty} \delta(n) z^{-n}$

$$= 1 + 0 + 0 + 0 \cdots = 1.$$

EXERCISES

1. Prove that $Z(n^3) = \dfrac{z^3 + 4z^2 + z}{(z-1)^4}$

2. Prove that $Z(n^4) = \dfrac{z^4 + 11z^3 + 11z^2 + z}{(z-1)^5}$

3. Let u_n, v_n and w_n be sequences and a, b, c are constants. Prove that $Z(au_n + bv_n + cw_n) = aZ(u_n) + bZ(v_n) + cZ(w_n)$.

4. Find the Z-transform of the following:

(i) $\dfrac{1}{n+1}$ (ii) a^{n+3} (iii) $(n-1)^2$

(iv) $n^2 a^n$ (v) $a^n \cos n\theta$ (vi) $\sinh n\theta$

(vii) $2^n u(n-z)$ (viii) $nu(n-1)$ (ix) $\left[\left(\dfrac{1}{2}\right)^n - \left(-\dfrac{1}{2}\right)^n\right] u(n)$

(x) $n^2 u(n)$ (xi) $2^n \cdot \sin\left(\dfrac{n\pi}{4}\right) u(n)$ (xii) $u(n) = \begin{cases} \left(\dfrac{1}{2}\right)^n, & 0 \le n \le 10 \\ 0, & \text{else where} \end{cases}$

5. Prove that $Z\left(\dfrac{1}{(n+2)(n-1)}\right) = (z^2 - z)\log\left(1 - \dfrac{1}{z}\right) + z$

6. Using $Z(n^2) = \dfrac{z^2 + 2}{(z-1)^3}$, show that $Z(n+1)^2 = \dfrac{z^3 + z^2}{(z-1)^3}$

7. Using $Z(pc_n) = \left(1 + \dfrac{1}{z}\right) p$, show that $Z(pc_n a^n) = \left(1 + \dfrac{a}{z}\right) p$.

8. Using $Z(n) = \dfrac{z}{(z-1)^2}$, show that $Z(n \cos n\theta) = \dfrac{(z^3 + z)\cos\theta - 2z^2}{(z^2 - 2z\cos\theta + 1)^2}$

9. If $Z(u_n) = \frac{z}{z-1} + \frac{z}{z^2+1}$, find $Z(u_n + 2)$.

10. If $Z(u_n) = \frac{2z^2+5z+14}{(z-1)^4}$, find u_2 and u_3

11. If $Z(u_n) = \frac{2z^2+3z+4}{(z-3)^3}$, find u_1 and u_2

Inverse Z-transforms

Given $\phi(z)$, we consider the problem of determining the sequence u_n such that $Z(u_n) = \phi(z)$. Then u_n is the inverse *Z*-transform of $\phi(z)$. The process of finding inverse *Z*-transforms is illustrated in the following examples.

(K) EXAMPLES

1. Find the inverse *Z*-transform of $\frac{z}{z^2+11z+24}$

Solution: Let $\phi(z) = \frac{z}{z^2+11z+24}$

$$= \frac{z}{(z+8)(z+3)}$$

$$\therefore \quad \frac{\phi(z)}{z} = \frac{1}{(z+8)(z+3)}$$

$$= \frac{1}{5}\left[\frac{1}{(z+3)} - \frac{1}{(z+8)}\right] \qquad \text{(By partial fractions)}$$

Hence $\phi(z) = \frac{1}{5}\left[\frac{z}{z+3} - \frac{z}{z+8}\right]$

$$\therefore \quad u_n = Z^{-1}(\bar{u}(z))$$

$$= \frac{1}{5}\left[Z^{-1}\left(\frac{z}{z+3}\right) - Z^{-1}\left(\frac{z}{z+8}\right)\right]$$

$$= \frac{1}{5}\left[(-3)^n - (-8)^n\right]. \qquad \left(\because Z(a^n) = \frac{z}{z-a}\right)$$

$$\therefore \quad Z^{-1}\left[\frac{z}{z^2+11z+24}\right]=\frac{1}{5}\left[(-3)^n-(-8)^n\right].$$

2. Find the inverse Z-transform of $\dfrac{2z^2+3z}{(z+2)(z-4)}$.

***Solution*:** Let $\phi(z)=\dfrac{2z^2+3z}{(z+2)(z-4)}$

$$\therefore \quad \frac{\phi(z)}{z}=\frac{2z+3}{(z+2)(z-4)}=\frac{1}{6(z+2)}+\frac{11}{6(z-4)} \qquad \text{(by partial fractions)}$$

$$\therefore \quad \phi(z)=\frac{1}{6}\left(\frac{z}{z+2}\right)+\frac{11}{6}\left(\frac{z}{z-4}\right)$$

$$\therefore \quad Z^{-1}[\phi(z)]=\frac{1}{6}z^{-1}\left(\frac{z}{z+2}\right)+\frac{11}{6}z^{-1}\left(\frac{z}{z-4}\right)=\frac{1}{6}(-2)^n+\frac{11}{6}(4^n)$$

3. Find $Z^{-1}[\phi(z)]$, where $\phi(z)=\dfrac{1}{(z-2)}$

***Solution*:** By actual division in an elegant form, we have

$$\begin{array}{l} z-2\,\big)\; 1 \;\; \left(\dfrac{1}{z}+\dfrac{2}{z^2}+\dfrac{4}{z^3}\right. \\ \quad\; \underline{-\left(1-\dfrac{2}{z}\right)} \\ \qquad\;\; \dfrac{2}{z} \\ \quad\; \underline{-\left(\dfrac{2}{z}-\dfrac{4}{z^2}\right)} \\ \qquad\;\; \dfrac{4}{z^2} \\ \quad\; \underline{-\left(\dfrac{4}{z^2}-\dfrac{8}{z^3}\right)} \\ \qquad\;\; \dfrac{8}{z^3} \end{array}$$

$$\therefore \quad \frac{1}{(z-2)} = \frac{1}{z} + \frac{2}{z^2} + \frac{4}{z^3} + \frac{8}{z^4} + \cdots$$

$$= z^{-1} + 2z^{-2} + 4z^{-3} + 8z^{-4} + \cdots + 2^{k-1} z^{-k}$$

$$= \{2^{k-1}\} z^{-k}$$

$$\therefore \quad z^{-1}\left\{\frac{1}{z-2}\right\} = \left\{2^{k-1}\right\}$$

For $|z| < 2$

$$\begin{array}{l} -2+z \big) \; 1 \; \left(-\frac{1}{2} - \frac{z}{4} - \frac{z^2}{8} - \frac{z^3}{16} \right. \\ \quad \underline{-\left(1 - \frac{z}{2}\right)} \\ \quad \quad \frac{z}{2} \\ \quad \underline{-\left(\frac{z}{2} - \frac{z^2}{4}\right)} \\ \quad \quad \frac{z^2}{4} \\ \quad \underline{-\left(\frac{z^2}{4} - \frac{z^3}{8}\right)} \\ \quad \quad \frac{z^3}{8} \\ \quad \underline{-\left(\frac{z^3}{8} - \frac{z^4}{16}\right)} \\ \quad \quad \frac{z^4}{16} \end{array}$$

$$\therefore \quad \frac{1}{(z-2)} = -\frac{1}{2} - \frac{z}{4} - \frac{z^2}{8} - \frac{z^3}{16} - \cdots$$

$$= -\frac{1}{2}\left[1 + \frac{z}{2} + \frac{z^2}{2^2} + \frac{z^3}{2^3} + \cdots + \frac{z^k}{2^k} + \cdots\right]$$

$$= -\frac{1}{2} - \frac{z}{2^2} - \frac{z^2}{2^3} - \frac{z^3}{2^4} - \cdots - \frac{z^k}{2^{k+1}} \cdots$$

$$= \{2^{-k-1}\}z^k$$

$$z^{-1}\left\{\frac{1}{z-2}\right\} = \left\{2^{-k-1}\right\}$$

4. Find $Z^{-1}\{\phi(z)\}$ where $\phi(z) = \dfrac{z}{(z-1)(z-0.5)}$

Solution: $\dfrac{z}{(z-1)(z-0.5)} = \dfrac{z}{z^2 - 1.5z + 0.5}$

$$= 0 \cdot z^0 + 1.0z^{-1} + 1.5z^{-2} + 1.75z^{-3}$$

(by actual division method)

$\therefore$ the coefficient giving by $f(n)$ are: 0, 1, 1.5, 1.75,$\cdots$

5. Evaluate $z^{-1}\left\{\dfrac{z}{(z-1)^2}\right\}$

Solution: $\dfrac{Z}{(z-1)^2} = \dfrac{2}{z^2 - 2z + 1}$

$$= 0 \cdot z^0 + z^{-1} + 2z^{-2} + 3z^{-3} + 4z^{-4} + \cdots$$

(by actual division method)

$$= \sum_{n=0}^{\infty} n \cdot z^{-n}$$

Inverse *Z*-transform by partial fractions

6. Find $z^{-1}\left\{\dfrac{4z^2 - 2z}{z^3 - 5z^2 + 8z - 4}\right\}$

Solution: Let $\phi(z) = \dfrac{4z^2 - 2z}{z^3 - 5z^2 + 8z - 4}$

$$= \frac{2z(2z-1)}{(z-1)(z-2)^2}$$

Let us resolve into partial fractions

$$\frac{2z-1}{(z-1)(z-2)^2}=\frac{A}{z-1}+\frac{B}{z-2}+\frac{C}{(z-2)^2}$$

$\Rightarrow \quad (2z-1)=A(z-2)^2+B(z-1)(z-2)+C(z-1)$

Put $z = 2$, we get $3 = C$;

Put $z = 1$, we get $A = 1$

Put $z = 0$, $-1=4A+2B-C$

$$\therefore \quad B=-1$$

Thus, $Z^{-1}\left[\dfrac{4z^2-2z}{(z-1)(z-2)^2}\right]$

$$=Z^{-1}\left[2z\cdot\frac{1}{z-1}\right]+Z^{-1}\left[2z\cdot\frac{-1}{z-2}\right]+Z^{-1}\left[2z\cdot\frac{-1}{(z-1)^2}\right]$$

$$=2\cdot 1-2\cdot 2^n+3\cdot Z^{-1}\left[\frac{2z}{(z-2)^2}\right]$$

$$=2-2\cdot 2^n+3\cdot Z^{-1}\left[-z\cdot\frac{d}{dz}\left(\frac{z}{z-2}\right)\right]$$

$$=2-2^{n+1}+3n\cdot 2^n \qquad \left(\because Z[\{n\cdot f(t)\}]=-z\frac{d}{dz}(\phi(z)) \text{ multiplication by 'n' rule}\right)$$

7. Evaluate $Z^{-1}\{(z-5)^{-3}\};\ |z|>5$ and hence find the region of convergence.

Solution: Let $(z-5)^{-3}=\dfrac{1}{(z-5)^3}=\dfrac{1}{z^3}\dfrac{1}{\left(z-\dfrac{5}{z}\right)^3}$

$\left(\text{Since } \left|\dfrac{5}{z}\right|<1 \text{ or } |z|>5\right)$, we have

$$\frac{1}{(z-5)^3}=\frac{1}{z^3}\left[1+3\left(\frac{5}{z}\right)+\frac{3\cdot 4}{1\cdot 2}\left(\frac{5}{z}\right)^2+\frac{3\cdot 4\cdot 5}{1\cdot 2\cdot 3}\left(\frac{5}{z}\right)^3+\frac{3\cdot 4\cdot 5\cdot 6}{1\cdot 2\cdot 3\cdot 6}\left(\frac{5}{z}\right)^4+\cdots\right]$$

$$=\frac{1}{2}\cdot\frac{1}{z^3}\left[1\cdot 2\left(\frac{5}{z}\right)^0+2\cdot 3\left(\frac{5}{z}\right)^1+3\cdot 4\left(\frac{5}{z}\right)^2+4\cdot 5\left(\frac{5}{z}\right)^3+5\cdot 6\left(\frac{5}{z}\right)^4+\cdots\right]$$

$$=\frac{1}{2}\sum_{m=0}^{\infty}(m+1)(m+2)5^m\cdot z^{-m-3};\ (\text{put } n=m+3)$$

$$\phi(z)=(z-5)^{-3}$$

$$=\frac{1}{2}\sum_{n=0}^{\infty}(n-1)(n-2)5^{n-3}\cdot z^{-n}$$

$$=\sum_{n=0}^{\infty}f(n)z^{-n}$$

$$\therefore \quad Z^{-1}[\phi(z)]=Z^{-1}\left[(z-5)^{-3}\right]=f(n)$$

$$=\frac{1}{2}(n-1)(n-2)5^{n-3} \qquad \text{for } n\geq 3$$

$$=0 \qquad \text{for } n<3$$

The region of convergence is the region outside the circle $|z|=5$; with centre at origin and of radius 5.

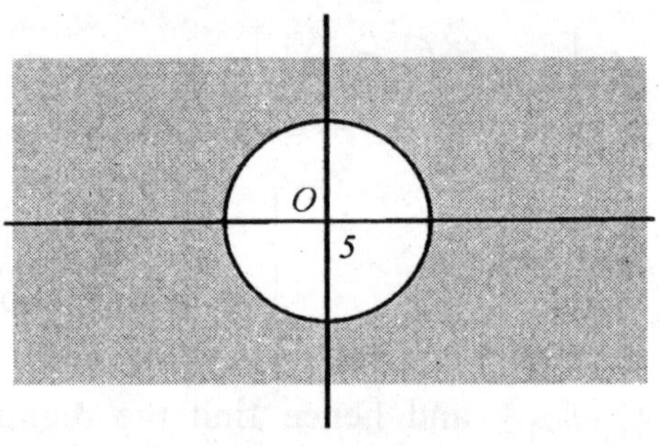

8. Evaluate the inverse Z-transform of the following:

(i) $\dfrac{z}{(z-a)^2}$ (ii) $z\log\left(\dfrac{z}{z+1}\right)$ (iii) $\dfrac{z^2}{(z-1)(z-3)}$

(iv) $\dfrac{z(2z+3)}{(z+2)(z-4)}$ (v) $\dfrac{z^3-20z}{(z+2)^3(z-4)}$

Solution: (i) $\dfrac{z}{(z-a)^2}=\dfrac{1}{2}\cdot\dfrac{1}{\left(1-\dfrac{a}{z}\right)^2}=\dfrac{1}{2}\left(1-\dfrac{a}{z}\right)^{-2}$

$$= \frac{1}{2}\left\{1 + 2\left(\frac{a}{z}\right) + 3\left(\frac{a}{z}\right)^2 + 4\left(\frac{a}{z}\right)^3 + \cdots\right\}$$

$$= \frac{1}{2}\sum_{n=1}^{\infty} \frac{na^{n-1}}{z^{n-1}} = \sum_{n=0}^{\infty} \left(na^{n-1}\right) z^{-n} = z\left\{n \cdot a^{n-1}\right\}$$

$$\therefore \quad Z^{-1}\left\{\frac{z}{(z-a)^2}\right\} = n \cdot a^{n-1}$$

(ii) $$\log\left(\frac{z}{z+1}\right) = \log\left(\frac{1}{1+\frac{1}{z}}\right) = -\log\left(1 + \frac{1}{z}\right)$$

$$= -\left[\frac{1}{z} - \frac{\left(\frac{1}{z}\right)^2}{2} + \frac{\left(\frac{1}{z}\right)^3}{3} - + \cdots\right]$$

$$= -\sum_{n=0}^{\infty} \frac{(-1)^n}{n+1}\left(\frac{1}{z}\right)^{n+1} = \sum_{n=0}^{\infty} \frac{(-1)^{n+1}}{n+1} \cdot z^{-(n+1)}$$

$$\therefore \quad Z\log\left(\frac{z}{z+1}\right) = \sum_{n=0}^{\infty} \frac{(-1)^{n+1}}{n+1} z^{-n}$$

$$= z\left\{\frac{(-1)^{n+1}}{n+1}\right\}$$

$$\Rightarrow \quad Z^{-1}\left\{z\log\left(\frac{z}{z+1}\right)\right\} = \frac{(-1)^{n+1}}{n+1}$$

(iii) Let $\phi(z) = \dfrac{z^2}{(z-1)(z+3)}$

$$\therefore \quad \frac{\phi(z)}{z} = \frac{z}{(z-1)(z+3)}$$

$$= \frac{1}{a}\left[\frac{1}{z-1} + \frac{3}{z+3}\right]$$

so that $\phi(z) = \dfrac{1}{a} \cdot \dfrac{z}{z-1} + \dfrac{3}{a} \cdot \dfrac{z}{z+3}$

$$= \frac{1}{a} Z\{1^n\} + \frac{3}{4} Z\{(-3)^n\}$$

$$= \frac{1}{a} Z\{1^n + 3(-3)^n\}$$

$$= \frac{1}{a} Z\{1 + (-1)^n + 3^{n+1}\}$$

Hence $Z^{-1}\{\phi(z)\} = \frac{1}{a}\{1 + (-1)^n + 3^{n+1}\}$

(iv) Let $\phi(z) = \dfrac{z(2z+3)}{(z+2)(z-4)}$

$$\therefore \quad \frac{\phi(z)}{2} = \frac{2z+3}{(z+2)(z-4)}$$

$$= \frac{1}{6}\left(\frac{1}{z+2} + \frac{11}{z-4}\right) \qquad \text{(by partial fractions)}$$

$$\Rightarrow \quad \phi(z) = \frac{1}{6}\left(\frac{z}{z+2} + 11 \cdot \frac{z}{z-4}\right)$$

$$= \frac{1}{6} Z\left\{(-2)^n + 11(4^n)\right\}$$

(v) Let $\phi(z) = \dfrac{z^3 - 20z}{(z-2)^3(z-4)}$, so that

$$\frac{\phi(z)}{z} = \frac{z^2 - 20}{(z-2)^3(z-4)} = \frac{Az^2 + Bz + C}{(z-2)^3} + \frac{D}{z-4}$$

$$\Rightarrow \quad z^2 - 20 = (Az^2 + Bz + C)(z-4) + D(z-2)^3$$

Putting $z = 4, 2, 0, 1,$ in this, we get

$$D = -\frac{1}{2}; \quad 4A + 2B + C = 8,$$

$$C + 2D = 5; \quad 3(A + B + C) + D = 19$$

On solving these we get

$$A = \frac{1}{2}; \; B = 0; \; C = 6, \; D = -\frac{1}{2}$$

Thus,

$$\frac{\phi(z)}{z} = \frac{1}{2}\left[\frac{z^2+12}{(z-2)^3} - \frac{1}{(z-4)}\right]$$

$$= \frac{1}{2}\left[\frac{(z-2)^2+4z+8}{(z-2)^3} - \frac{1}{(z-4)}\right]$$

or $$\phi(z) = \frac{1}{2}\cdot\frac{z}{z-2} + \frac{2z(z+2)}{(z-2)^3} - \frac{1}{z}\cdot\frac{z}{z-4}$$

$$= \frac{1}{2}Z\{2^n\} + Z\{2^n\cdot n^2\} - \frac{1}{2}Z(4^n)$$

$\therefore$ $$Z^{-1}\{\phi(z)\} = \frac{1}{2}(2^n) + 2^n n^2 - \frac{1}{2}(4^n)$$

$$= 2^{n-1} + 2^n.n^2 - 2^{2n-1}$$

$$\left(\because\ Z(a^n) = \frac{z}{z-a};\ Z(a^n\cdot n^2) = \frac{9z(z+a)}{(z-a)^3}\right)$$

Convolution theorem

If $Z^{-1}\{\phi_1(z)\} = f_1(n)$ and

$Z^{-1}\{\phi_2(z)\} = f_2(n)$, then

$Z^{-1}\{\phi_1(z)\cdot\phi_2(z)\}$

$= f_1(n) * f_2(n) =$ convolution of $f_1(n)$ and $f_2(n)$

$$= \sum_{m=0}^{\infty} f_1(m)f_2(n-m)$$

Proof: $$\phi_1(z)\cdot\phi_2(z) = \left[\sum_{n=0}^{\infty} f_1(n)z^{-n}\right]\left[\sum_{n=0}^{\infty} f_2(n)z^{-n}\right]$$

$$= \left[f_1(0) + \frac{f_1(1)}{z} + \frac{f_1(2)}{z^2} + \cdots + \frac{f_1(n)}{z^n} + \cdots\right]$$

$$\times\left[f_2(0) + \frac{f_2(1)}{z} + \frac{f_2(2)}{z^2} + \cdots + \frac{f_2(n)}{z^n}\cdots\right]$$

Collecting the coefficient of z^{-n},

$$= \sum_{n=0}^{\infty}[f_1(0)f_2(n)+f_1(1)f_2(n-1)+f_1(2)f_2(n-2)+\cdots+f_1(n)f_2(0)]z^{-n}$$

$$= Z(f_1(0)f_2(n)+f_1(1)f_2(n-1)+\cdots+f_1(n)f_2(0), \text{ by definition}$$

$$= Z\left(\sum_{m=0}^{\infty} f_1(m)f_2(n-m)\right)$$

$$\Rightarrow \quad Z^{-1}[\phi_1(z)\phi_2(z)] = \sum_{m=0}^{\infty} f_1(m)f_2(n-m)$$

9. Using convolution theorem, evaluate

$$\mathsf{L}^{-1}\left[\frac{z^2}{(z-1)(z-3)}\right]$$

Solution: We know that $Z^{-1}\left[\frac{z}{z-1}\right]=1^n$ and $Z^{-1}\left[\frac{z}{z-3}\right]=3^n$

Now, $Z^{-1}\left[\frac{z^2}{(z-1)(z-3)}\right] = Z^{-1}\left[\frac{z}{z-1}\cdot\frac{z}{z-3}\right]$

$$= 1^n * 3^n \qquad \text{(by convolution theorem)}$$

$$= \sum_{m=0}^{n} 1^m 3^{n-m}$$

$$= 3^n + 3^{n-1} + 3^{n-2} + 3^{n-3} + \cdots + 3' + 1$$

$$= 1 + 3 + \cdots + 3^n$$

$$= \frac{3^{n+1}-1}{2}$$

$$= \frac{1}{2}(3^{n+1}-1)$$

10. Prove that $Z^{-1}\left[\left(\frac{z}{z-a}\right)^2\right] = (n+1)a^n$

Solution: We know that $Z^{-1}\left[\frac{z}{z-a}\right]=a^n$

$$Z^{-1}\left[\left(\frac{z}{z-a}\right)^2\right]=Z^{-1}\left[\frac{z}{z-a}\cdot\frac{z}{z-a}\right]$$

$$=a^n * a^n$$

$$=\sum_{m=0}^{n} a^m \cdot a^{n-m}$$

$$=a^n+a^n+\cdots+a^n,\ [n+1 \text{ terms}]$$

$$=(n+1)a^n.$$

11. Find $Z^{-1}\left[\frac{1}{\left(1-\frac{1}{2}z^{-1}\right)\left(1-\frac{1}{4}z^{-1}\right)}\right]$

Solution: Let $\phi_1(z)=\frac{1}{1-\frac{1}{2}\cdot z^{-1}}$ and $\phi_2(z)=\frac{1}{1-\frac{1}{4}\cdot z^{-1}}$

so that $f(n)=Z^{-1}\{\phi_1(z)\}=2^{-n}f(n)$

$$g(n)=Z^{-1}\{\phi_2(z)\}=4^{-n}f(n)$$

Using convolution theorem we obtain:

$$Z^{-1}\left[\frac{1}{\left(1-\frac{1}{2}\cdot z^{-1}\right)\left(1-\frac{1}{4}\cdot z^{-1}\right)}\right]=\sum_{m=0}^{n}\left[2^{-(n-m)}\right]\cdot f(n-m)\left[4^{-m}+(m)\right]$$

$$=\left(2^{-n}\sum_{m=0}^{n}2^{-m}\right)f(n)$$

$$=2^{-n}\left(\frac{2^{-n-1}}{2^{-1}-1}\right)\cdot f(n)$$

$$=\left[2\left(\frac{1}{2}\right)^n-\left(\frac{1}{a}\right)^n\right]f(n).$$

12. Find $Z^{-1}\left[\dfrac{z^2}{(z-a)(z-b)}\right],\ a \neq b$

Solution: We know that

$$Z\{a^n\} = \frac{z}{z-a} \qquad [1]$$

$$\frac{z^2}{(z-a)(z-b)} = \frac{z}{(z-a)} \cdot \frac{z}{(z-b)}$$

$$= Z\{a^n\} \cdot Z(b^n)$$

$$= Z\{f_1(n)\} \cdot Z\{f_2(n)\}$$

where $f_1(n) = a^n;\ f_2(n) = b^n$

$$= \phi_1(z) \cdot \phi_2(z) = Z\left\{\sum_{m=0}^{n} f_1(m) f_2(n-m)\right\} \qquad \text{(by convolution theorem)}$$

$$= Z\left\{\sum_{m=0}^{n} a^m \cdot b^{n-m}\right\}$$

$$= Z\left\{b^n \sum_{m=0}^{n} \left(\frac{a}{b}\right)^m\right\} \qquad [2]$$

$$= Z\left\{b^n \cdot \frac{\left(\frac{a}{b}\right)^{n+1} - 1}{\left(\frac{a}{b}\right) - 1}\right\}, \text{ if } a \neq b$$

$$= Z\left\{\frac{a^{n+1} - b^{n+1}}{a-1}\right\}$$

$$\Rightarrow \quad Z^{-1}\left[\frac{z^2}{(z-a)(z-b)}\right] = \left\{\frac{a^{n+1} - b^{n+1}}{a-1}\right\} \quad (a \neq b)$$

13. A sequence $\{f(n)\}$ satisfies the equation $f(n+2) - 4f(n) = n-1$. with $f(0) = 1;\ f(1) = 2$ Find $f(n)$ by using Z-transform.

Solution: $Z\{f(n+2)\} - 4Z\{f(n)\} = Z[n] - Z(1)$

$$= \frac{z}{(z-1)^2} - \frac{z}{(z-1)}$$

Let us evaluate $Z\{f(n)\} = \phi(z)$ and using shifting rule, we get

$$Z^2\left\{\phi(z) - f(0) - f(1)\cdot z^{-1}\right\} - 4\phi(z) = \frac{z(2-z)}{(z-1)^2}$$

$$\Rightarrow \quad (z^2-4)\phi(z) = f(0)z^2 + f(1)z + \frac{z(2-z)}{(z-1)^2}$$

$$= z(z+2) - \frac{z(z-2)}{(z-1)^2} \qquad (\because\ f(0)=1;\ f(1)=2)$$

$$\Rightarrow \quad \frac{\phi(z)}{z} = \frac{1}{(z-2)} - \frac{1}{(z-1)^2(z+2)} \qquad [1]$$

by partial fractions, we find $\dfrac{1}{(z-1)^2(z+2)} = \dfrac{1}{9}\cdot\dfrac{1}{z+2} - \dfrac{1}{9}\cdot\dfrac{1}{z-1} + \dfrac{1}{3}\cdot\dfrac{1}{(z-1)^2}$

Using these in [1], we set

$$\phi(z) = \frac{z}{Z-2} - \frac{1}{9}\cdot\frac{1}{z+2} + \frac{1}{9}\cdot\frac{1}{z-1} - \frac{1}{3}\cdot\frac{z}{(z-1)^2}$$

$$= Z\left[z^n\right] - \frac{1}{9}Z\left\{(-2)^n\right\} + \frac{1}{9}Z(1) - \frac{1}{3}Z(n)$$

$$\Rightarrow \quad f(n) = Z^{-1}\left[\phi(z)\right]$$

$$f(n) = 2^n - \frac{1}{9}(-2)^n + \frac{1}{9} - \frac{1}{3}n$$

14. Evaluate $Z^{-1}\left\{\log(1+az^{-1}\right\},\ |z| > |a|$

Solution: We have $\dfrac{d}{dz}[\phi(z)] = \dfrac{-9z^{-2}}{1+az^{-1}}$

where $\phi(z) = \log(1+az^{-1})$

Using differential rule, we get

$$n\cdot f(n) = Z^{-1}\left[-z\cdot\frac{d}{dz}\{\phi(z)\}\right]$$

$$= Z^{-1}\left[\frac{-az^{-1}}{1+az^{-1}}\right]$$

$$= (-1)^{n+1} \cdot a^n f(n-1)$$

$$\Rightarrow \quad f(n) = (-1)^{n+1} \cdot \frac{a^n}{n} f(n-1)$$

EXERCISES

1. Evaluate inverse Z-transforms of the following.

(a) $\dfrac{3z^2+2}{(5z-1)(5z-2)}$ (b) $\dfrac{8z-z^3}{(4-z)^3}$ (c) $\dfrac{z^2-20z}{(z-2)^3(z-4)}$

(d) $\dfrac{z}{(z-1)(z-2)}$ (e) $\dfrac{8z^2}{(2z-1)(4z-1)}$ (f) $\dfrac{z}{(z+1)^2(z-1)}$

(g) $\dfrac{10z}{(z-1)(z^2+1)}$ (h) $\dfrac{z^2+z}{(z-1)(z^2+1)}$

2. Find (a) $Z^{-1}\{(1-az^{-1})^{-2}\}$

(b) $Z^{-1}\left\{\dfrac{az^{-1}}{(1-az^{-1})^2}\right\}$

3. Use convolution to find $f(n)$

if $f_1(n) = a^n \cdot f(n)$; $f_2(n) = f(n)$

4. If $\phi(z) = \dfrac{z^2+3z+5}{(z+1)^3}$ find $f(1)$ and $f(2)$.

1.4 Difference Equations and their solutions by Z-transform

An expressed relation between an independent variable x and one or more dependent variables or functions f, g, ..., and any successive differences of f, g, etc., as $\Delta f(x) = f(x+h) - f(x)$, $\Delta^2 f(x) = f(x+2h) - 2f(x+h) + f(x)$, etc., or equivalently, the results of any successive applications of the operator E, where $Ef(x) = f(x+h)$.

The order of a difference (or exponent of the highest power of E and the *degree* is the highest power of E), and the degree is the highest power to which the highest difference is involved. A difference equation is linear if it is of the first degree with respect to all of the quantities $f(x)$, $\Delta f(x)$, $\Delta^2 f(x)$, etc., or $f(x)$, $Ef(x)$, etc. The equation $f(x+1) = xf(x)$ is a linear difference equation.

Solution of difference equations by Z-transforms

Since *Z*-transforms act on sequences (u_n), it can be employed to solve difference equations. Using the relations

$$Z(u_{n+1}) = Z\left[\phi(z) - u_0\right]$$
$$Z(u_{n+2}) = Z^2\left[\phi(z) - u_0 - u_1 z^{-1}\right]$$

the given difference equation can be converted to an equation of the form $\overline{u}(z) = \phi(z)$ by taking *Z*-transforms and then by applying inverse *Z*-transforms, we obtain the required solution.

Definition: An equation which expresses a relation between the independent variable, the dependent variable and the successive differences of the dependent variable is called a difference equation.

For example,

$$\Delta^3 u(n) - 4\Delta u(n) + 7u(n) = n^2 + \cos n + 7 \qquad [1]$$

$\Delta^2 u(n) - 2\Delta u(n) + u(n) = 0$ are difference equations.

Order of $u(n+3) - 5u(n+2) + 7u(n+1) + u(n) = 10n$ is $(n+3)(n) = 3$.

Order of $u(n+3) - 5u(n+2) + u(n+1) = 0$ is $(n+3) - (n+1) = 2$ and not 3.

The *degree* of a difference equation is written in a form free from Δ's in the highest power of the *u*'s.

For example, $u(n+1)u^5(n+2) - u(u+1)u(n) + u^2(n+3) = \cos n$ is order 5 and of order 3.

Working rule

(1) Take *Z*-transform on both sides of the given difference equation.

(2) Use given conditions and solve for $\phi(z)$.

(3) Apply partial fractions method.

(4) Take inverse *Z*-transform on both sides which result in the given sequence.

(L) Examples

1. Solve $4u(n) - u(n+2) = 0$ using *Z*-transform. (Given: $u(0) = 0$, $u(1) = 2$)

Solution: Given: $4u(n)-u(n+2)=0$ [1]

Let $Z\{u(n)\}=\phi(z)$

Taking Z-transforms in [1] on both the sides

we have $4Z\{u(n)\}-Z\{u(n+2)\}=0$

$$\therefore \quad 4\phi(z)-z^2\left[\phi(z)-u(0)-u(1)Z^{-1}\right]=0$$

or, $\quad 4\phi(z)-z^2\phi(z)+2z \qquad (\because u(0)=0;\ u(1)=2)$

$$\Rightarrow \quad \phi(z)[z^2-4]=2z$$

$$\therefore \quad \phi(z)=\frac{2z}{z^2-4}$$

$$\therefore \quad \frac{\phi(z)}{z}=\frac{2}{z^2-4}$$

$$=\left[\frac{2}{(z+2)(z-2)}\right]$$

$$=-\frac{1}{2(z+2)}+\frac{1}{2(z-2)} \qquad \text{(by partial fractions)}$$

Let $\dfrac{2}{(z+2)(z-2)}=\dfrac{A}{(z+2)}+\dfrac{B}{(z-2)}$

$$\therefore \quad A=\left[\frac{2}{(z-2)}\right]_{z=-2}=-\frac{2}{4}=-\frac{1}{2}$$

$$B=\left[\frac{2}{(z+2)}\right]_{z=2}=\frac{2}{4}=\frac{1}{2}$$

$$\therefore \quad \frac{2}{(z+2)(z-2)}=-\frac{1}{2(z+2)}+\frac{1}{2(z-2)}$$

$$\therefore \quad \phi(z)=-\frac{z}{2(z+2)}+\frac{z}{2(z-2)}$$

Taking inverse Z-transform, we get

$$Z^{-1}\{\phi(z)\}=-\frac{1}{2}Z^{-1}\left[\frac{z}{(z+2)}\right]+\frac{1}{2}Z^{-1}\left[\frac{z}{z-2}\right]$$

$$=-\frac{1}{2}(-2)^n+\frac{1}{2}(2)^n$$

$$= (-2)^{-1}(-2)^n + (2^{-1})2^n$$

$\therefore$ $\boxed{u(n) = (-2)^{n-1} + 2^{n-1}}$ $\left(\because Z^{-1}[\phi(z)] = u(n)\right)$

2. Using Z-transform solve

$$u(n+2) - 4u(n+1) + 3u(n) = 5^n$$

Solution: Given $u(n+2) - 4u(n+1) + 3u(n) = 5^n$ [1]

Let $Z\{u(n)\} = \phi(z)$

Taking Z-transforms in [1] on both sides we have

$$Z\{u(n+2)\} - 4Z\{u(n+1)\} + 3Z\{u(n)\} = Z\{5^n\}$$

$$\Rightarrow \quad z^2\left[\phi(z) - u(0) - u(1)z^{-1}\right] - 4z[\phi(z) - u(0)] + 3\phi(z) = \frac{z}{z-5}$$

$$\Rightarrow \quad \phi(z)\left[z^2 - 4z + 3\right] - u(0)[z^2 - 4z] - 4(1)z = \frac{z}{z-5}$$

$$\Rightarrow \quad \frac{\phi(z)}{z} = \frac{1}{(z-5)(z-1)(z-3)} + \frac{u(0)(z-4) + u(1)}{(z-1)(z-3)}$$

By partial fractions $\dfrac{\phi(z)}{z} = \left[\dfrac{A}{(z-1)} + \dfrac{B}{(z-3)} + \dfrac{C}{(z-5)}\right] + \left[\dfrac{D}{(z-1)} + \dfrac{E}{(z-3)}\right]$

$$\phi(z) = \left[\frac{1}{8}\frac{z}{z-1} - \frac{1}{4}\frac{z}{z-3} + \frac{1}{8}\frac{z}{z-5}\right] + \left[\frac{(u(1) - 3u(0))z}{z(z-1)} + \frac{3z(u(0) - u(1))}{2(z-3)}\right]$$

Taking inverse Z-transform on both sides, we get

$$u(n) = Z^{-1}[\phi(z)] = \frac{1}{8}\cdot 1 - \frac{1}{4}3^n + \frac{1}{8}5^n + \left(\frac{u(1) - 3u(0)}{2}\right)1 + \frac{3(u(0) - u(1))}{2}3^n$$

$u(n) = A + B\,3^n + \dfrac{1}{8}5^n$ is the solution.

where $A = \dfrac{1}{8} + \dfrac{u(1) - 3u(0)}{2}$,

$$B = \frac{3(u(0) - u(1))}{2} - \frac{1}{4}.$$

3. Solve $u(n+2)-2u(n+1)+u(n)=2^n$ with $u(0)=2;\ u(1)=1,$ using Z-transform.

Solution: Given $u(n+2)-2u(n+1)+u(n)=2^n$ [1]

Let $Z\{u(n)\}=\phi(z)$

Taking Z-transform in [1] on both sides we have

$$Z\{u(n+2)\}-2Z\{u(n+1)\}+Z\{u(n)\}=Z\{2^n\}$$

$$\therefore \quad z^2\left[\phi(z)-u(0)-u(1)z^{-1}\right]-2z\left[\phi(z)-u(0)\right]+\phi(z)=\frac{z}{z-2}$$

Since $u(0)=2;\ u(1)=2,$ we get

$$z^2\left[\phi(z)-2-z^{-1}\right]-2z\phi(z)+4z+\phi(z)=\frac{z}{z-2}$$

$$\Rightarrow \quad \phi(z)\left[z^2-2z+1\right]=\frac{z}{z-2}+2z^2+z-4z=\frac{z+(2z^2-3z)(z-2)}{(z-2)}$$

$$=\frac{2z^3-7z^2+7z}{z-2}$$

$$\therefore \quad \phi(z)=\frac{2z^3-7z^2+7z}{(z-2)(z^2-2z+1)}$$

$$\text{or,} \quad \phi(z)=\frac{2z^3-7z^2+7z}{(z-2)(z-1)^2}$$

$$\therefore \quad \frac{\phi(z)}{z}=\frac{2z^2-7z+7}{(z-2)(z-1)^2}$$

$$=\frac{1}{(z-2)}+\frac{1}{(z-1)}-\frac{2}{(z-1)^2} \qquad \text{(by partial fractions)}$$

$$\therefore \quad \phi(z)=\frac{z}{(z-2)}+\frac{z}{(z-1)}-\frac{2z}{(z-1)^2}$$

Taking inverse Z-transform we get

$$Z^{-1}[\phi(z)]=Z^{-1}\left[\frac{1}{(z-2)}\right]+Z^{-1}\left[\frac{1}{z-1}\right]-2Z^{-1}\left[\frac{2}{(z-1)^2}\right]$$

$$\therefore \quad \boxed{u(n)=2^n+1-2n}$$

4. Solve $u(n+2)+2u(n+1)+u(n)=n$ with $u(0)=u(1)=0$ using *Z*-transform.

Solution: Given $u(n+2)+2u(n+1)+u(n)=n$ [1]

Let $Z\{u(n)\}=\phi(z)$

Taking *Z*-transform of [1] on both sides,

we have

$$Z\{u(n+2)\}+2Z\{u(n+1)\}+Z\{u(n)\}=Z\{n\}$$

$$\Rightarrow \quad z^2\left[\phi(z)-u(0)-u(1)z^{-1}\right]+2z[\phi(z)-u(0)]+\phi(z)=\frac{z}{(z-2)^2}$$

putting $u(0)=u(1)=0$ solving

$$\phi(z)=\frac{z}{(z-1)^2(z+1)^2} \qquad \text{(by partial fractions)}$$

$$\frac{1}{(z-1)^2(z+1)^2}=\frac{A}{(z-1)}+\frac{B}{(z-1)^2}+\frac{C}{(z+1)}+\frac{D}{(z+1)^2}$$

we get $A=-\frac{1}{4},\ B=C=D=\frac{1}{4}$

$$\therefore \quad \phi(z)=\frac{1}{4}\left[-\frac{z}{(z-1)}+\frac{z}{(z-1)^2}+\frac{z}{(z+1)}+\frac{z}{(z+1)^2}\right]$$

Taking inverse *Z*-transform on either side

$$u(n)=\frac{1}{4}\left[(-1)^n+n+(-1)^n-n(-1)^n\right]$$

$$\left(\frac{n-1}{4}\right)\left[1-(-1)^n\right].$$

EXERCISES

Using *Z*-transform solve the following difference equations.

1. $u(n+2)-4u(n+1)+4u(n)=0$ where $u(0)=0;\ u(1)=0$ $\left[\textit{Ans.}\ \ 2^n(1-n)\right]$

2. $u(n+2)-4u(n+1)+3u(n)=0$ where $u(0)=2;\ u(1)=4$ $\left[\textit{Ans.}\ \ 1+3^n\right]$

3. $u(n+2)-3u(n+1)+2u(n)=0$ where $u(0)=0;\ u(1)=1$ $\left[\textit{Ans.}\ \ 2^n-1\right]$

4. $u(n+2)-2\cos\alpha\, u(n+1)+u(n)=0$ where $u(0)=0;\ u(1)=1$ $\left[\textit{Ans.}\ \ \dfrac{\sin\alpha n}{\sin\alpha}\right]$

5. $u(n+2)-u(n)=2^n$ where $u(0)=0;\ u(1)=1$ $\left[\textit{Ans.}\ \ \dfrac{1}{3}\left[2^n-(-1)\right]\right]$

6. $u(n+1)-2u(n)=1,$ where $u(0)=0.$ $\left[\textit{Ans.}\ \ 2^n-1\right]$

7. $u(n+1)+4(n)=1,$ where $u(0)=0$ $\left[\textit{Ans.}\ \ \dfrac{1}{2}-\dfrac{1}{2}(-1)^n\right]$

8. $u(n+2)+2u(n+1)+u(n)=n$ given $u(0)=0;\ u(1)=0$

$$\left[\textit{Ans.}\ \ \left(\frac{n-1}{4}\right)\left[1-(-1)^n\right]\right]$$

9. $u(n+2)+6u(n+1)+9u(n)=2^n,$ if $u(0)=0;\ u(1)=0$

$$\left[\textit{Ans.}\ \ \frac{1}{25}2^n-\frac{1}{25}(-3)^n-\frac{1}{5}n(-3)^{n-1}\right]$$

10. $u(n+2)-5u(n+1)+6u(n)=5^n,$ if $u(0)=0;\ u(1)=0$ $\left[\textit{Ans.}\ \ \dfrac{2^n}{3}-\dfrac{3^n}{2}+\dfrac{5^n}{6}\right]$

11. $u(n+2)-5u(n+1)+6u(n)=4^n,$ given $u(0)=0;\ u(1)=1$ $\left[\textit{Ans.}\ \ \dfrac{1}{2}(4^n-2^n)\right]$

12. $u(n)=u(n-1)+u(n-2),$ given $u(0)=1;\ u(1)=1$

$$\left[\textit{Ans.}\ \ u(n)=\frac{1}{\sqrt{5}}\left(\frac{1}{2}\right)^{n+1}\left[\left(1+\sqrt{5}\right)^{n+1}-\left(1-\sqrt{5}\right)^{n+1}\right]\right]$$

CHAPTER 2
Integral Transforms

Introduction

In this chapter, we introduce different integral transforms which are very useful in obtaining the solution of linear differential equations, partial differential equations, boundary value problems and also in different concepts of engineering field. If we consider a function $f(t)$, $(a < t < b)$, then the integral transform of $f(t)$ is defined as $I[f(t)] = \phi(s) = \int_a^b f(t) \cdot k(s,t) dt$ [1]

where $k(s,t)$ is called the *kernel* of the transform. Depending on the various types of kernels, i.e, $k(s,t)$, or $k(\alpha,t)$, and (a,b). We obtain various integral transforms, i.e., Laplace transforms, Fourier transforms, Hankel transforms, Mellin transforms, Laguerre transforms, Legendre transforms, etc. Keeping in mind the scope of the book, the discussions will be restricted to Fourier transforms (F.T.), Hankel transforms (H.T.) and Mellin transforms (M.T.) which are widely used in solving problems in science and engineering. We obtain different integral transforms with a different choice of the *kernel,* as you will see in the following examples.

Example 1. If by denoting the kernel $k(s,t)$ and choosing the same as

$$k(s,t) = \begin{cases} e^{-st}, & \text{for } t \geq 0 \\ 0, & \text{for } t < 0 \end{cases}$$

then the transform $I[f(t)]$ in Equation [1] is known as Laplace transform (L.T.) and is denoted by $L[f(t)]$ and is defined as:

$$L[f(t)] = \phi(s) = \int_0^\infty e^{-st} \cdot f(t) dt \qquad [2]$$

Example 2. If we choose

$$k(\alpha,t) = \begin{cases} t^{\alpha-1}, & \text{if } t \geq 0 \\ 0, & \text{otherwise} \end{cases}$$

then the corresponding transform is called Mellin's transform and is given by

$$M(\alpha) = \int_0^{\infty} f(t) t^{\alpha-1} dt \qquad [3]$$

Example 3. Let $k(\alpha, x) = \frac{1}{\sqrt{2\pi}} e^{i\alpha x}, \ -\infty < x < \infty$. The corresponding transform is called thc Fourier transform and is given by

$$F(\alpha) = F\{f(x)\} = \frac{1}{\sqrt{2\pi}} \int_{-\infty}^{\infty} f(x) e^{i\alpha x} dx$$

Example 4. If we choose $k(\alpha, t) = \begin{cases} \sqrt{\frac{2}{\pi}} \sin \alpha t, & t \geq 0 \\ 0, & \text{otherwise} \end{cases}$

The corresponding transform is called the Fourier sine transform and is given by

$$F_s(\alpha) = F_s\{f(t)\} = \sqrt{\frac{2}{\pi}} \int_0^{\infty} f(t) \sin \alpha t \, dt$$

Example 5. If we choose

$$k(\alpha, t) = \begin{cases} \sqrt{\frac{2}{\pi}} \cos \alpha t, & \text{if } t \geq 0 \\ 0, & \text{otherwise} \end{cases}$$

The corresponding transform is called the Fourier cosine transform and is given by

$$F_c(\alpha) = F_c\{f(t)\} = \sqrt{\frac{2}{\pi}} \int_0^{\infty} f(t) \cos \alpha t \, dt.$$

2.1 Fourier Transform

Several engineering problems lead to ordinary or partial differential equations which have to be solved under various conditions. Integral transform enables us to solve some of these equations.

If a function defined is periodic in a finite internal $(-l, +l)$, then it can be expressed in Fourier series. By extending this concept, non-periodic function defined in $(-\infty < x < \infty)$ can be expressed as Fourier integral. Since in practice periodic functions are fairly rare.

We have the complex form of Fourier integral representation of $f(x)$ as:

$$f(x)=\frac{1}{2\pi}\int_{-\infty}^{\infty}\int_{-\infty}^{\infty}f(t)\cdot e^{i\alpha(t-x)}dtd\alpha$$

$$=\frac{1}{2\pi}\int_{-\infty}^{\infty}e^{i\alpha x}\left[\int_{-\infty}^{\infty}f(t)\cdot e^{i\alpha t}dt\right]d\alpha \qquad [1]$$

if $\quad F(\alpha)=\frac{1}{\sqrt{2\pi}}\left[\int_{-\infty}^{\infty}f(t)\cdot e^{i\alpha t}dt\right]$ [2]

then [1] becomes

$$f(x)=\frac{1}{\sqrt{2\pi}}\int_{-\infty}^{\infty}F(\alpha)\cdot e^{-i\alpha t}d\alpha \qquad [3]$$

In the above [2] is called the Fourier transform of $f(x)$ or $f(t)$ and [3] is called the inverse Fourier transform of $F(\alpha)$.

We note that the Fourier Transform $F(\alpha)$ is a frequency domain representation of a function $f(x)$, that characterises the frequency of the decomposed cosinusoids and sinusoids and is equal to the number of cycles per unit of *x*. If a function or waveform is not periodic, then the Fourier transform of the function will be a continuous function of frequency. The graph of $(\alpha,|F(\alpha)|)$ is called the amplitude Spectrum of $f(t)$.

We have the Fourier sine integral representation of $f(x)$ as

$$f(x)=\frac{2}{\pi}\int_0^{\infty}\sin dx\left[\int_0^{\infty}f(t)\sin\alpha t dt\right]d\alpha$$

Let $\quad F_s(\alpha)=\sqrt{\frac{2}{\pi}}\int_0^{\infty}f(t)\sin\alpha t\,dt$ [4]

then $\quad f(x)=\sqrt{\frac{2}{\pi}}\int_0^{\infty}F_s(\alpha)(d)\sin\alpha x\,d\alpha$ [5]

In the above [4] is called the Fourier sine transform of $f(x)$ and [5] is called the inverse Fourier sine transform of $F_s(\alpha)$.

Again, we have the Fourier cosine integral representation of $f(x)$ as

$$f(x)=\frac{2}{\pi}\int_0^{\infty}\cos\alpha x\left[\int_0^{\infty}f(t)\cos\alpha t\,dt\right]d\alpha$$

Let $\quad F_c(\alpha)=\sqrt{\frac{2}{\pi}}\int_0^{\infty}f(t)\cdot\cos\alpha t\,dt$ [6]

then $\quad f(x)=\sqrt{\frac{2}{\pi}}\int_0^{\infty}F_c(\alpha)\cos\alpha x\,d\alpha$ [7]

In the above, [6] is called the Fourier cosine transform of $f(x)$ and [7] is called the inverse Fourier cosine transform of $F_c(\alpha)$

Infinite Fourier transform or complex Fourier transform

Definition: Let $f(x)$ be a function defined in $(-\infty,\infty)$ and piece-wise continuous in each finite partial interval and absolutely integrable in $(-\infty,\infty)$. Then the complex Fourier transform of $f(x)$ is defined by

$$F(\alpha)=F\{f(x)\}=\frac{1}{\sqrt{2\pi}}\int_{-\infty}^{\infty} f(x)e^{i\alpha x}dx \qquad [1]$$

The function $f(x)$ is called the inverse Fourier transform of $F(\alpha)$

i.e., $f(x)=F^{-1}[F(\alpha)]$

$$=\frac{1}{\sqrt{2\pi}}\int_{-\infty}^{\infty} F(\alpha)e^{-i\alpha x}d\alpha$$

$$\Rightarrow \quad f(x)=\frac{1}{\sqrt{2\pi}}\int_{-\infty}^{\infty} F(\alpha)e^{-i\alpha x}d\alpha \qquad [2]$$

Note: The complex Fourier transform or simply Fourier transform can also be defined in some different forms as shown below.

1. $\begin{Bmatrix} F(\alpha)=\int_{-\infty}^{\infty} e^{i\alpha x} f(x)dx\cdot \\ f(x)=\frac{1}{2\pi}\int_{-\infty}^{\infty} F(\alpha)e^{i\alpha x}dx \end{Bmatrix}$	2. $\begin{Bmatrix} F(\alpha)=\int_{-\infty}^{\infty} e^{i\alpha x} f(x)dx \\ f(x)=\frac{1}{2\pi}\int_{-\infty}^{\infty} e^{i\alpha x}F(\alpha)d\alpha \end{Bmatrix}$
3. $F(\alpha)=\frac{1}{\sqrt{2\pi}}\int_{-\infty}^{\infty} e^{-i\alpha x} f(x)dx; f(x)=\frac{1}{\sqrt{2\pi}}\int_{-\infty}^{\infty} e^{-i\alpha x}F(\alpha)d\alpha$	

(A) EXAMPLES

1. Find $F\{f(x)\}$, where $f(x)=\begin{cases}1, \text{for} \lvert x\rvert < a \\ 0, \text{for} \lvert x\rvert > a\end{cases}$

***Solution*:** We have, by definition of Fourier transform

$$F(\alpha)=F\{f(x)\}=\frac{1}{\sqrt{2\pi}}\int_{-\infty}^{\infty} f(x)e^{i\alpha x}dx \qquad [1]$$

Here, $f(x) = 0$ for $|x| > a \Rightarrow$ for $x > a$, and $x < -a$

$f(x) = 1$ for $|x| < a \Rightarrow (-a < x < a)$

$$\therefore \quad F\{f(x)\} = \frac{1}{\sqrt{2\pi}} \int_{-a}^{a} 1 \cdot e^{i\alpha x} dx = \frac{1}{\sqrt{2\pi}} \left[\frac{e^{i\alpha x}}{+i\alpha} \right]_{-a}^{a}, \text{ for } \alpha \neq 0$$

$$= \frac{1}{\sqrt{2\pi}} \cdot \frac{1}{i\alpha} (e^{i\alpha a} - e^{-i\alpha a})$$

$$= \frac{1}{\sqrt{2\pi}} \cdot \frac{1}{i\alpha} (2i \sin a\alpha)$$

$$\therefore \quad F\{f(x)\} = F(\alpha) = \sqrt{\frac{2}{\pi}} \cdot \frac{\sin \alpha a}{\alpha} \qquad [2]$$

For $\alpha = 0$, we have, $F(0) = \frac{1}{\sqrt{2}} \int_{-a}^{a} 1 \cdot e^{0} dx = \frac{1}{\sqrt{2\pi}} (x)_{-a}^{a} = \sqrt{\frac{2}{\pi}} a.$

$$\therefore \quad F(\alpha) = \sqrt{\frac{2}{\pi}} a, \text{ when } \alpha = 0.$$

we note that $\lim_{\alpha \to 0} \sqrt{\frac{2}{\pi}} \frac{\sin a\alpha}{\alpha} = \sqrt{\frac{2}{\pi}} a \left(\lim_{a\alpha \to 0} \frac{\sin a\alpha}{a\alpha} \right) = \sqrt{\frac{2}{\pi}}$

$$\left(\because \lim_{\theta \to 0} \frac{\sin \theta}{\theta} = 1 \right)$$

Thus, we have $F\{f(x)\} = \sqrt{\frac{2}{\pi}} \cdot \frac{\sin a\alpha}{\alpha}$, for all α.

2. Find the Fourier transform of functions

$$f(x) = x, \text{ if } |x| \leq a$$
$$= 0, \text{ if } |x| > a$$

Solution: By definition of Fourier transform,

$$f(\alpha) = F\{f(x)\} = \frac{1}{\sqrt{2\pi}} \int_{-\infty}^{\infty} f(x) e^{i\alpha x} dx$$

$$= \frac{1}{\sqrt{2\pi}} \int_{-a}^{a} f(x) e^{i\alpha x} dx$$

$$(\therefore f(x) = 0 \text{ for } |x| > a$$

$$= \frac{1}{\sqrt{2\pi}} \int_{-a}^{a} x \cdot e^{i\alpha x} dx$$

$$(\therefore f(x) = x \text{ for } |x| \leq a)$$

$$= \frac{1}{\sqrt{2\pi}} \left[x(\frac{e^{i\alpha x}}{i\alpha}) - 1 \cdot \frac{e^{i\alpha x}}{(i\alpha)^2} \right]_{-a}^{a}, \text{ (by parts)}$$

$$= \frac{1}{\sqrt{2\pi}} \left\{ \frac{1}{i\alpha} [xe^{i\alpha x}]_{-a}^{a} + \frac{1}{\alpha^2} [e^{i\alpha x}]_{-a}^{a} \right\} \qquad (\because i^2 = -1)$$

$$= \frac{1}{\sqrt{2\pi}} \cdot \left\{ \frac{1}{i\alpha} (ae^{i\alpha x} + ae^{-i\alpha a}) + \frac{1}{\alpha^2} (e^{i\alpha a} e^{-i\alpha a}) \right\}$$

$$= \frac{1}{\sqrt{2\pi}} \left\{ \frac{1}{i\alpha} (a \cdot 2\cos a\alpha) + \frac{1}{\alpha^2} (2i \sin a\alpha) \right\},$$

$$(\because e^{i\theta} = \cos\theta + i\sin\theta; e^{-i\theta} = \cos\theta - i\sin\theta$$

$$\frac{1}{i} = -i; \qquad (\because i^2 = -1)$$

Thus $$F\{f(x)\} = F(\alpha) = \frac{1}{\sqrt{2\pi}} \left\{ \frac{-i}{\alpha} (2a\cos a\alpha) + i \cdot \frac{2}{\alpha^2} (\sin a\alpha) \right\}$$

$$F\{\alpha\} = \sqrt{\frac{2}{\pi}}\, i(\sin a\alpha - a\alpha \cos \alpha)$$

3. Find the Fourier transforms of the function

$$f(x) = \begin{cases} 0, \text{if } x \leq a \\ 1, \text{if } a < x < b \\ 0, \text{if } x \geq b \end{cases}$$

Solution: Given: $$f(x) = \begin{cases} 0, \text{if } x \leq a \\ 1, \text{if } a < x < b \\ 0, \text{if } x \geq b \end{cases} \qquad [1]$$

$$F\{f(x)\} = \frac{1}{\sqrt{2\pi}} \int_{-\infty}^{\infty} f(x) e^{i\alpha x} dx$$

$$= \frac{1}{\sqrt{2\pi}} \int_a^b 1 \cdot e^{i\alpha x} dx$$

$$= \frac{1}{\sqrt{2\pi}} \int_a^b 1 \cdot e^{i\alpha x} dx = \frac{1}{\sqrt{2\pi}} \left[\frac{e^{i\alpha x}}{i\alpha} \right]_a^b$$

$$F\{f(x)\} = \frac{1}{\sqrt{2\pi}} \frac{1}{i\alpha} \cdot (e^{i\alpha b} - e^{i\alpha a})$$

$$= -\frac{i}{\alpha\sqrt{2\pi}} (e^{i\alpha b} - e^{i\alpha a})$$

$$\therefore \quad F\{f(x)\} = -\frac{i}{\alpha\sqrt{2\pi}} (e^{i\alpha b} - e^{i\alpha a})$$

4. Find the Fourier transform of the following functions,

$$f(x) = \begin{cases} \sin x, \text{for } 0 < x < \pi \\ 0, \text{otherwise} \end{cases}$$

Solution: By definition, we have

$$F\{f(x)\} = \frac{1}{\sqrt{2\pi}} \int_{-\infty}^{\infty} f(x) e^{i\alpha x}$$

$$= \frac{1}{\sqrt{2\pi}} \int_0^{\pi} (\sin \pi) e^{i\alpha x} dx$$

$$= \frac{1}{\sqrt{2\pi}} \left[\frac{e^{i\alpha\pi}(i\alpha \sin x - \cos x)}{(i\alpha)^2 + 1} \right]_0^{\pi}$$

$$= \frac{1}{\sqrt{2\pi}} \frac{1}{(1-\alpha^2)} \left\{ e^{i\alpha\pi} \cdot (-\cos \pi) - (-\cos 0) \right\}$$

$$= \frac{1}{\sqrt{2\pi}} \cdot \frac{1}{(1-\alpha^2} \{(\cos \alpha\pi + i \sin \alpha\pi) + 1\}$$

$$= \frac{(1 + \cos \alpha\pi) + i \sin \alpha\pi}{\sqrt{2\pi}(1-\alpha^2)}$$

5. Find the Fourier transform of $f(x)=\begin{cases} a^2-x^2, \text{if} |x|<a \\ 0, \quad \text{if} |x|>a \end{cases}$

Hence evaluate

(i) $\int_0^\infty \frac{\sin x - x\cos x}{x^3}dx$

(ii) $\int_0^\infty \frac{\sin x - x\cos x}{x^3}\cdot \cos\frac{x}{2}dx = \frac{3\pi}{16}$

Solution: By definition,

$$F(\alpha)=F\{f(x)\}=\frac{1}{\sqrt{2\pi}}\int_{-\infty}^{\infty} f(x)e^{i\alpha x}dx$$

$$=\frac{1}{\sqrt{2\pi}}\int_{-a}^{a} f(x)e^{i\alpha x}dx \qquad (\because f(x)=0, \text{ if} |x|>a$$

$$=\frac{1}{\sqrt{2\pi}}\int_{-a}^{a} (a^2-x^2)e^{i\alpha x}dx \left(\because f(x)=(a^2-x^2)\text{if}|x|<a\right)$$

$$=\frac{1}{\sqrt{2\pi}}\left[(a^2-x^2)\frac{e^{i\alpha x}}{i\alpha}-(-2x)\frac{e^{i\alpha x}}{(i\alpha)^2}+(-2)\frac{e^{i\alpha x}}{(i\alpha)^3}\right]_{-a}^{a}$$

(by parts)

$$=\frac{2}{\sqrt{2\pi}}\left[-\frac{a}{\alpha^2}(e^{i\alpha a}+e^{i\alpha a})-\frac{i}{\alpha^3}(e^{i\alpha a}-e^{-i\alpha a})\right]$$

$$=\frac{2}{\sqrt{2\pi}}\left[-\frac{a}{\alpha^2}(2\cos\alpha a)-\frac{i}{\alpha^3}(2i\sin\alpha a)\right]$$

$$=\frac{2}{\alpha^3}\cdot\frac{\sqrt{2}}{\sqrt{\pi}}(\sin\alpha a-\alpha a\cos\alpha a)$$

Hence $\quad F(\alpha)=\frac{2}{\alpha^3}\cdot\frac{\sqrt{2}}{\sqrt{\pi}}(\sin\alpha a-\alpha a\cos\alpha a) \qquad [1]$

To prove the result: $\quad \int_0^\infty \frac{\sin x - x\cos x}{x^3}dx=\frac{\pi}{4}$

By using inversion formula, substituting for $F(\alpha)$ from [1] in this formula, we get

$$\left\{\begin{aligned} f(x) &= \frac{1}{\sqrt{2\pi}}\int_{-\infty}^{\infty}\frac{2\sqrt{2}}{\sqrt{\pi}}\cdot\frac{1}{\alpha^3}(\sin\alpha a-\alpha a\cos\alpha a)e^{-i\alpha x}dx \\ &= \frac{2}{\pi}\int_{-\infty}^{\infty}\frac{1}{\alpha^3}(\sin\alpha a-\alpha a\cos\alpha a)\cos\alpha x\,d\alpha \\ &\quad -i\frac{2}{\pi}\int_{-\infty}^{\infty}\frac{1}{\alpha^3}(\sin\alpha a-\alpha a\cos\alpha a)\sin\alpha x\,d\alpha \end{aligned}\right\} \quad [2]$$

The integrand in the second integral in the RHS of expression [2] is an odd function of α, and hence this integral vanishes. Also, the integrand is the first integral in an even function.

Hence [2] gives

$$\int_0^{\infty}\frac{1}{\alpha^3}(\sin\alpha a-\alpha a\cos\alpha a)\cos\alpha x d\alpha = \frac{\pi}{4}f(x) \quad [3]$$

$$\int_0^{\infty}\frac{1}{\alpha^3}(\sin\alpha-\cos\alpha)d\alpha = \frac{\pi}{4}f(0)$$

Since $f(0) = a^2 = 1$, when $a = 1$, we get

$$\int_0^{\infty}\frac{1}{\alpha^3}(\sin\alpha-\alpha\cos\alpha)\,d\alpha = \frac{\pi}{4} \quad [4]$$

or $\displaystyle\int_0^{\infty}\frac{(\sin x - x\cos x)}{x^3}dx = \frac{\pi}{4}$ which proves the required result.

Next, taking $a = 1,\ x = \dfrac{1}{2}$ in [3], we get

$$\int_0^{\infty}\frac{1}{\alpha^3}(\sin\alpha-\alpha\cos\alpha)\cos\left(\frac{\alpha}{2}\right)d\alpha = \frac{\pi}{4}+\left(\frac{1}{2}\right) \quad [5]$$

(Since $f\left(\dfrac{1}{2}\right) = \left(a^2-\dfrac{1}{4}\right) = \dfrac{3}{4}$ when a = 1)

Hence, we obtain

$$\int_0^{\infty}\frac{1}{\alpha^3}(\sin\alpha-\alpha\cos\alpha)\cos\left(\frac{\alpha}{2}\right)d\alpha = \frac{3\pi}{16} \quad [6]$$

which leads to the required result:

$$\int_0^{\infty}\frac{(\sin x - x\cos x)}{x^3}\cos\left(\frac{x}{2}\right)dx = \frac{3\pi}{16}$$

6. Find the Fourier transform of $f(x)=\begin{cases}1-|x| & \text{for}|x|\le 1\\ 0 & \text{for}|x|>1\end{cases}$

Hence deduce that $\int_0^{\infty}\frac{\sin^2 t}{t^2}dt=\frac{\pi}{2}$

Solution: Since $F(\alpha)=\frac{1}{\sqrt{2\pi}}\int_{-\infty}^{\infty}f(x)e^{i\alpha x}dx$

$$=\frac{1}{\sqrt{2\pi}}\int_{-1}^{1}(1-|x|)e^{i\alpha x}dx$$

$$=\frac{1}{\sqrt{2\pi}}\left\{\int_{-1}^{0}(1+x)e^{i\alpha x}dx+\int_{0}^{1}(1-x)e^{i\alpha x}dx\right\}$$

$$=\frac{1}{\sqrt{2\pi}}\left\{\left[(1+x)\frac{e^{i\alpha x}}{i\alpha}-\frac{e^{i\alpha x}}{(i\alpha)^2}\right]_{-1}^{0}+\left[(1-x)\frac{e^{i\alpha x}}{i\alpha}+\frac{e^{i\alpha x}}{(i\alpha)^2}\right]_{0}^{1}\right\}$$

(by parts)

$$=\frac{1}{\sqrt{2\pi}}\left\{\frac{2}{\alpha^2}-\frac{1}{\alpha^2}(e^{i\alpha}+e^{-i\alpha})\right\}=\frac{1}{\sqrt{2\pi}}\left\{\frac{2-2\cos\alpha}{\alpha^2}\right\}$$

$$=\sqrt{\frac{2}{\pi}}\cdot\frac{1-\cos\alpha}{\alpha^2} \qquad [1]$$

Now, by using the inversion formula and substituting for $F(\alpha)$ from [1] into this formula, we get

$$f(x)=\frac{1}{\sqrt{2\pi}}\int_{-\infty}^{\infty}F(\alpha)e^{-i\alpha x}d\alpha$$

$$=\frac{1}{\sqrt{2\pi}}\int_{-\infty}^{\infty}\sqrt{\frac{2}{\pi}}\cdot\frac{1-\cos\alpha}{\alpha^2}e^{-i\alpha x}d\alpha \text{ z}$$

$$=\frac{1}{\pi}\int_{-\infty}^{\infty}\frac{1-\cos\alpha}{\alpha^2}(\cos\alpha x-i\sin\alpha x)d\alpha$$

$$=\frac{1}{\pi}\int_{-\infty}^{\infty}\frac{1-\cos\alpha}{\alpha^2}\cdot\cos\alpha x d\alpha-\frac{i}{\pi}\int_{-\infty}^{\infty}\frac{1-\cos\alpha}{\alpha^2}\sin\alpha x \qquad [2]$$

The second term of Equation [2] contains the integrand which is an odd function of α; hence this integral vanishes. Accordingly, [2] yields

$$\int_{-\infty}^{\infty}\frac{1-\cos\alpha}{\alpha^2}\cos\alpha x d\alpha=\pi f(x) \qquad [3]$$

for $x = 0$, this yields

$$\int_{-\infty}^{\infty} \frac{1-\cos\alpha}{\alpha^2} dx = \pi f(0) = \pi,$$

($\because f(0) = 1$, by the definition of $f(x)$)

Hence, $\displaystyle\int_0^{\infty} \frac{1-\cos\alpha}{\alpha^2} d\alpha = \frac{\pi}{2}$

or $\displaystyle\int_0^{\infty} \frac{\sin^2 t}{t^2}\, dt = \frac{\pi}{2}, \left(\text{setting } t = \frac{\alpha}{2}\right)$

7. Find the Fourier transform of $e^{-a^2x^2}$, where a is a constant. Also deduce that $e^{\frac{-x^2}{2}}$ is self reciprocal under Fourier transform.

Solution: $\displaystyle F(\alpha) = \frac{1}{\sqrt{2\pi}} \int_{-\infty}^{\infty} e^{-a^2x^2} \cdot e^{i\alpha x} dx$

$$= \frac{1}{\sqrt{2\pi}} \int_{-\infty}^{\infty} e^{-(a^2x^2 - i\alpha x)} dx \qquad [1]$$

Note: $\displaystyle -a^2x^2 + i\alpha x - \frac{i^2\alpha^2}{4a^2} + \frac{i^2\alpha^2}{4a^2}$

$$= -\left[a^2x^2 - i\alpha x + \left(\frac{i\alpha}{2a}\right)^2\right] - \frac{\alpha^2}{4a^2}$$

$$= -\left(ax - \frac{i\alpha}{2a}\right)^2 - \frac{\alpha^2}{4a^2}$$

$$= \frac{e^{-\left(\frac{\alpha}{2a}\right)^2}}{\sqrt{2\pi}} \int_{-\infty}^{\infty} e^{-\left(ax - \frac{i\alpha}{2a}\right)^2} dx$$

Putting $y = \left(ax - \dfrac{i\alpha}{2a}\right)$: $dy = a\,dx$ in this, we obtain

$$F(\alpha) = \frac{e^{-\left(\frac{\alpha}{2a}\right)^2}}{a\sqrt{2\pi}} \int_{-\infty}^{\infty} e^{-y^2} dy$$

$$= 2\frac{e^{-\left(\frac{\alpha}{2a}\right)^2}}{a\sqrt{2\pi}}\int_0^{\infty} e^{-y^2}\,dy$$

$$= 2\cdot\frac{e^{-\left(\frac{\alpha}{2a}\right)^2}}{a\sqrt{2\pi}}\left\{\frac{1}{2}\Gamma\left(\frac{1}{2}\right)\right\}$$

$$= \frac{e^{-\left(\frac{\alpha}{2a}\right)^2}}{a\sqrt{2\pi}}\left(\sqrt{\pi}\right) = \frac{1}{\sqrt{2}}\cdot\frac{e^{-\left(\frac{\alpha}{2a}\right)^2}}{a}$$

This is the Fourier transform of the given function.

Note: In the above example, if we take $a = \frac{1}{\sqrt{2}}$, we find $f(x) = e^{-\frac{x^2}{2}}$ and $\hat{f}(\alpha) = e^{-\frac{\alpha^2}{2}}$. Thus, in this case, $\hat{f}(\alpha) = f(\alpha)$, and we say that $f(x)$ is *self-reciprocal* with respect to Fourier transform.

8. Find the complex Fourier transform of $f(x) = \cos ax^2$

***Solution*:** By using the Formula from the definition,

We obtain: $F(\alpha) = \frac{1}{\sqrt{2\pi}}\int_{-\infty}^{\infty}(\cos at^2)e^{i\alpha t}\,dt$

Which may be written as

$$F(\alpha) = \frac{1}{\sqrt{2\pi}}\int_{-\infty}^{\infty}\left[\frac{e^{iat^2} + e^{-iat^2}}{2}\right]e^{i\alpha t}\,dt$$

$$= \frac{1}{2\sqrt{2\pi}}\int_{-\infty}^{\infty}\left[e^{ia(t+\alpha/2a)^2 - i\alpha^2/4a} + e^{-ia(t-\alpha/2a)^2 + i\alpha^2/4a}\right]dt$$

$$= \frac{1}{2\sqrt{2\pi}}\int_{-\infty}^{\infty}\left[e^{-i(\alpha^2/4a)}\int_{-\infty}^{\infty} e^{ia(t+\alpha/2a)^2}\,dt\right.$$

$$\left. + e^{i(\alpha^2/4a)}\int_{-\infty}^{\infty} e^{-ia(t-\alpha/2a)^2}\,dt\right]$$

Setting $u=\sqrt{a}\left(t+\frac{\alpha}{2a}\right)$ and $v=\sqrt{a}\left(t-\frac{\alpha}{2a}\right)$, we obtain

$$F(\alpha)=\frac{1}{2\sqrt{2\pi a}}\left[e^{-i\left(\frac{\alpha^2}{4a}\right)}\int_{-\infty}^{\infty}e^{iu^2}du+e^{i\left(\frac{\alpha^2}{4a}\right)}\int_{-\infty}^{\infty}e^{-iv^2}dv\right]$$

$$=\frac{1}{2\sqrt{2\pi a}}\left[\left\{\cos\frac{\alpha^2}{4a}-i\sin\left(\frac{\alpha^2}{4a}\right)\right\}\int_{-\infty}^{\infty}(\cos u^2+i\sin u^2)du\right.$$

$$\left.+\left\{\cos\left(\frac{\alpha^2}{4a}\right)+i\sin\left(\frac{\alpha^2}{4a}\right)\right\}\int_{-\infty}^{\infty}(\cos v^2-i\sin v^2)dv\right]$$

Using the results

$$\int_{-\infty}^{\infty}\cos u^2du=\int_{-\infty}^{\infty}\sin u^2du=\sqrt{\frac{\pi}{2}},$$

we obtain

$$F(\alpha)=\frac{1}{2\sqrt{2\pi a}}\left[\left\{\cos\left(\frac{\alpha^2}{4a}\right)-i\sin\left(\frac{\alpha^2}{4a}\right)\right\}(1+i)\sqrt{\frac{\pi}{2}}\right.$$

$$\left.+\left\{\cos\left(\frac{\alpha^2}{4a}\right)-i\sin\left(\frac{\alpha^2}{4a}\right)\right\}[\ (1-i)\sqrt{\frac{\pi}{2}}\right.$$

which simplifies to

$$F(\alpha)=\frac{1}{2\sqrt{a}}\left[\cos\left(\frac{\alpha^2}{4a}\right)+\sin\left(\frac{\alpha^2}{4a}\right)\right]$$

This may be put in the alternative form

$$F(\alpha)=\frac{1}{\sqrt{2a}}\cos\left(\frac{\alpha^2}{4a}-\frac{\pi}{4}\right)$$

This is the complex Fourier transform of $f(x)=\cos ax^2$.

9. Show that the complex Fourier transform of $f(x)=xe^{-a|x|}$ where a is a positive constant and $-\infty<x<\infty$ is

$$\hat{f}(x)=\sqrt{\frac{2}{\pi}}\frac{2i\alpha a}{(a^2+\alpha^2)^2}$$

Solution: For the given $f(x)$,

$$\hat{f}(x)=\frac{1}{\sqrt{2\pi}}\int_{-\infty}^{\infty} xe^{-a|x|}e^{i\alpha x}dx$$

$$=\frac{1}{\sqrt{2\pi}}\left[\int_{-\infty}^{0} xe^{a|x|}e^{i\alpha x}dx+\int_{0}^{\infty} xe^{-a|x|}e^{i\alpha x}dx\right]$$

Using the fact that $|x| = x \text{ for } x \geq 0 \text{ and } |x| = -x \text{ for } x < 0,$ we get

$$\hat{f}(x)=\frac{1}{\sqrt{2\pi}}\left[\int_{-\infty}^{0} xe^{(a+i\alpha)x}dx+\int_{0}^{\infty} xe^{(a-i\alpha)x}dx\right]$$

Integration by parts yields:

$$=\frac{1}{\sqrt{2\pi}}\left\{\left[\frac{xe^{(a+i\alpha)x}}{a+i\alpha}\right]_{-\infty}^{0}-\left[\frac{e^{(a+i\alpha)x}}{(a+i\alpha)^2}\right]_{-\infty}^{0}\right.$$

$$\left.+\left[\frac{xe^{-(a-i\alpha)x}}{-(a-i\alpha)}\right]_{0}^{\infty}-\left[\frac{e^{-(a-i\alpha)x}}{(a-i\alpha)^2}\right]_{0}^{\infty}\right\}$$

$$=\frac{1}{\sqrt{2\pi}}\left[\frac{-1}{(a+i\alpha)^2}+\frac{1}{(a-i\alpha)^2}\right]$$

$$=\sqrt{\frac{2}{\pi}}\frac{2ia\alpha}{(a^2+\alpha^2)^2},\text{ as required.}$$

10. Show that $F\{\delta(x-a)\}=\frac{1}{\sqrt{2\pi}}e^{i\alpha a}$ where δ is the Dirac Delta function.

Solution: We know that δ function possesses the property that

$$\int_{-\infty}^{\infty}\varphi(x)\delta(x-a)dx=\varphi(a)$$

for any differentiable function $\varphi(x)$. Hence, in particular, we have

$$\int_{-\infty}^{\infty}e^{i\alpha x}\delta(x-a)\,dx=e^{i\alpha a}$$

Therefore,

$$F\{\delta(x-a)\}=\frac{1}{\sqrt{2\pi}}\int_{-\infty}^{\infty}e^{i\alpha x}\delta(x-a)\,dx$$

$$=\frac{1}{\sqrt{2\pi}}e^{i\alpha a}$$

Table 2.1: Infinite (or complex) Fourier transform

$f(x)$	$F(\alpha)$
$e^{-a^2x^2}$	$\frac{1}{a\sqrt{2}}e^{-\left(\frac{\alpha}{2a}\right)^2}$
$\frac{a}{a^2+x^2}, a>0$	$\sqrt{\frac{\pi}{2}}e^{-\lvert\alpha\rvert a}$
$\delta(x-a)$	$\frac{1}{\sqrt{2\pi}}e^{i\alpha a}$
$\cos ax^2$	$\frac{1}{\sqrt{2a}}\cos\left(\frac{\alpha^2}{4a}-\frac{\pi}{4}\right)$
$\sin ax^2$	$\frac{1}{\sqrt{2a}}\sin\left(\frac{\alpha^2}{4a}+\frac{\pi}{4}\right)$
$\frac{\sin ax}{x}$	$\sqrt{\frac{\pi}{2}}$ for $\lvert\alpha\rvert<a$, 0 for $\lvert\alpha\rvert>a$
$\frac{1}{\lvert x\rvert}, x\neq 0$	$\frac{1}{\lvert\alpha\rvert}$
$e^{-a\lvert x\rvert}$	$\sqrt{\frac{2}{\pi}}\frac{a}{a^2+\alpha^2}$
$xe^{-a\lvert x\rvert}$	$\sqrt{\frac{2}{\pi}}\frac{2ia\alpha}{(a^2+\alpha^2)^2}$
$\lvert x\rvert e^{-a\lvert x\rvert}$	$\sqrt{\frac{2}{\pi}}\frac{a^2-\alpha^2}{(a^2+\alpha^2)^2}$

Relationship between Fourier and Laplace Transforms

Given a function $f(x)$, suppose we define a new function $g(x)$ as

$$g(x)=\begin{cases}\sqrt{2\pi}e^{-\beta x}f(x), & \text{for } x>0\\ 0, & \text{for } x\leq 0\end{cases} \qquad [1]$$

where β is a real parameter.

Then

$$F\{g(x)\} = \frac{1}{\sqrt{2\pi}}\int_{-\infty}^{\infty} g(t)e^{i\alpha t}dt$$

$$= \frac{1}{\sqrt{2\pi}}\left[\int_{-\infty}^{\infty} 0\cdot e^{i\alpha t}dt + \int_{0}^{\infty}\sqrt{2\pi}f(t)e^{-\beta t}e^{i\alpha t}dt\right]$$

$$= \int_{0}^{\infty} e^{-st}f(t)dt, (\text{where } s = \beta - i\alpha) \qquad [2]$$

Thus, the Laplace transform of $f(x)$ is identical with the Fourier transform of $g(x)$ defined by [1].

Properties of Fourier Transform

Linearity property

For any two functions $f(x)$ and $g(x)$ (whose Fourier transform exists) and any two constants a and b.

$$F\{af(x)+bg(x)\} = aF\{f(x)\}+bF\{g(x)\} \qquad [3]$$

$$F^{-1}\{aF(\alpha)+bG(\alpha)\} = aF^{-1}\{F(\alpha)\}+bF^{-1}\{G(\alpha)\} \qquad [4]$$

Proof: We have, by the definition,

$$F\{f(x)\} = F(\alpha) = \frac{1}{\sqrt{2\pi}}\int_{-\infty}^{\infty} f(t)e^{i\alpha t}dt$$

$$F\{g(x)\} = G(\alpha) = \frac{1}{\sqrt{2\pi}}\int_{-\infty}^{\infty} g(t)e^{i\alpha t}dt$$

Therefore

$$F\{af(x)\}+bF\{g(x)\} = \frac{1}{\sqrt{2\pi}}\int_{-\infty}^{\infty}\{af(t)+bg(t)\}e^{i\alpha t}dt$$

$$= F\{af(x)+bg(x)\}.$$

This proves result [3]. The proof of result [4] is similar.

Change of scale property

If $F(\alpha)$ is the complex Fourier transform of $f(x)$, then for any nonzero real constant a, the complex Fourier transform of $f(ax)$ is $\frac{1}{|a|}F\left(\frac{\alpha}{a}\right)$.

Proof: By the definition of complex Fourier transform, we have

$$F\{f(ax)\}=\frac{1}{\sqrt{2\pi}}\int_{-\infty}^{\infty} f(at)e^{i\alpha t}dt \qquad [1]$$

Suppose a>0. Let us set *at=y.* Then [1] becomes

$$F\{f(ax)\}=\frac{1}{a}\frac{1}{\sqrt{2\pi}}\int_{\infty}^{-\infty} f(y)e^{i\left(\frac{\alpha}{a}\right)y}dy$$

$$=-\frac{1}{a}F\left(\frac{\alpha}{a}\right) \qquad [2]$$

On the other hand, suppose a<0. Then, if we set again $at = y$, [1] becomes

$$F\{f(ax)\}=\frac{1}{a}\frac{1}{\sqrt{2\pi}}\int_{\infty}^{-\infty} f(y)e^{i\left(\frac{\alpha}{a}\right)y}dy$$

$$=-\frac{1}{a}F\left(\frac{\alpha}{a}\right) \qquad [3]$$

Since $|a|=a$ for a>0 and $|a|=-a$ for a<0, expressions [2] and [3] show that

$$F\{f(dx)\}=\frac{1}{|a|}F\left(\frac{\alpha}{a}\right) \qquad [4]$$

as required.

Shift property

For any real constant *a,*

(i) $F\{f(x-a)\}=e^{i\alpha a}F(\alpha)$ [1]

(ii) $F\{e^{iax}f(x)\}=F(\alpha+a)$ [2]

Proof: (i) We have

$$F(\alpha)=\frac{1}{\sqrt{2\pi}}\int_{-\infty}^{\infty} f(y)e^{i\alpha y}dy=F\{f(x)\}$$

Therefore

$$e^{i\alpha a}F(\alpha)=\frac{1}{\sqrt{2\pi}}\int_{-\infty}^{\infty} f(y)e^{i\alpha(a+y)}dy$$

$$=\frac{1}{\sqrt{2\pi}}\int_{-\infty}^{\infty} f(t-a)e^{i\alpha t}dt,\ \text{on setting } t=a+y$$

$$=F\{f(x-a)\}$$

(ii) Next,

$$F(\alpha+a)=\frac{1}{\sqrt{2\pi}}\int_{-\infty}^{\infty} f(t)e^{i(\alpha+a)t}\,dt$$

$$=\frac{1}{\sqrt{2\pi}}\int_{-\infty}^{\infty} g(t)e^{i\alpha t}\,dt, \qquad \text{where } g(t)=f(t)e^{iat}$$

$$=F\{g(x)\}$$

$$=F\left\{f(x)e^{iax}\right\}$$

This completes the proof.

Note: The results [1] and [2] can be put in the following equivalent forms:

$$\left\{\begin{aligned} &F^{-1}\left\{e^{i\alpha a}F(\alpha)\right\}=f(x-a) && [1]\\ &F^{-1}\{F(\alpha+a)\}=e^{iax}f(x) && [2]\end{aligned}\right\}$$

11. If $f(x)$ is a complex-valued function, show that

(i) $F\left\{\overline{f(x)}\right\}=\overline{F(-\alpha)}$

(ii) $F\left\{\overline{f(-x)}\right\}=\overline{F(\alpha)}$

where an over bar denotes the complex conjugate, as usual.

Solution: (i) We have $F(-\alpha)=\frac{1}{\sqrt{2\pi}}\int_{-\infty}^{\infty} f(t)e^{i(-\alpha)t}\,dt$

Therefore

$$\overline{F(-\alpha)}=\frac{1}{\sqrt{2\pi}}\int_{-\infty}^{\infty} \overline{f(t)}e^{i\alpha t}\,dt$$

$$=F\left\{\overline{f(x)}\right\}$$

(ii) Next, $\overline{F(\alpha)}=\frac{1}{\sqrt{2\pi}}\int_{-\infty}^{\infty} \overline{f(t)}e^{i\alpha t}\,dt$

Setting $y=-t$, this becomes

$$\overline{F(\alpha)}=-\frac{1}{\sqrt{2\pi}}\int_{+\infty}^{-\infty} \overline{f(-y)}e^{i\alpha y}\,dy$$

$$=\frac{1}{\sqrt{2\pi}}\int_{-\infty}^{\infty} \overline{f(-y)}e^{i\alpha y}\,dy$$

$$=F\left\{\overline{f(-x)}\right\}$$

12. If $F\{f(x)\} = F(\alpha)$, show that $F\{f(x)\}\cos ax = \frac{1}{2}[F(\alpha + a) + F(\alpha - a)]$ where a is a real constant.

Solution: We note that $f(x)\cos ax = \frac{1}{2} f(x)\left\{e^{iax} + e^{-iax}\right\}$

Therefore, by linearity and shift properties, we find

$$F\{f(x)\cos ax\} = \frac{1}{2}\left[F\left\{f(x)e^{iax}\right\} + F\left\{f(x)e^{-iax}\right\}\right]$$

$$= \frac{1}{2}[F(\alpha + a) + F(\alpha - a)]$$

13. If a is a nonzero real constant and b is any real constant, show that

$$F\{f(ax - b)\} = \frac{e^{i\alpha b/a}}{|a|} F\left(\frac{\alpha}{a}\right)$$

Solution: We have $F\{f(ax - b)\} = \frac{1}{\sqrt{2\pi}} \int_{-\infty}^{\infty} f(ay) e^{i\alpha\left(\frac{y+b}{a}\right)} dy$

(setting $ax - b = ay$)

$$= e^{\frac{i\alpha b}{a}} F\{f(ax)\}$$

$$= \frac{e^{\frac{i\alpha b}{a}}}{|a|} F\left(\frac{\alpha}{a}\right).$$

on using the change of scale property.

Transform of the Derivative and Derivative of the Transform

(1) Transform of the nth derivative

If $F(\alpha)$ is the Fourier transform of $f(x)$, then the transform of the nth derivative $f^{(n)}(x)$ is $(-i\alpha)^n F(\alpha)$, provided that the first $(n-1)$derivatives of $f(x)$ tend to zero as $x \to \pm\infty$, i.e.,

$$F\left\{f^{(n)}(x)\right\} = (-i\alpha)^n F\{f(x)\} \qquad [1]$$

when $f^{(r)}(x) \to 0$ as $x \to \pm\infty$ for $r = 0, 1, 2, \ldots n-1$

Proof: By definition, we have

$$F\{f'(x)\} = \frac{1}{\sqrt{2\pi}}\int_{-\infty}^{\infty} f'(t)e^{i\alpha t}\,dt$$

$$F\{f'(x)\} = \frac{1}{\sqrt{2\pi}}\left\{\left[f(t)e^{i\alpha t}\right]_{-\infty}^{\infty} - \int_{-\infty}^{\infty}(i\alpha)f(t)e^{i\alpha t}\,dt\right\} \text{ (by parts)}$$

$$= -i\alpha F\{f(x)\}$$

provided $f(x)\to 0$ as $x\to\pm\infty$, result [1] is true for $n=1$. Let us assume that the result [1] is valid for $n=k$, i.e.,

$$F\left\{f^{(k)}(x)\right\} = (-i\alpha)^k F\{f(x)\}$$

Then

$$F\left\{f^{(k+1)}(x)\right\} = \frac{1}{\sqrt{2\pi}}\int_{-\infty}^{\infty} f^{(k+1)}(t)e^{i\alpha t}\,dt$$

Integrating by parts, we obtain

$$F\left\{f^{(k+1)}(x)\right\} = \frac{1}{\sqrt{2\pi}}\left\{\left[f^{(k)}(t)e^{i\alpha t}\right]_{-\infty}^{\infty} - \int_{-\infty}^{\infty}(i\alpha)f^{(k)}(t)e^{i\alpha t}\,dt\right\}$$

$$= (-i\alpha)F\left\{f^{(k)}(x)\right\}, \text{provided } f^{(k)}(x)\to 0 \text{ as } x\to\pm\infty$$

$$= (-i\alpha)(-i\alpha)^k F\{f(x)\}$$

$$= (-i\alpha)^{k+1} F\{f(x)\}$$

Thus, the statement is true for $n=k+1$. Hence, by the principle of mathematical induction, the result [1] is valid for all positive integral values of *n*.

$$F\left\{\overset{(n)}{f(x)}\right\} = (-i\alpha)^n F\{f(x)\}$$

(2) n[th] derivative of the Transform

If $F(\alpha)$ is the Fourier transform of $f(x)$, then

$$\frac{d^n}{d\alpha^n}\{F(\alpha)\} = i^n F\left\{x^n f(x)\right\} \qquad [2]$$

Proof: By definition, we have

$$F(\alpha) = \frac{1}{\sqrt{2\pi}}\int_{-\infty}^{\infty} f(t)e^{i\alpha t}\,dt$$

$$\therefore \quad F(\alpha) = \frac{1}{\sqrt{2\pi}} \frac{d}{d\alpha}\left\{\int_{-\infty}^{\infty} f(t) e^{i\alpha t} dt\right\}$$

$$= \frac{1}{\sqrt{2\pi}} \int_{-\infty}^{\infty} f(t) \frac{\partial}{\partial a}\left\{e^{i\alpha t}\right\} dt$$

$$= \frac{1}{\sqrt{2\pi}} \int_{-\infty}^{\infty} f(t) \text{i} t\, e^{i\alpha t} dt$$

$$= \text{i}F\{xf(x)\}$$

This shows that the result [2] is valid for $n = l$. We suppose that the result is valid for $n = k$, i.e.,

$$\hat{f}^{(k)}(\alpha) = \text{i}^k F\left\{x^k f(x)\right\}$$

Then

$$\hat{f}^{(k+1)}(\alpha) = \frac{d}{d\alpha}\left\{\hat{f}^{(k)}(\alpha)\right\} = \text{i}^k \frac{d}{d\alpha}\left[F\left\{x^k f(x)\right\}\right]$$

$$= \text{i}^k \frac{d}{d\alpha}\left[\frac{1}{\sqrt{2\pi}} \int_{-\infty}^{\infty} t^k f(t) e^{i\alpha t} dt\right]$$

$$= \text{i}^k \frac{1}{\sqrt{2\pi}} \int_{-\infty}^{\infty} t^k f(t) \text{i} t\, e^{i\alpha t} dt$$

$$= \text{i}^{k+1} F\left\{x^{k+1} f(x)\right\}.$$

Thus, the result os true for $n = k+1$. Hence, by the principle of mathematical induction, the result [2] is true for all integral values of n.

14. Find the inverse complex Fourier transform of $F(\alpha) = e^{|\alpha| a}, a > 0$

Solution: For the given F(α)

$$f(x) = \frac{1}{\sqrt{2\pi}} \int_{-\infty}^{\infty} e^{|\alpha| a} e^{-i\alpha x} d\alpha$$

$$= \frac{1}{\sqrt{2\pi}}\left\{\left[\frac{e^{(a-\text{i}x)\alpha}}{a-\text{i}x}\right]_{-\infty}^{0} + \left[\frac{e^{-(a+\text{i}x)\alpha}}{-(a+ix)}\right]_{0}^{\infty}\right\}$$

$$= \frac{1}{\sqrt{2\pi}}\left\{\frac{1}{a-\text{i}x} + \frac{1}{a+\text{i}x}\right\}$$

$$= \sqrt{\frac{2}{\pi}} \frac{a}{a^2 + x^2}$$

This is the inverse complex Fourier transform of the given F(α)

15. If $\int f(x)dx \to 0$ as $x \to \pm\infty$, show that $F\left\{\int f(x)\,dx\right\} = \frac{\mathrm{i}}{\alpha} F(\alpha)$

***Solution*:** Let $g(x) = \int f(x)dx$. Then

$$F(\alpha) = F\{f(x)\} = F\{g'(x)\}$$

$$= (-\mathrm{i}\alpha)F\{g(x)\},$$

Using [1] provided $g(x) \to 0$ as $x \to \pm\infty$. Hence

$$F\left\{\int f(x)dx\right\} = F\{g(x)\} = \frac{1}{(-\mathrm{i}\alpha} F(\alpha) = \frac{\mathrm{i}}{\alpha} F(\alpha)$$

16. Prove that $F\left\{x^k f^{(n)}(x)\right\} = (-\mathrm{i})^{n+k} \frac{d^k}{d\alpha^k}\left\{\alpha^n F(\alpha)\right\}$

***Solution*:** Let $g(x) = f^{(n)}(x)$. Then

$$F\left\{x^k \overset{(n)}{f}(x)\right\} = F\left\{x^k g(x)\right\}$$

$$= \left(\frac{1}{\mathrm{i}^k}\right) \frac{d^k}{d\alpha^k}\{G(\alpha)\} \qquad \left(\because \frac{d^n}{d\alpha^n}\{F(\alpha)\} = \mathrm{i}^n F\left\{x^n f(x)\right\}\right)$$

$$= (-\mathrm{i})^k \frac{d^k}{d\alpha^k}\left[F\left\{f^{(n)}(x)\right\}\right]$$

$$= (-\mathrm{i})^k \frac{d^k}{d\alpha^k}\left[(-\mathrm{i}\alpha)^n F\{f(x)\}\right]$$

$$\left(\text{Using } F\left\{f^{(n)}(x)\right\} = (-\mathrm{i}\alpha)^n F\{f(x)\}\right)$$

$$= (-\mathrm{i})^{k+n} \frac{d^k}{d\alpha^k}\left[\alpha^n F(\alpha)\right]$$

17. Find the Fourier transforms of the following function

(i) $x \cos x^2$ (ii) $x^2 e^{-x^2}$

Solution: (i) We note that

$$F\left\{\sin x^2\right\} = \frac{1}{\sqrt{2}} \sin\left(\frac{\alpha^2}{4} + \frac{\pi}{4}\right) \qquad [1]$$

Let $f(x) = \sin x^2$;

$$\therefore \quad f'(x) = 2x \cos x^2$$

Hence, $F\left\{x \cos x^2\right\} = \frac{1}{2} F\{f'(x)\}$

$$= \frac{1}{2}(-\mathrm{i}\alpha) F\left\{\sin x^2\right\}$$

$$\left(\because F\left\{f^{(n)}(x)\right\} = (\mathrm{i}\alpha)^n F\{f(x)\}\right)$$

$$= -\frac{\mathrm{i}\alpha}{2} \cdot \frac{1}{\sqrt{2}} \sin\left(\frac{\alpha^2}{4} + \frac{\pi}{4}\right) \qquad \text{(Using [1])}$$

$$= -\frac{\mathrm{i}\alpha}{4}\left(\sin\frac{\alpha^2}{4} + \cos\frac{\alpha^2}{4}\right)$$

$$\therefore \quad F\left\{x \cos x^2\right\} = -\frac{\mathrm{i}\alpha}{4}\left(\sin\frac{\alpha^2}{4} + \cos\frac{\alpha^2}{4}\right.$$

(ii) We know that

$$F\left\{e^{-x^2}\right\} = \frac{1}{\sqrt{2}} e^{-\alpha^2/4}$$

Now, by using $\frac{d^n}{d\alpha^n}\{F(\alpha)\} = \mathrm{i}^n F\left\{x^n f(x)\right\}$

We have, $F\left\{x^2 e^{-x^2}\right\} = \frac{1}{\mathrm{i}^2} \frac{d^2}{d\alpha^2}\left\{\frac{1}{\sqrt{2}} e^{-\alpha^2/4}\right\}$

$$= \frac{1}{2\sqrt{2}}\left(1 - \frac{\alpha^2}{2}\right) e^{-\alpha^2/4}$$

Convolution theorem and parseval's identity

Convolution theorem

Definition: The convolution of two functions $f(x)$ and $g(x)$ is defined as

$$(f * g)(x) = \frac{1}{\sqrt{2\pi}} \int_{-\infty}^{\infty} f(x-t) g(t)\, dt \qquad [1]$$

Evidently, $(f * g)(x)$ is a function of x putting $x - t = u$ in [1], we see that

$$(f * g) = \frac{1}{\sqrt{2\pi}} \int_{-\infty}^{\infty} f(u)\, g(x-u)\, du$$

$$= \frac{1}{\sqrt{2\pi}} \int_{-\infty}^{\infty} g(x-t) f(t)\, dt \qquad [2]$$

Clearly, RHS is $g * f \; = \; g * f$

Thus, $f * g = g * f$

Valid Properties:

(i) (f * g) * v = f * (g * v) - Associative

(ii) f * (g_1+g_2) = (f * g_1) + (f * g_2) - Distributive

(iii) f * 0 = 0 * f = 0 - Commutative

The following theorem (known as the convolution theorem) provides a simple formula for the Fourier transform of ($f * g$).

Theorem (I): The Fourier transform of the convolution of $f(x)$ and $g(x)$ is the product of their Fourier transforms.

i.e., $F\big(f(x) * g(x)\big) = F(\alpha)\cdot G(\alpha) = F\big[f(x)\big]\cdot F\big[g(x)\big]$ [3]

Proof: $F\{f * g\} = \frac{1}{\sqrt{2\pi}} \int_{-\infty}^{\infty} (f * g)(x) e^{i\alpha x} dx$

(by the definition of Fourier transform)

Now $F\{f * g\} = \frac{1}{\sqrt{2\pi}} \int_{-\infty}^{\infty} \left[\frac{1}{\sqrt{2\pi}} \int_{-\infty}^{\infty} f(x-t) g(t) dt \right] e^{i\alpha x} dx$

$= \frac{1}{2\pi} \int_{-\infty}^{\infty} \int_{-\infty}^{\infty} f(x-t) g(t) e^{i\alpha x} dt dx$ (Using Equation [1])

Changing the variables of integration in the double integral from *x* and *t* to *y* and *u*, where $u = x - t,\; u = t^2$

We obtain $F\{f*g\}=\frac{1}{2\pi}\int_{-\infty}^{\infty}\int_{-\infty}^{\infty}f(y)g(u)e^{i\alpha(y+u)}dudt$

$$=\left\{\frac{1}{\sqrt{2\pi}}\int_{-\infty}^{\infty}f(y)e^{i\alpha y}dy\right\}\left\{\frac{1}{\sqrt{2\pi}}\int_{-\infty}^{\infty}g(u)e^{i\alpha u}du\right\}$$

$$=F(\alpha)G(\alpha)$$

Hence, $F\{f*g\}=F(\alpha)G(\alpha)=F\{f(x)\}\cdot F\{g(x)\}$

or equivalently, $f*g=F^{-1}\left[F\{f(x)F\{g(x)\}\right]$

$$=F^{-1}\{F(\alpha)G(\alpha)\}$$

Parseval's identity for Fourier Transforms

We now obtain an important relation called the Parseval's identity which is a direct consequence of the convolution theorem.

Theorem (II): Let F(α) be the Fourier transform of $f(x)$. Then

$$\int_{-\infty}^{\infty}\left|f(x)^2\right|dx=\int_{-\infty}^{\infty}\left|F(\alpha)\right|^2 d\alpha.$$

Proof: For any two functions *f*(*x*) and *g*(*x*) we have by the convolution theorem (Theorem [I])

$$F(f*g)=F(\alpha)G(\alpha)$$

$$\Rightarrow \quad f*g=F^{-1}[F(\alpha)G(\alpha)]$$

$$\therefore \quad \frac{1}{\sqrt{2\pi}}\int_{-\infty}^{\infty}f(t)g(x-t)dt=\frac{1}{\sqrt{2\pi}}\int_{-\infty}^{\infty}F(\alpha)G(\alpha)e^{-i\alpha x}d\alpha$$

Putting *x* = *0*, we get

$$\int_{-\infty}^{\infty}f(t)g(-t)dt=\int_{-\infty}^{\infty}F(\alpha)G(\alpha)d\alpha$$

Now taking $g(t)=\overline{f(-t)}$, we get

$$\int_{-\infty}^{\infty}f(t)\overline{f(t)}\,dt=\int_{-\infty}^{\infty}F(\alpha)\overline{G(\alpha)}\,d\alpha \qquad \left(\begin{array}{l}\because F\left[\overline{f(-x)}\right]=\overline{F(\alpha)}\\ F\left[f(-x)\right]=F(-\alpha)\end{array}\right)$$

Hence, $\int_{-\infty}^{\infty}|f(x)|^2\,dx=\int_{-\infty}^{\infty}|F(\alpha)|^2\,d\alpha$

This relation is known as the Parseval's identity for complex Fourier transform.

Fourier sine and Fourier cosine transforms

Definition 1: The (infinite) Fourier sine transform. The infinite Fourier sine transform (IFST) of *f(x)*, $(0 < x < \infty)$, is denoted by $F_s(\alpha)$ or $F_s\{(f(x)\}$ and is defined as:

$$F_s(\alpha) = \sqrt{\frac{2}{\pi}} \int_0^\infty f(x) \sin \alpha x \, dx \qquad [1]$$

The inverse formula for infinite Fourier sine transform is given by

$$f(x) = F_s^{-1}\{F_s(\alpha)\} = \sqrt{\frac{2}{\pi}} \int_0^\infty F_s(\alpha) \sin \alpha x \, d\alpha \sqrt{a^2 + b^2} \qquad [2]$$

IFST-pair

$$F_s(\alpha) = \sqrt{\frac{2}{\pi}} \int_0^\infty f(x) \sin \alpha x \, dx$$

$$f(x) = F_s^{-1}\{F_s(\alpha)\} = \sqrt{\frac{2}{\pi}} \int_0^\infty F_s(\alpha) \sin \alpha x \, d\alpha$$

Definition 2: The (Infinite) Fourier cosine transform (IFCT). The infinite Fourier cosine transform of $f(x), 0 < x < \infty$, is denoted by $F_c(\alpha)$ or $F_c\{f(x)\}$ and is defined as:

$$F_c(\alpha) = \sqrt{\frac{2}{\pi}} \int_0^\infty f(x) \cos \alpha x \, dx \qquad [1]$$

The inverse formula for infinite Fourier cosine transform is given by:

$$F_c^{-1}\{F_c(\alpha)\} = f(x) = \sqrt{\frac{2}{\pi}} \int_0^\infty F_c(\alpha) \cos \alpha x \, d\alpha \qquad [2]$$

IFCT-pair

$$F_c(\alpha) = \sqrt{\frac{2}{\pi}} \int_0^\infty f(x) \cos \alpha x \, dx$$

$$f(x) = F_c^{-1}\{F_c(\alpha)\} \cos \alpha x \, d\alpha$$

Remark: We may also take $\frac{2}{\pi}$ or 1 instead of $\sqrt{\frac{2}{\pi}}$ as the coefficient of the integrals.

Now, let us prove some basic properties of Fourier sine and cosine transforms

P$_1$: **(Linearity property)** if a and b are constants, then for two functions $f(x)$ and $g(x)$,

(i) $F_s\{af(x) + bg(x)\} = aF_s\{f(x)\} + bF_s\{g(x)\}$ [1]

(ii) $F_s^{-1}\{aF_s(\alpha)+bG_s(\alpha)\}=af(x)+bg(x)$ [2]

Proof: (i) The definition of Fourier transform yields

$$F_s\{af(x)+bg(x)\}=\sqrt{\frac{2}{\pi}}\int_0^\infty\{af(x)+bg(x)\}\sin\alpha x\,dx$$

$$=a\left\{\sqrt{\frac{2}{\pi}}\int_0^\infty g(x)\sin\alpha x\,dx\right\}$$

$$+b\left\{\sqrt{\frac{2}{\pi}}\int_0^\infty g(x)\sin\alpha x\,dx\right\}$$

$$=aF_s(\alpha)+bG_s(\alpha)$$

or $F_s\{af(x)+bg(x)\}=aF_s(\alpha)+bG_s(\alpha)$ which is the required result [1]

(ii) Therefore, $af(x)+bg(x)=F_s^{-1}[aF_s(\alpha)+bG_s(\alpha)]$

This proves result [2]

Generalization:

(1) $F_s[C_1\cdot f_1+C_2\cdot f_2+...]$

$=C_1F_3|f_1|+C_2\cdot F_s|f_2|+...$

The proof of this is similar to that of infinite complex Fourier transforms. Similarly, this result is true for cosine transforms.

Hence, if $C_1\cdot C_2...C_n$ are all constants then

(2) $F_c[C_1f_1+C_2f_2+...]$

$=C_1F_c|f_1|+C_2F_c|f_2|+...$

P$_2$: (Change of scale property)

If a is a constant, and

(i) $F_c\{f(x)\}=F_c(\alpha)$, then $F_c\{f(ax)\}=\frac{1}{\alpha}F_c\left(\frac{\alpha}{a}\right)$

(ii) $F_s\{f(x)\}=F_s(\alpha)$, then $F_s\{f(ax)\}=\frac{1}{\alpha}F_s\left(\frac{\alpha}{a}\right)$

Proof: (i) By definition, we have

$$F_c\{f(ax)\} = \sqrt{\frac{2}{\pi}}\int_0^\infty f(ax)\cdot\cos\alpha x\,dx \qquad [1]$$

$$\left[\begin{array}{l}\left(\text{put } ax = u; x = \frac{1}{a}u; dx = \frac{1}{a}du\right)\\ \text{Limits: } u \text{ varies from } 0 \text{ to } \infty\end{array}\right]$$

$$\therefore \quad F_c[f(ax)] = \sqrt{\frac{2}{\pi}}\int_0^\infty f(u)\cos\left(\frac{\alpha u}{a}\right)\cdot\frac{1}{a}du$$

$$= \frac{1}{a}\cdot\sqrt{\frac{2}{\pi}}\int_0^\infty f(u)\cdot\cos\left(\frac{\alpha}{a}\right)u\cdot du \qquad [2]$$

But $F_c(\alpha) = F_c[f(x)] = \sqrt{\frac{2}{\pi}}\int_0^\infty f(x)\cos\alpha x\,dx$

$$\therefore \quad \sqrt{\frac{2}{\pi}}\int_0^\infty f(u)\cos\left(\frac{\alpha}{a}\right)u\cdot du = F_c\left(\frac{\alpha}{a}\right) \qquad [3]$$

$$\Rightarrow \quad F_c[f(ax)] = \frac{1}{a}\cdot F_c\left(\frac{\alpha}{a}\right)$$

Proof: (ii) By definition, we have

$$F_s\{f(ax)\} = \sqrt{\frac{2}{\pi}}\int_0^\infty f(x)\sin\alpha x\,dx \qquad [1]$$

$$\left[\begin{array}{l}\left(\text{put } ax = u; dx = \frac{du}{a}\right)\\ \text{Limits: } u \text{ varies from } 0 \text{ to } \infty\end{array}\right]$$

$$\therefore \quad F_s[f(x)] = \sqrt{\frac{2}{\pi}}\int_0^\infty f(u)\sin\left(\alpha\cdot\frac{u}{a}\right)\cdot\frac{du}{a}$$

$$= \frac{1}{a}\int_0^\infty f(u)\sin\left\{\left(\frac{\alpha}{a}\right)u\right\}du$$

$$= \frac{1}{a}F_s\left(\frac{\alpha}{a}\right) \qquad \left(\text{on replacing } \alpha \text{ by } \frac{\alpha}{a} \text{ in [1]}\right)$$

$$\therefore \quad F_s[f(ax)] = \frac{1}{a}\cdot F_s\left(\frac{\alpha}{a}\right)$$

P_3: (Shifting property)

If '*a*' is a constant and

$$F_c[f(x)] = F_c(\alpha); F_s[f(x)] = F_s(\alpha).$$

then,

$$\begin{cases} F_c[f(x)\cos ax] = \frac{1}{2}[F_c(\alpha+a)+F_c(\alpha-a)] \\ F_s[f(x)\cos ax] = \frac{1}{2}[F_s(\alpha+a)+F_s(\alpha-a)] \\ F_c[f(x)\sin ax] = \frac{1}{2}[F_s(\alpha+a)-F_s(\alpha-a)] \\ F_c[f(x)\sin ax] = \frac{1}{2}[F_c(\alpha-a)-F_c(\alpha+a)] \end{cases}$$

Proof: By definition, we have

$$F_c[f(x)\cos ax] = \sqrt{\frac{2}{\pi}}\int_0^\infty f(x)\cos ax \cdot \cos\alpha x\, dx$$

$$= \sqrt{\frac{2}{\pi}}\int_0^\infty f(x)\cdot\frac{1}{2}\{\cos(\alpha+a)x+\cos(\alpha-a)x\}\cdot dx$$

$\because$ $2C_A C_B = C_{A+B} + C_{A-B}$

$[(\text{put } A = \alpha x : B = ax)]$

$$\therefore \quad F_c[f(x)\cos ax] = \frac{1}{2}\left\{\sqrt{\frac{2}{\pi}}\int_0^\infty f(x)\cos(\alpha+a)x\,dx + \sqrt{\frac{2}{\pi}}\int_0^\infty f(x)\cos(\alpha-a)x\,dx\right\} \qquad [1]$$

But, $\sqrt{\frac{2}{\pi}}\int_0^\infty f(x)\cos\alpha x\,dx = F_c(\alpha)$

$\therefore$ $\sqrt{\frac{2}{\pi}}\int_0^\infty f(x)\cos(\alpha+a)x\,dx = F_c(\alpha-\alpha)$

and $\sqrt{\frac{2}{\pi}}\int_0^\infty f(x)\cos(\alpha-a)x\,dx = F_c(\alpha-\alpha)$

putting in [1], we get

$$F_c\left[f(x)\cos ax) = \frac{1}{2}[F_c(\alpha+a)+F_c(\alpha-a)]\right]$$

Again, by definitions:

$$F_s[f(x)\cos ax] = \sqrt{\frac{2}{\pi}}\int_0^\infty f(x)\cos\alpha x\,dx$$

$$= \sqrt{\frac{2}{\pi}}\int_0^\infty f(x)\cdot\frac{1}{2}\{\sin(\alpha+a)x + \sin(\alpha-a)\}dx \quad [1]$$

$\because \quad 2S_AC_B = S_{A+B} + S_{A-B}$

with $\quad A = \alpha x; \qquad B = ax$

$$\therefore \quad F_s[f(x)\cos ax] = +\left\{\sqrt{\frac{2}{\pi}}\int_0^\infty f(x)\sin(\alpha+a)x\,dx + \sqrt{\frac{2}{\pi}}\int_0^\infty f(x)\sin(\alpha-a)x\,dx\right\} \quad [2]$$

Since $\sqrt{\frac{2}{\pi}}\int_0^\infty f(x)\sin(\alpha x)dx = F_s(\alpha)$

$$\therefore \quad \sqrt{\frac{2}{\pi}}\int_0^\infty f(x)\sin(\alpha+a)x\,dx = F_s(\alpha+a)$$

and $\sqrt{\frac{2}{\pi}}\int_0^\infty f(x)\sin(\alpha-a)x\,dx = F_s(\alpha-a)$

Substituting in [1] we get

$$F_s[f(x)\cos ax] = \frac{1}{2}[F_s(\alpha+a) + F_s(\alpha-a)]$$

Similarly, we can prove the other results.

Parseval's Identities

The following identities (known as Parseval's Identities) for Fourier sine and cosine transforms are true:

(i) $\int_0^\infty F_s(\alpha)G_s(\alpha)\cos\alpha t\,d\alpha = \frac{1}{2}\int_0^\infty g(x)[f(x+t)+f(x-t)dx]$ [1]

(ii) $\int_0^\infty F_s(\alpha)G_s(\alpha)\,d\alpha = \int_0^\infty f(x)g(x)dx$ [2]

(iii) $\int_0^\infty |F_s(\alpha)|^2\,d\alpha = \int_0^\infty |f(x)|^2 dx$ [3]

Proof: Let $F_s\{f(x)\} = F_s(\alpha)$

$F_s\{g(x)\} = G_s(\alpha)$

Then $\int_0^\infty F_s(\alpha)G_s(\alpha)\cos\alpha t\,d\alpha$

$$= \int_0^\infty F_s(\alpha)\left\{\sqrt{\frac{2}{\pi}}\int_0^\infty g(x)\sin\alpha x\,dx\right\}\cos\alpha t\,d\alpha$$

$$= \sqrt{\frac{2}{\pi}}\int_0^\infty g(x)\left\{\int_0^\infty F_s(\alpha)\sin\alpha x\cos\alpha t\,dx\right\}\cos\alpha t\,d\alpha$$

$$= \frac{1}{2}\int_0^\infty g(x)\left[\sqrt{\frac{2}{\pi}}\int_0^\infty F_s(\alpha)\{\sin\alpha(x+t)+\sin\alpha(x-t)\}\cdot d\alpha\right]dx$$

$$= \frac{1}{2}\int_0^\infty g(x)[f(x+t)+f(x-t)]dx$$

on using the inversion Formula
This proves the result [1]
For t = 0, the result [1] becomes

$\int_0^\infty F_s(\alpha)G_s(\alpha)d\alpha = \int_0^\infty f(x)g(x)dx$ which is the required result [2]

Suppose, $g(x) = \overline{f(x)}$

then $G_s(\alpha) = \overline{F_s(\alpha)}$, and [2] becomes

$\int_0^\infty |F_s(\alpha)|^2\,d\alpha = \int_0^\infty |f(x)|^2\,dx$ which is the required identity [3]

Similarly, for Fourier cosine transform, the Parseval's identities are true:

(i) $\int_0^\infty F_c(\alpha)\cos\alpha x\,d\alpha = \frac{1}{2}\int_0^\infty g(t)\{f(x+t)+f(x-t)\}dt$ [4]

(ii) $\int_0^\infty F_c(\alpha)G_c(\alpha)d\alpha = \int_0^\infty f(x)g(x)dx$ [5]

(iii) $\int_0^\infty |F_c(\alpha)|^2\,d\alpha = \int_0^\infty |f(x)|^2\,dx$ [6]

18. Verify the convolution theorem for the functions $f(x) = g(x) = e^{-x^2}$

Solution: Given: $f(x) = g(x) = e^{-x^2}$

We note that $F(\alpha) = G(\alpha) = \frac{1}{\sqrt{2}} e^{\frac{-\alpha^2}{4}}$

So that $\qquad F(\alpha) = G(\alpha) = \frac{1}{2} e^{\frac{-\alpha^2}{2}}$ [1]

Also, $\qquad f * g = \frac{1}{\sqrt{2\pi}} \int_{-\infty}^{\infty} f(x-t) g(t) dt$

$$= \frac{1}{\sqrt{2\pi}} \int_{-\infty}^{\infty} e(x-t)^2 + t^2 dt$$

$$= \frac{1}{\sqrt{2\pi}} \int_{-\infty}^{\infty} e^{-\{(x-t)^2 + t^2\}} dt$$

$$= \frac{1}{2\sqrt{2}} e^{\frac{-x^2}{2}}$$

Therefore

$$F\{f * g\} = \frac{1}{\sqrt{2\pi}} \int_{-\infty}^{\infty} \frac{1}{2\sqrt{2}} e^{\frac{-x^2}{2}} e^{i\alpha x} dx$$

$$= \frac{1}{4\sqrt{\pi}} \int_0^{\infty} e^{\frac{-x^2}{2}} e^{i\alpha x} dx$$

$$= \frac{2}{2\sqrt{\pi}} \int_0^{\infty} e^{-\frac{x^2}{2}} \cos \alpha x \, dx$$

$$= \frac{1}{2} e^{\frac{-\alpha^2}{2}}$$

These results verify the convolution theorem.

19. By using convolution theorem, find the inverse Fourier transform.

$$\hat{h}(\alpha) = \frac{1}{(1+\alpha^2)^2}$$

Solution: If we let $F(\alpha) = \frac{1}{1+\alpha^2}; G(\alpha) = \frac{1}{1+\alpha^2}$ [1]

then, $\hat{h}(\alpha) = F(\alpha) G(\alpha)$

$$= F\{f * g\} \quad \text{by convolution theorem}$$

Hence, the inverse Fourier transform $h(x)$ of $\hat{h}(\alpha)$ is

$$h(x) = F^{-1}\{\hat{h}(\alpha)\} = f * g$$

$$= \frac{1}{\sqrt{2\pi}} \int_{-\infty}^{\infty} f(x-t)\, g(t)\, dt \qquad [2]$$

Let us recall the result

$$F\{e^{-|x|}\} = \sqrt{\frac{2}{\pi}} \frac{1}{(1+\alpha^2)}$$

Therefore, $F(\alpha) = G(\alpha) = \dfrac{1}{\alpha^2 + 1} = \sqrt{\dfrac{\pi}{2}} F\{e^{-|x|}\}$

Using this in (ii) we get

$$h(x) = \frac{1}{2}\sqrt{\frac{\pi}{2}} \int_{-\infty}^{\infty} e^{-\{|x-t|+|t|\}} dt.$$

This is the inverse Fourier transform of the given function.

20. Verify the Parseval's identity for the function

$$f(x) = \begin{cases} 1 & \text{for } |x| < a \\ 0 & \text{for } |x| > a \end{cases}$$

Solution: For the given function, we recall that

$$F(\alpha) = \sqrt{\frac{2}{\pi}} \frac{\sin a\alpha}{\alpha}$$

Therefore, we have to verify that

$$\int_{-a}^{a} |1|^2 dt = \int_{-\infty}^{\infty} \frac{2}{\pi} \left| \frac{\sin a\alpha}{\alpha} \right|^2 d\alpha$$

or $\quad 2a = \dfrac{4}{\pi} \displaystyle\int_0^{\infty} \frac{\sin^2 a\alpha}{\alpha^2} d\alpha$

or $\quad \displaystyle\int_0^{\infty} \frac{\sin^2 ax}{x^2} dx = \frac{\pi a}{2}$

which is true. Thus, for the given function the Parseval's Identity is verified.

21. Employing the Parseval's identity to the function

$$f(x)=\begin{cases}1-|x|, & |x|<1\\ 0, & |x|>1\end{cases}$$

Show that

$$\int_0^{\infty}\frac{(1-\cos x)^2}{x^4}dx=\frac{\pi}{6}$$

Solution: For the given function $f(x)$, we have

$$F(\alpha)=\sqrt{\frac{2}{\pi}}\frac{1}{\alpha^2}(1-\cos\alpha)$$

Therefore, Parseval's identity gives

$$\int_{-1}^{+1}\left|1-|t|\right|^2 dt=\frac{2}{\pi}\int_{-\infty}^{\infty}\left|\frac{1}{\alpha^2}(1-\cos\alpha)\right|^2 d\alpha$$

or $$\int_{-\infty}^{\infty}\frac{(1-\cos\alpha)^2}{\alpha^4}d\alpha=\frac{\pi}{2}\int_{-1}^{+1}(1-|t|)^2 dt$$

$$=\pi\int_0^1(1-t)^2 dt=\frac{\pi}{3}$$

Therefore $$\int_0^{\infty}\frac{(1-\cos\alpha)^2}{\alpha^4}d\alpha=\frac{\pi}{6}$$

This is essentially the required result.

22. Find the Fourier transform of

$$f(x)=\begin{cases}1-|x|, & \text{if } |x|<1\\ 0, & \text{if } |x|>1\end{cases}$$

Hence, deduce that

$$\int_0^{\infty}\left(\frac{\sin t}{t}\right)^4 dt=\frac{\pi}{3}$$

Solution: $$F(\alpha)=\frac{1}{\sqrt{2\pi}}\int_{-\infty}^{\infty}f(x)e^{i\alpha x}dx$$

$$=\frac{1}{\sqrt{2\pi}}\int_{-1}^{1}(1-|x|)e^{i\alpha x}dx$$

$$= \frac{1}{\sqrt{2\pi}}\left[\int_{-1}^{1}(1-|x|)\cos\alpha x\,dx + i\int_{-1}^{1}(1-|x|)\sin\alpha x\,dx\right]$$

$$= \frac{1}{\sqrt{2\pi}}\int_{-1}^{1}(1-|x|)\cos\alpha x\,dx$$

$$\left[\because (1-|x|)\sin\alpha x \text{ is odd function}\right]$$

$$= \frac{2}{\sqrt{2\pi}}\int_{0}^{1}(1-x)\cos\alpha x\,dx$$

$$\left[\because 1-|x|\alpha x \text{ is even and } 1-|x| = 1-x \text{ is } x\in(0,1)\right]$$

$$= \frac{2}{\sqrt{2\pi}}\int_{0}^{1}(1-x)\,d\left(\frac{\sin\alpha x}{\alpha}\right)$$

$$= \sqrt{\frac{2}{\pi}}\left[(1-x)\left(\frac{\sin\alpha x}{\alpha}\right) - (-1)\left(\frac{-\cos\alpha x}{\alpha^2}\right)\right]_0^1$$

$$= \sqrt{\frac{2}{\pi}}\left[\left(\frac{-\cos\alpha}{\alpha^2}\right) - \left(\frac{-1}{\alpha^2}\right)\right]$$

$$\therefore F(\alpha) = \sqrt{\frac{2}{\pi}}\left(\frac{1-\cos\alpha}{\alpha^2}\right)$$

$$|F(\alpha)|^2 = \frac{2}{\pi}\left(\frac{1-\cos\alpha}{\alpha^2}\right)$$

Using Parseval's identity

$$\int_{-\infty}^{\infty}|F(\alpha)|^2\,d\alpha = \int_{-\infty}^{\infty}|f(x)|^2\,dx, \text{ we have}$$

$$\frac{2}{\pi}\int_{-\infty}^{\infty}\left(\frac{1-\cos\alpha}{\alpha^2}\right)^2 d\alpha = \int_{-\infty}^{\infty}(1-|x|)^2\,dx$$

$$= 2\int_0^1 (1-x)^2\,dx$$

$$= 2\left[-\frac{(1-x)^3}{3}\right]_0^1$$

$$= 2\left[0-\left(-\frac{1}{3}\right)\right] = \frac{2}{3}$$

$$\therefore \quad \int_{-\infty}^{\infty}\left(\frac{1-\cos\alpha}{\alpha^2}\right)^2 d\alpha = \frac{\pi}{3}$$

$$\therefore \quad 2\int_0^{\infty}\left(\frac{2\sin^2\left(\frac{\alpha}{2}\right)}{\alpha^2}\right)^2 d\alpha = \frac{\pi}{3}$$

$$\left(\because \quad \left(\frac{2sin^2\left(\frac{\alpha}{2}\right)}{\alpha^2}\right)^2 \text{ is even}\right)$$

$$\Rightarrow \quad \int_0^{\infty}\frac{\sin^4 t}{2^4 t^4}(2dt) = \frac{\pi}{24} \quad \left(\text{putting } t = \frac{\alpha}{2}\right)$$

$$\therefore \quad \int_0^{\infty}\left(\frac{\sin t}{t}\right)^4 = \frac{\pi}{3}$$

23. Find $F_c e^{-ax}; F_s e^{ax}$

Solution: $F_c e^{-ax} = \sqrt{\frac{2}{\pi}}\int_0^{\infty} e^{-ax}\cos\alpha x\,dx = \sqrt{\frac{2}{\pi}}\cdot\frac{\mathrm{a}}{a^2+\alpha^2}$

$$F_c e^{-ax} = \sqrt{\frac{2}{\pi}}\int_0^{\infty} e^{-ax}\sin\alpha x\,dx = \sqrt{\frac{2}{\pi}}\cdot\frac{\mathrm{a}}{a^2+\alpha^2}$$

24. Prove that (i) $F_c[xf(x)] = \frac{d}{d\alpha}\{F_s\}$ (ii) $F_s[xf(x)] = \frac{-dF_c}{d\alpha}$

Proof: (i) $F_s(\alpha) = F_s[f(x)] = \sqrt{\frac{2}{\pi}}\int_0^{\infty} f(x)\sin\alpha x\,dx$

Hence $\frac{d}{d\alpha}\{F_s(\alpha)\} = \sqrt{\frac{2}{\pi}}\int_0^{\infty} x\,f(x)\cos\alpha x\,dx$

$$= F_c[x\,f(x)]$$

Similarly (ii) $\frac{d}{d\alpha}\{F_c(\alpha)\} = \sqrt{\frac{2}{\pi}}\int_0^\infty x f(x)(-\sin\alpha x)dx$

$$= \sqrt{\frac{2}{\pi}}\int_0^\infty x f(x)\sin\alpha x dx = -F_s[x f(x)]$$

$$\Rightarrow \qquad F_s[x f(x)] = -\frac{d}{d\alpha}\{F_c\}$$

25. Find: (i) $F_c\left\{xe^{-ax}\right\}$ (ii) $F_s\left\{xe^{-ax}\right\}$

Solution: (i) $F_c\left\{xe^{-ax}\right\} = F_c[x f(x)] = \frac{d}{d\alpha}\left\{F_s\left[e^{-ax}\right]\right\}$

$$= \frac{d}{d\alpha}\left\{\sqrt{\frac{2}{\pi}}\frac{\alpha}{a^2+\alpha^2}\right\}$$

$$\because \quad F_c[x f(x)] = \frac{d}{d\alpha}\{F_s\}$$

$$= \sqrt{\frac{2}{\pi}}\frac{(a^2-\alpha^2)}{(a^2+\alpha^2)^2}$$

$$\because \quad F_c\left[xe^{-ax}\right] = \sqrt{\frac{2}{\pi}}\frac{(a^2+\alpha^2)}{(a^2+\alpha^2)^2}$$

(ii) $F_s\left[xe^{-ax}\right] = -\frac{d}{d\alpha}F_c e^{-ax}$

$$= -\frac{d}{d\alpha}\sqrt{\frac{2}{\pi}}\frac{a}{a^2+\alpha^2}$$

$$= \sqrt{\frac{2}{\pi}}\cdot\frac{2a\alpha}{(a^2+\alpha^2)^2}$$

$$F_s\left[xe^{-ax}\right] = \sqrt{\frac{2}{\pi}}\cdot\frac{2a\alpha}{(a^2+\alpha^2)}$$

Note: $F_s[x f(x)] = -\frac{d}{d\alpha}F_c$

26. Find (i) $F_c\left\{e^{-ax}\cos ax\right\}$ (ii) $F_c\left\{e^{-ax}\sin ax\right\}$

***Solution*:** Since $F_c e^{-ax} = \sqrt{\frac{2}{\pi}} \cdot \frac{a}{a^2+\alpha^2}$ [see solved example 23]

Now replace a by (1-i) a, we get

$$F_c\left\{e^{-ax}\cos ax + ie^{-ax}\sin ax\right\} = \sqrt{\frac{2}{\pi}}.\frac{a(1-i)}{-2a^2 i+\alpha^2}$$

$$= \sqrt{\frac{2}{\pi}}.\frac{a(\alpha^2+2a^2)+ia(2a^2-\alpha^2)}{\alpha^2+4a^4}$$

Equate Re and Im, we get

$$\text{(i) } F_c\left\{e^{-ax}\cos ax\right\} = \sqrt{\frac{2}{\pi}}.\frac{a(\alpha^2+2a^2)}{\alpha^4+4a^4} \qquad [1]$$

$$\text{(ii) } F_c\left\{e^{-ax}\sin ax\right\} = \sqrt{\frac{2}{\pi}}.\frac{a(2a^2+\alpha^2)}{\alpha^4+4a^4} \qquad [2]$$

Hence [1] + [2] gives

$$F_c\left[e^{-ax}(\cos ax + \sin ax\right] = \sqrt{\frac{2}{\pi}}\frac{4a^3}{\alpha^4+4a^4}$$

If we get $4a^2 = k^4$, we find:

$$F_c\left\{\frac{1}{x^4+k^4}\right\} = \sqrt{\frac{\pi}{2}}\frac{2\sqrt{2}}{4k^3}e^{-\frac{ka}{\sqrt{2}}}\left(\cos\frac{k\alpha}{\sqrt{2}} + \sin\frac{k\alpha}{\sqrt{2}}\right)$$

$$= \sqrt{\frac{2}{\pi}}\frac{1}{k^3}e^{-\frac{k\alpha}{\sqrt{2}}}\sin\left(\frac{k\alpha}{\sqrt{2}} + \frac{\pi}{\sqrt{2}}\right)$$

Similarly [1] – [2] gives

$$F_c\left[e^{-ax}(\cos ax - \sin ax\right] = \sqrt{\frac{2}{\pi}}\frac{2a\alpha^2}{\alpha^2+4a^4}$$

If we get $4a^4 = k^4$, we find:

$$F_c\left\{\frac{x^2}{x^4+k^4}\right\} = \sqrt{\frac{\pi}{2}}\frac{1}{\sqrt{2}k}e^{-\frac{k\alpha}{\sqrt{2}}}\left(\cos\frac{k\alpha}{\sqrt{2}} - \sin\frac{k\alpha}{\sqrt{2}}\right)$$

$$= \sqrt{\frac{\pi}{2}} \cdot \frac{1}{k} \cdot e^{-\frac{k\alpha}{\sqrt{2}}}\cos\left(\frac{\pi}{4} + \frac{k\alpha}{\sqrt{2}}\right)$$

27. Find the cosine transform of a function x which is unity for $0 < x < a$ and zero for $x \geq a$. What is the function whose cosine transform is $\left(\dfrac{\sin a\alpha}{\alpha}\right)$?

Solution: Given: $f(x) = \begin{cases} 1, & 0 < x < a \\ 0, & x \geq a \end{cases}$ [1]

$$\therefore \quad F_c\{f(x)\} = \sqrt{\frac{2}{\pi}} \int_0^\infty f(x) \cos \alpha x \, dx$$

$$= \int_0^a f(x) \cos \alpha x \, dx + \int_a^\infty f(x) \cos \alpha x \, dx$$

$$= \int_0^a \cos \alpha x \, dx + 0. \qquad \text{(by [1])}$$

$$= \left[\frac{\sin \alpha x}{\alpha}\right]_0^a = \left(\frac{\sin \alpha x}{\alpha}\right)$$

Let $f(x)$ be such that

$$F_c\{f(x)\} = \left(\frac{\sin \alpha x}{\alpha}\right)$$

Now, by inversion formula, we get

$$f(x) = \sqrt{\frac{2}{\pi}} \int_0^\infty \frac{\sin \alpha x}{\alpha} \cos \alpha x \, dx$$

$$= \frac{1}{\sqrt{\pi}} \int_0^\infty \frac{\sin(a+x)\alpha + \sin(a-x)\alpha}{\alpha} dx$$

$$= \frac{1}{\sqrt{\pi}} \left[\int_0^\infty \frac{\sin(a+x)\alpha}{\alpha} d\alpha + \int_0^\infty \frac{\sin(a-x)\alpha}{\alpha} d\alpha\right]$$

$$= \begin{cases} \left(\dfrac{1}{\sqrt{\pi}}\right)\left(\dfrac{\pi}{2} + \dfrac{\pi}{2}\right) \cdot x < a \\ \left(\dfrac{1}{\sqrt{\pi}}\right)\left(\dfrac{\pi}{2} - \dfrac{\pi}{2}\right) \cdot x > a \end{cases}$$

$$= \begin{cases} \sqrt{\pi}, x < a \\ 0, \quad x > a \end{cases}$$

Note: If the coefficient of integral in the inversion formula is $2/\pi$, then, we get,

$$f(x) = \left\{ \begin{array}{l} 1, x < a \\ 0, x < a \end{array} \right)$$

28. Find $F_s\left\{x^{m-1}\right\}$ and $F_c\left\{x^{m-1}\right\}(0<m<1)$

Deduce that $\frac{1}{\sqrt{x}}$ is self-reciprocal under Fourier sine and cosine transforms.

***Solution*:** We know that from Gamma functions

$$\frac{\Gamma(m)}{a}=\int_0^{\infty}e^{-ax}x^{m-1}dx,\quad m>0$$

(put $a=\mathrm{i}\alpha$), we get

$$\int_0^{\infty}e^{-ax}x^{m-1}dx=\frac{\Gamma(m)}{(\mathrm{i}\alpha)^m}=\frac{(-\mathrm{i})^m\Gamma(m)}{\alpha^m}$$

$$=\frac{e^{-\mathrm{i}\frac{\pi m}{2}}\Gamma(m)}{\alpha^m}.\ \text{since}\ =e^{-\mathrm{i}\frac{\pi \mathrm{i}}{2}}=-\mathrm{i}$$

Equating *Re* and I_m parts, we get

$$\int_0^{\infty}x^{m-1}\cos\alpha x\,dx=\frac{\Gamma(m)\cos\frac{\pi m}{2}}{\alpha^m}$$

and $$\int_0^{\infty}x^{m-1}\sin\alpha x\,dx=\frac{\Gamma(m)\sin\frac{\pi m}{2}}{\alpha^m}$$

Hence

$$F_c\left\{x^{m-1}\right\}=\sqrt{\frac{2}{\pi}}\cdot\frac{\Gamma(m)\cos\frac{\pi m}{2}}{\alpha^m}$$

$$F_s\left\{x^{m-1}\right\}=\sqrt{\frac{2}{\pi}}\cdot\frac{\Gamma(m)\sin\frac{\pi m}{2}}{\alpha^m}$$

with $m=\frac{1}{2}$,

$$F_c\frac{1}{\sqrt{x}}=\frac{1}{\sqrt{\alpha}}\qquad\left(\sqrt{\frac{1}{2}}=\sqrt{\pi}\right)$$

29. Find $F_c\left\{e^{-x^2}\right\}$

***Solution*:** $F_c\left\{e^{-x^2}\right\}=\int_0^\infty e^{-x^2}\cos\alpha x\,dx=I$, say [1]

$$\frac{dI}{d\alpha}=\int_0^\infty e^{-x^2}(-x\sin\alpha x)\,dx$$

$$=\frac{1}{2}\int_0^\infty \sin(\alpha x)\left(-2xe^{-x^2}\right)dx$$

$$=\frac{1}{2}\left[\left\{\sin\alpha x\cdot e^{-x^2}\right\}_0^\infty-\alpha\int_0^\infty \cos\alpha x e^{-x^2}\,dx\right]$$

Integrating by parts

if $-x^2=u$; so that $-2xdx=du$

$\therefore$ $\dfrac{dI}{d\alpha}=-\dfrac{1}{2}\alpha I$ or $\dfrac{dI}{I}=-\dfrac{1}{2}\alpha\,d\alpha$

Integrating $\log I=-\dfrac{1}{2}\alpha^2+\log c$ or $\cdot I=ce^{-\frac{\alpha^2}{4}}$ [2]

For $\alpha=0$, [1] and [2] give

$$c=\int_0^\infty e^{-x^2}dx=\sqrt{\frac{\pi}{2}}$$

$\therefore$ From [1] and [2]

$$F_c\left\{e^{-x^2}\right\}=I=\sqrt{\frac{\pi}{2}}e^{-\frac{\alpha^2}{4}}$$

30. Find Fourier sine transform of $\left(\dfrac{1}{x}\right)e^{-ax}$

***Solution*:** $F_s\left\{\dfrac{e^{-ax}}{x}\right\}=\int_0^\infty \dfrac{e^{-ax}}{x}x\cos\alpha x\,dx=I$, say [1]

$\therefore$ $\dfrac{dI}{d\alpha}=\int_0^\infty \dfrac{e^{-ax}}{x}x\cos\alpha x\,dx$

$$=\int_0^\infty e^{-ax}\cos\alpha x\,dx$$

or $\dfrac{dI}{d\alpha}=\dfrac{a}{a^2+\alpha^2}$

or $\quad dl = \dfrac{a\,d\alpha}{a^2+\alpha^2}$

Integrating $l = \tan^{-1}\left(\dfrac{\alpha}{a}\right) + c$ [2]

For $\alpha = 0$, [1] and [2] given

$c + \tan^{-1}(0) = 0 \qquad \therefore\ c = 0$

$$\therefore\quad F\left\{e^{-\frac{ax}{x}}\right\} = l = \tan^{-1}\left(\frac{\alpha}{a}\right)$$

31. Find $F_c\left\{\dfrac{1}{1+x^2}\right\}$ and $F_s\left\{\dfrac{1}{1+x^2}\right\}$

Solution: $F_c\left\{\dfrac{1}{1+x^2}\right\} = \displaystyle\int_0^\infty \frac{\cos\alpha x}{1+x^2}dx = l$, say [1]

$$\therefore\quad \frac{dl}{d\alpha} = \int_0^\infty \frac{-x\sin\alpha x}{1+x^2}dx = -\int_0^\infty \frac{x\sin\alpha x}{1+x^2}dx \qquad [2]$$

$$\frac{dl}{d\alpha} = -\int_0^\infty \frac{x^2\sin\alpha x}{x(1+x^2)}dx = -\int_0^\infty \frac{[(1+x^2)-1]\sin\alpha x}{x(1+x^2)}dx$$

or $$\frac{dl}{d\alpha} = -\int_0^\infty \frac{\sin\alpha x}{x}dx + \int_0^\infty \frac{\sin\alpha x}{x(1+x^2)}dx$$

$$\therefore\quad \frac{dl}{d\alpha} = \left(-\frac{\pi}{2}\right) + \int_0^\infty \frac{\sin\alpha x}{x(1+x^2)}dx \qquad [3]$$

$$\left[\because \int_0^\infty \frac{\sin\alpha x}{x} = \frac{\pi}{2}\right]$$

From [3], $\dfrac{d^2l}{d\alpha^2} = \displaystyle\int_0^\infty \frac{x\cos\alpha x}{x(1+x^2)}dx = \int_0^\infty \frac{\cos\alpha x}{1+x^2}dx = l$

or $\quad \dfrac{d^2l}{d\alpha^2} - l = 0$

or $(D^2 - l)l = 0$, where $D = \dfrac{d}{d\alpha}$

Whose solution is $l = Ae^{-\alpha} + Be^{\alpha}$ [4]

From [4], $\dfrac{dl}{d\alpha} = -Ae^{-\alpha} + Be\alpha$ [5]

For $\alpha = 0$, [1] and [4] give

$$-A + B = \int_0^\infty \frac{dx}{1+x^2} = \frac{\pi}{2} \qquad [6]$$

For $\alpha = 0$, [3] and [5] give

$$-A + B = \frac{\pi}{2} \qquad [7]$$

Solving [6] and [7],

$$A = \frac{\pi}{2} \cdot B = 0$$

So [1] and [4], we have

$$F_c\left\{\frac{1}{(1+x^2)}\right\} = l = \left(\frac{\pi}{2}\right)e^{-\alpha}$$

and
$$F_s\left\{\frac{1}{(1+x^2)}\right\} = \int_0^\infty \frac{x \sin \alpha x}{1+x^2} dx,$$

$$= -\frac{dl}{d\alpha},$$

$$= \left(\frac{\pi}{2}\right)e^{-\alpha}$$

From [5] and with $A = \left(\dfrac{\pi}{2}\right)$: and $B = 0$

32. Find the Fourier sine transforms of the following functions:

(i) $\dfrac{\cos x}{\sqrt{x}}$ (ii) $\dfrac{\sin^2 x}{x^{\frac{2}{3}}}$

***Solution*:** (i) We know that

$$F_s\left[\frac{1}{\sqrt{x}}\right] = \frac{1}{\sqrt{x}} \qquad [1]$$

Therefore, the shift property $F_s\left[f(x)\cos ax\right]=\frac{1}{2}\left[F_s(\alpha+a)+F_s(\alpha-a)\right]$, yields

$$F_s\left[\frac{\cos}{\sqrt{x}}x\right]=\frac{1}{2}\left[\frac{1}{\sqrt{\alpha+1}}+\frac{1}{\sqrt{\alpha-1}}\right]$$

(ii) We know that

$$F_s\left[x^{m-1}\right]=\sqrt{\frac{2}{\pi}}\frac{\lceil(m)}{\alpha^m}\sin\left(\frac{m\pi}{2}\right)$$

Here, $\frac{1}{x^{2/3}}=x^{\left(\frac{1}{3}\right)-1}$, We have

$$F_s\left\{\frac{1}{x^{2/3}}\right\}=\sqrt{\frac{2}{\pi}}\frac{\Gamma\left(\frac{1}{3}\right)}{\alpha^{1/3}}\sin\frac{\pi}{6}=\frac{1}{\sqrt{2\pi}}\frac{\Gamma\left(\frac{1}{3}\right)}{\alpha^{1/3}}$$

Therefore,

$$F_s\left\{\frac{\sin^2 x}{x^{2/3}}\right\}=F_s\left\{\frac{1-\cos 2x}{2x^{2/3}}\right\}$$

$$=\frac{1}{2}\left[F_s\left\{\frac{1}{x^{2/3}}\right\}-F_s\left\{\frac{\cos 2x}{x^{2/3}}\right\}\right]$$

$$=\frac{1}{2}\left[\frac{1}{\sqrt{2\pi}}\frac{\Gamma\left(\frac{1}{2}\right)}{\alpha^{1/3}}-\frac{1}{2}\frac{1}{\sqrt{2\pi}}\Gamma\left(\frac{1}{3}\right)\left\{\frac{1}{(\alpha)+2)^{1/3}}+\frac{1}{(\alpha-2)^{1/3}}\right\}\right]$$

$$=\frac{1}{2\sqrt{2\pi}}\Gamma\left(\frac{1}{3}\right)\left[\frac{1}{\alpha^{1/3}}-\frac{1}{2}\left\{\frac{1}{(\alpha+2)^{1/3}}+\frac{1}{(\alpha-2)^{1/3}}\right\}\right]$$

33. Show that $F_s\left\{xe^{-\left(\frac{1}{2}\right)x^2}\right\}=\alpha e^{-\left(\frac{1}{2}\right)\alpha^2}$ (i.e., the function $xe^{-(1/2)x^2}$ is self-reciprocal w.r.t. Fourier sine transform).

Solution: Let us recall that $f(x)=e^{-(1/2)x^2}$ is self-reciprocal w.r.t. complex Fourier transform, i.e.,

$$\frac{1}{\sqrt{2\pi}}\int_{-\infty}^{\infty}e^{-(1/2)x^2}e^{i\alpha x}dx=e^{-(1/2)\alpha^2}$$

Differentiating both sides w.r.t. α, we obtain

$$\frac{i}{\sqrt{2\pi}}\int_{-\infty}^{\infty} xe^{(-1/2)x^2} e^{i\alpha x}\,dx = -\alpha e^{-(1/2)\alpha^2}$$

or
$$\alpha e^{(-1/2)\alpha^2} = \frac{-i}{\sqrt{2\pi}}\int_{-\infty}^{\infty} xe^{(-1/2)x^2} \sin\alpha x\,dx$$

$$= \frac{2}{\sqrt{2\pi}}\int_{0}^{\infty} xe^{(-1/2)x^2} \sin\alpha x\,dx$$

$$= F_s\left\{xe^{-(1/2)x^2}\right\}.$$

34. Find the Fourier sine transform of

$$f(x) = \begin{cases} x, & 0<x<1 \\ a-x, & 1<x<a \\ 0, & x>a \end{cases}$$

Solution: By the definition: we have

$$F_s\{f(x)\} = \sqrt{\frac{2}{\pi}}\int_0^{\infty} f(x)\sin\alpha x\,dx$$

$$= \sqrt{\frac{2}{\pi}}\left\{\int_0^1 x\sin\alpha x\,dx + \int_1^0 (a-x)\sin\alpha x\,dx\right\}$$

Integration by parts yields

$$F_s\{f(x)\} = \sqrt{\frac{2}{\pi}}\left\{\left[-\frac{1}{\alpha}x\cos\alpha x + \frac{1}{\alpha^2}\sin\alpha x\right]_0^1\right.$$

$$\left. + \left[-\frac{(a-x)}{\alpha}\cos\alpha x - \frac{\sin\alpha x}{\alpha^2}\right]_1^a\right\}$$

$$= \sqrt{\frac{2}{\pi}}\frac{1}{\alpha^2}\{(a-2\}\alpha\cos\alpha + 2\sin\alpha - \sin a\alpha$$

35. Using the Fourier sine transform of e^{-ax}, a>0, show that

$$\int_0^{\infty}\frac{x\sin kx}{a^2+x^2}dx = \frac{\pi}{2}e^{-ak}, \mathrm{K}>0$$

Hence obtain the Fourier sine transform of $\dfrac{x}{a^2+x^2}$

Solution: We recall

$$F_s\left\{e^{-ax}\right\}=\sqrt{\frac{2}{\pi}}\frac{\alpha}{a^2+\alpha^2}$$

Therefore, the inversion formula gives

$$e^{-ax}=\sqrt{\frac{2}{\pi}}\int_0^\infty\left\{\sqrt{\frac{2}{\pi}}\frac{\alpha}{a^2+\alpha^2}\right\}\sin\alpha x\,d\alpha$$

or $$\int_0^\infty\frac{\alpha}{a^2+\alpha^2}\sin\alpha x\,d\alpha=\frac{\pi}{2}e^{-ax}$$

Changing the symbols x and α to k and x respectively, we get

$$\int_0^\infty\frac{x\sin kx}{a^2+x^2}dx=\frac{\pi}{2}e^{-ak} \qquad \text{...(i)}$$

Next, the Fourier sine transform of $\frac{x}{1+x^2}$ is

$$F_s\left\{\frac{x}{a^2+x^2}\right\}=\sqrt{\frac{2}{\pi}}\int_0^\infty\frac{x}{a^2+x^2}\sin\alpha x\,dx$$

$$=\sqrt{\frac{2}{\pi}}\frac{\pi}{2}e^{-\alpha a},\text{by. (i)}$$

$$=\sqrt{\frac{\pi}{2}}e^{-\alpha a} \qquad \text{...(ii)}$$

36. Show that $F_s\left\{\frac{1}{x}\right\}=\sqrt{\frac{\pi}{2}}\operatorname{sgn}\alpha$

Solution: We know that $F_s\{f(x)\}=\begin{cases}(\alpha), & \alpha>0\\(-\alpha), & \alpha<0\end{cases}$

$$\operatorname{sgn}\alpha=\begin{cases}1, & \alpha>0\\-1, & \alpha<0\end{cases}$$

$$F_s\left\{\frac{1}{x}\right\}=\sqrt{\frac{2}{\pi}}\int_0^\infty\frac{1}{x}\sin\alpha x\,dx$$

For $\alpha>0$, we have

$$\int_0^\infty\frac{\sin\alpha x}{x}dx=\frac{\pi}{2}$$

Therefore, for $\alpha > 0$,

$$F_s\left\{\frac{1}{x}\right\} = \sqrt{\frac{2}{\pi}} \cdot \frac{\pi}{2} = \sqrt{\frac{\pi}{2}}$$

Hence, by the result $F_s\{f(x)\} = \phi(|\alpha|)\operatorname{sgn}\alpha$, we have for $\alpha > 0$, and $\alpha < 0$,

$$F_s\left\{\frac{1}{x}\right\} = \sqrt{\frac{\pi}{2}}\operatorname{sgn}\alpha$$

Table 2.2: Fourier sine transforms

f(x)	$F_s(\alpha)$
e^{-x}	$\sqrt{\frac{2}{\pi}}\frac{1}{1+\alpha^2}$
$xe^{-\left(\frac{1}{2}\right)x^2}$	$\alpha e^{-\left(\frac{1}{2}\right)\alpha^2}$
$x^{p-1}, 0 < p < 1$	$\sqrt{\frac{2}{\pi}}\frac{\Gamma(p)}{\alpha^p}\sin\frac{p\pi}{2}$
$\frac{e^{-\frac{a}{\sqrt{x}}}}{\sqrt{x}}$	$\frac{1}{\sqrt{\alpha}}\left[\cos\sqrt{2a\alpha} - \sin\sqrt{2a\alpha}\right]$
$\frac{\sin x}{x}$	$\frac{1}{\sqrt{2\pi}}\log\frac{\lvert\alpha+1\rvert}{\lvert\alpha-1\rvert}$

37. Find $f(x)$ given that $F_s(\alpha) = \alpha^n e^{-a\alpha}$ where α is a positive constant and n is a positive integer. By using the Fourier sine inversion formula, we get

$$f(x) = \sqrt{\frac{2}{\pi}}\int_0^\infty \alpha^n e^{-a\alpha}\sin\alpha x\, d\alpha \qquad [1]$$

***Solution*:** Let us recall that

$$\int_0^\infty e^{-a\alpha}\sin\alpha x\, d\alpha = \frac{x}{a^2 + x^2}$$

Differentiating both sides w.r.t. *a*, we obtain

$$\int_0^\infty (-\alpha)^n e^{-a\alpha} \sin \alpha x \, d\alpha = \frac{d^n}{da^n}\left\{\frac{x}{x^2+a^2}\right\}$$

Setting $x = a\tan\theta$ in the RHS, we obtain

$$\int_0^\infty (-\alpha)^n e^{-a\alpha} \sin \alpha x \, dx = (-1)^n \frac{n!}{a^n+1}(\cos\theta)^{n+1} \sin(n+1)\theta$$

$$= (-1)^n \frac{n!}{a^{n+1}} \sin(n+2)\theta \frac{a^n+1}{(a^2+x^2)^{\left(\frac{n+1}{2}\right)}}$$

or

$$\int_0^\infty \alpha^n e^{-a\alpha} \sin \alpha x \, d\alpha = \frac{n! \sin(n+1)\theta}{(a^2+x^2)^{\left(\frac{n+1}{2}\right)}} \qquad [2]$$

Therefore [1] gives

$$f(x) = \sqrt{\frac{2}{\pi}} \frac{n! \sin(n+1)\theta}{(a^2+x^2)^{\left(\frac{n+1}{2}\right)}} \qquad [3]$$

38. By using the Fourier sine transform of e^{-ax} and Parseval's Identity prove that

$$\int_0^\infty \frac{x^2 dx}{(x^2+a^2)(x^2+b^2)} = \frac{\pi}{2(a+b)}, a>0, b>0$$

Solution: Let $f(x) = e^{-ax}$, $g(x) = e^{-bx}$. Then

$$F_s(\alpha) = \sqrt{\frac{2}{\pi}} \cdot \frac{\alpha}{(a^2+\alpha^2)}$$

$$G_s(\alpha) = \sqrt{\frac{2}{\pi}} \frac{\alpha}{(b^2+\alpha^2)}.$$

Therefore

$$\int_0^\infty f(x) g(x) dx = \int_0^\infty e^{-(a+b)x} dx = \frac{1}{a+b} \qquad [1]$$

$$\int_0^\infty F_s(\alpha) G_s(\alpha) d\alpha = \frac{2}{\pi}\int_0^\infty \frac{\alpha^2}{(a^2+\alpha^2)(b^2+\alpha^2)} d\alpha \qquad [2]$$

By Parseval's identity the LHS of [1] and [2] are equal. Therefore,

$$\frac{2}{\pi}\int_0^\infty \frac{\alpha^2}{(a^2+\alpha^2)(b^2+\alpha^2)}d\alpha = \frac{1}{a+b}$$

or $$\int_0^\infty \frac{x^2}{(x^2+a^2)(x^2+b^2)}dx = \frac{\pi}{2(a+b)}.$$

Note: For $b = a$, we get

$$\int_0^\infty \frac{x^2}{(x^2+a^2)^2}dx = \frac{\pi}{4a}$$

39. Find the Fourier cosine transform of the function

$$f(x) = \begin{cases} \cos x, & 0 < x < a \\ 0, & x > a \end{cases}$$

***Solution*:** Using the definition of Fourier cosine transform, we obtain

$$F_c\{f(x)\} = \sqrt{\frac{2}{\pi}}\int_0^\infty f(x)\cos\alpha x\,dx$$

$$= \sqrt{\frac{2}{\pi}}\int_0^a \cos x\cos\alpha x\,dx$$

$$= \frac{1}{2}\sqrt{\frac{2}{\pi}}\int_0^a \{\cos(x+\alpha x)+\cos(x-\alpha x)\}\,dx$$

$$= \frac{1}{\sqrt{2\pi}}\left[\frac{\sin(1+\alpha)x}{1+\alpha}+\frac{\sin(1-\alpha)x}{1-\alpha}\right]_0^a$$

$$= \frac{1}{\sqrt{2\pi}}\left[\frac{\sin(1+\alpha)a}{1+\alpha}+\frac{\sin(1-\alpha)a}{1-\alpha}\right]$$

40. Find the Fourier cosine transform of $\dfrac{\cos 2x}{1+x^2}$

***Solution*:** Let $f(x) = \dfrac{1}{1+x^2}$ so that $\hat{f}_c(\alpha) = \sqrt{\dfrac{\pi}{2}}e^{-|\alpha|}$

Therefore, by the shift property,

$$F_c\left\{\frac{\cos 2x}{1+x^2}\right\} = F_c\{f(x)\cos 2x\}$$

$$=\frac{1}{2}\left\{\hat{f}_c(\alpha+a)+\hat{f}_c(\alpha-a)\right\}$$

$$=\frac{1}{2}\left\{\sqrt{\frac{\pi}{2}}e^{-|\alpha+a|}+\sqrt{\frac{\pi}{2}}e^{-|\alpha-a|}\right\}$$

$$=\frac{1}{2}\sqrt{\frac{\pi}{2}}\left\{e^{-|\alpha+a|}+e^{-|\alpha-a|}\right\}$$

41. Using the Fourier cosine transform of e^{-ax} and the Parseval's identity for Fourier cosine transforms, show that

***Solution*:** Let $f(x)=e^{-ax}, a>0$. Then

$$\int_0^\infty |f(x)|^2 = \int_0^\infty e^{-2ax}dx=\frac{1}{2a} \qquad [1]$$

and $$F(\alpha)=\sqrt{\frac{2}{\pi}}\frac{a}{\alpha^2+a^2}$$

so that $$\int_0^\infty |F(\alpha)|^2 d\alpha=\frac{2a^2}{\pi}\int_0^\infty \frac{d\alpha}{(\alpha^2+a^2)^2} \qquad [2]$$

Using [1] and [2], the Parseval's identity gives

$$\frac{2a^2}{\pi}\int_0^\infty \frac{d\alpha}{(\alpha^2+a^2)}=\frac{1}{2a}$$

$$\Rightarrow \quad \int_0^\infty \frac{d\alpha}{(a^2+x^2)}=\frac{\pi}{4a^3}.$$

Table 2.3: Fourier cosine transform

$f(x)$	$\hat{f}_c(\alpha)$
e^{-ax}	$\sqrt{\frac{2}{\pi}}\frac{a}{a^2+\alpha^2}$
$x^{p-1}, 0<p<1$	$\sqrt{\frac{2}{\pi}}\frac{\Gamma(p)}{\alpha^p}\cos\frac{p\pi}{2}$
$e^{-a^2x^2}$	$\frac{1}{\sqrt{2}}\frac{1}{\lvert a\rvert}e^{-\left(\frac{\alpha}{2a}\right)^2}$
$\cos\frac{x^2}{2}$	$\frac{1}{\sqrt{2}}\left\{\cos\frac{\alpha^2}{2}+\sin\frac{\alpha^2}{2}\right\}$
$\sin\frac{x^2}{2}$	$\frac{1}{\sqrt{2}}\left\{\cos\frac{\alpha^2}{2}-\sin\frac{\alpha^2}{2}\right\}$
$1,\ 0<x<a$ $0,\ x>a$	$\sqrt{\frac{2}{\pi}}\frac{\sin\alpha a}{\alpha}$

EXERCISES

Find the Fourier transform of following functions:

1. $f(x)-\begin{cases}x, & \lvert x\rvert\le a\\ 0, & \lvert x\rvert>a\end{cases}$ where a is a positive constant. $\left[Ans.\ \frac{2i}{\alpha^2}(a\alpha\cos\alpha a-\sin\alpha a)\right]$

2. $f(x)=\begin{cases}0, & -\infty<x<a\\ x, & a\le x\le b\\ 0, & x>b\end{cases}$ $\left[Ans.\ \frac{1}{\alpha}\left(ae^{i\alpha a}-be^{i\alpha b}\right)+\frac{1}{\alpha^2}\left(e^{i\alpha b}-e^{i\alpha a}\right)\right]$

3. $f(x)=xe^{-x},\ 0\le x<\infty$ $\left[Ans.\ \frac{1}{2\pi}\cdot\frac{(1+i\alpha)^2}{(1+\alpha^2)^2}\right]$

4. $f(x) - \begin{cases} \cos x, & \text{if } 0 < x < 1 \\ 0, & \text{otherwise} \end{cases}$

$$\left[Ans.\ \frac{1}{4\pi}\left[\frac{\sin[\alpha+1]}{\alpha+1}+\frac{\sin(\alpha-1)}{\alpha-1}+i\left\{\frac{1-\cos(\alpha+1)}{\alpha+1}+\frac{1-\cos(\alpha-1)}{\alpha-1}\right\}\right]\right]$$

5. $f(x)\begin{cases} x^2, & |x| < a, \text{ where } a \text{ is a positive constant} \\ 0, & |x| > a \end{cases}$

$$\left[Ans.\ \sqrt{\frac{2}{\pi}}\frac{1}{\alpha^3}\left[(a^2\alpha^2+1)\sin a\alpha + 2a\alpha\cos a\alpha\right]\right]$$

6. $f(x) = \begin{cases} e^{iwx}, & a < x < b \\ 0, & x < a,\ x > b \end{cases}$ $\left[Ans.\ -\frac{i}{\sqrt{2\pi}}\frac{1}{\alpha+w}\left[e^{-i(\alpha+w)b} - e^{i(\alpha+w)a}\right]\right]$

7. $f(x) = \begin{cases} 1-|x|, & |x| < 1 \\ 0, & |x| > 1. \end{cases}$ $\left[Ans.\ \sqrt{\frac{2}{\pi}}\frac{1}{\alpha^2}(1-\cos\alpha)\right]$

8. $f(x) = \begin{cases} \frac{1}{2\in}, & |x| < 1 \\ 0, & |x| > 1 \end{cases}$ $\left[Ans.\ \frac{1}{\sqrt{2\pi}}\cdot\frac{\sin\alpha}{\alpha\in}\right]$

9. $f(x) = \begin{cases} 1-\frac{x}{a}, & 0 < x < a \\ 1+\frac{x}{a}, & -a < x < 0 \\ 0, & \text{otherwise} \end{cases}$ $\left[Ans.\ \frac{2\sqrt{2}}{a\alpha^2\sqrt{\pi}}\sin^2\left(\frac{\alpha a}{2}\right)\right]$

10. $f(x) = \frac{x}{a^2+x^2},\ a > 0$ $\left[Ans.\ \frac{-i\alpha}{2a}\sqrt{\frac{\pi}{2}}e^{-a|x|}\right]$

11. $f(x) = \sin ax^2.$ $\left[Ans.\ \frac{1}{\sqrt{2a}\sin}\left(\frac{\pi}{4}+\frac{\alpha^2}{4a}\right)\right]$

12. $f(x) = e^{-x^2}\sin cx$ $\left[Ans.\ \frac{i}{\sqrt{2}}e^{-\left(\frac{\alpha^2+c^2}{4}\right)}\sinh\left(\frac{\alpha c}{2}\right)\right]$

13. $f(x) = \dfrac{\sin ax}{x}$ $\left[Ans.\ \sqrt{\dfrac{\pi}{2}} \text{ if } |\alpha| < a, 0 \text{ if } |\alpha| > a\right]$

14. $f(x) = e^{-a(x-b)^2}$ $\left[Ans.\ \dfrac{1}{\sqrt{2a}} e^{i\alpha b} e^{-\frac{\alpha^2}{4a}}\right]$

15. $f(x) = \dfrac{e^{-a|x|}}{\sqrt{|x|}}$ $\left[Ans.\ \dfrac{\left[(a^2+\alpha^2)^{\frac{1}{2}} + a\right]^{\frac{1}{2}}}{(a^2+\alpha^2)^{\frac{1}{2}}}\right]$

16. $f(x) - |x| e^{-a|x|}$ $\left[Ans.\ \sqrt{\dfrac{2}{\pi}} \cdot \dfrac{a^2-\alpha^2}{(a^2+\alpha^2)^2}\right]$

17. $f(x) - \dfrac{\cosh ax}{\cosh \pi x}, -\pi < a < \pi$ $\left[Ans.\ \sqrt{\dfrac{2}{\pi}} \dfrac{\cos\left(\dfrac{a}{2}\right)\cosh\left(\dfrac{\alpha}{2}\right)}{\cosh\alpha + \cos a}\right]$

18. $F(x) = \dfrac{\sinh ax}{\sinh \pi x}$ $\left[Ans.\ \dfrac{1}{\sqrt{2x}} \dfrac{\sin a}{\cosh\alpha + \cos a}\right]$

19. Show that $F\{f(-x)\} = F(-\alpha)$

20. Prove that $F\{f(x)\sin ax\} = \dfrac{1}{2i}[F(\alpha+a) - F(\alpha-a)]$

21. Prove the following:

(i) $F\left\{e^{-ax^2}\cos bx\right\} = \dfrac{1}{2a\sqrt{2}}\left[e^{-\frac{(\alpha+b)^2}{4a}} + e^{-\frac{(\alpha-b)^2}{4a}}\right]$

(ii) $F\left\{e^{-ax^2}\sin bx\right\} = \dfrac{i}{2a\sqrt{2}}\left[e^{-\frac{(\alpha-b)^2}{4a}} - e^{-\frac{(\alpha-b)^2}{4a}}\right]$

(iii) $F\left\{\dfrac{1}{x}\sin ax \cos bx\right\} = \begin{cases} \sqrt{\dfrac{\pi}{2}}, & |\alpha| < a \\ 0, & |\alpha| > a \end{cases}$

(iv) $F\left\{\frac{1}{x}\sin ax \sin bx\right\} = \begin{cases} -i\sqrt{\frac{\pi}{2}}, & |\alpha| < a \\ 0, & |\alpha| > a \end{cases}$

(v) $F\left\{\sin ax^2 \cos bx\right\} = \frac{1}{2\sqrt{2a}}\left[e^{-i\frac{(\alpha+b)^2}{4a}} + e^{-i\frac{(\alpha-b)^2}{4a}}\right]$

(vi) $F\left\{\cos ax^2 \cos bx\right\} = \frac{i}{2\sqrt{2a}}\left[e^{-i\frac{(\alpha+b)^2}{4a}} + e^{-i\frac{(\alpha-b)^2}{4a}}\right]$

(vii) $F\left\{\sin ax^2 \sin bx\right\} = \frac{1}{2\sqrt{2a}}\left[e^{-i\frac{(\alpha+b)^2}{4a}} - e^{-i\frac{(\alpha-b)^2}{4a}}\right]$

(viii) $F\left\{\frac{1}{|ax-b|}, x \neq \frac{b}{a} = \frac{1}{|\alpha|}e^{\frac{ib\alpha}{a}}\right.$

22. Find: (i) $F\left\{x \sin 4x^2\right\}$ (ii) $F\left\{x^2 e^{-4x^2}\right\}$

$\left[\textit{Ans.}\ \text{(i)}\ \frac{i\alpha}{64}\left[\cos\frac{\alpha^2}{16} + \sin\frac{\alpha^2}{16}\right];\quad \text{(ii)}\ \frac{1}{16\sqrt{2}}\left[1 - \frac{\alpha^2}{8}\right]e^{-\frac{\alpha^2}{16}}\right]$

23. By using Fourier transform, show the differential equation

(i) $f*(g*h) = (f*g)*h$ (ii) $f*(g+h) = (f*g)+(f*h)$

24. Find g(x) given that

$$\int_{-\infty}^{\infty} \frac{g(x)}{(x-1)^2 + a^2}\,dt = \frac{1}{x^2+b^2}, 0 < a < b.$$

$\left[\textit{Ans.}\ g(x) = \frac{a}{\pi b}\frac{(b-a)|\alpha|}{x^2+(b-a)^2}\right]$

25. $f(x) = g(x) = \begin{cases} 1-x^2, & |x| < 1 \\ 0, & |x| > 1 \end{cases}$ $\left[\textit{Ans.}\ \frac{4}{\alpha^3\sqrt{2x}}(\sin\alpha - \alpha\cos\alpha)\right]$

26. By employing the Parseval's identity to the function $f(x)$ defined by

$$f(x) = \begin{cases} 1-x^2, & |x| < 1 \\ 0, & |x| > 1 \end{cases}$$

Prove the following:

(i) $\int_0^\infty \frac{(1-\cos x)^2}{x^2}dx = \frac{\pi}{2}$ (ii) $\int_0^\infty \frac{\sin^4 x}{x^2}dx = \frac{\pi}{4}$

(iii) $\int_0^\infty \frac{(x\cos x - \sin x)^2}{x^6}dx = \frac{\pi}{15}$

27. Prove the following by using Parseval's identity

(i) $\int_0^\infty \frac{dx}{(x^2+a^2)^2} = \frac{\pi}{4a^3}$, $a > 0$ (ii) $\int_0^\infty \frac{x^2 dx}{(x^2+a^2)} = \frac{\pi}{4a}$, $a > 0$

28. Find the Fourier sine transforms of the following functions:

(i) $f(x) = \begin{cases} 1, & 0 \le x \le 1 \\ 0, & x > 1 \end{cases}$ (ii) $f(x) = \begin{cases} 0, & 0 \le x < a \\ x, & a \le x < b \\ 0, & x \ge b \end{cases}$

(iii) $f(x) = \begin{cases} \sin x, & 0 < x < a \\ 0, & x > a \end{cases}$ (iv) $xe^{-|x|}$

Ans. (i) $\sqrt{\frac{2}{\pi}}\frac{1-\cos\alpha}{\alpha}$

(ii) $\sqrt{\frac{2}{\pi}}\frac{1}{\alpha^2}\{(\sin b\alpha) - a(b\cos b\alpha - a\cos a\alpha\}$.

(iii) $\frac{1}{\sqrt{2\pi}}\frac{1}{\alpha^2-1}\left[\{\sin(\alpha-1)a + \sin(\alpha+1)a\} + \alpha\{\sin(\alpha-1)a - \sin(\alpha+1)a\}\right]$

(*iv*) $\sqrt{\frac{2}{\pi}}\frac{4\alpha}{(1+\alpha^2)^2}$

29. Prove the following.

(i) $F_s\left\{\frac{e^{ax}+e^{-ax}}{e^{\pi x}-e^{-\pi x}}\right\} = \frac{1}{\sqrt{2\pi}}\frac{\sin h\alpha}{\cos h\alpha + \cos a}$ (ii) $F_s\left\{\frac{1}{e^{\pi x}-e^{-\pi x}}\right\} = \frac{1}{2\sqrt{2\pi}}\tan h\frac{\alpha}{2}$

(iii) $F_s\{\cos e c h x\} = \sqrt{\frac{\pi}{2}}\tan h\left(\frac{\alpha\pi}{2}\right)$

(iv) $F_s\left\{\frac{e^{-ax}-e^{-bx}}{x}\right\} = \sqrt{\frac{2}{\pi}}\tan^{-1}\left[\frac{(b-a)\alpha}{ab+\alpha^2}\right]$, $a < b$

(v) $F_s\left\{\frac{x^2-a^2}{x(x^2+a^2)}\right\} = \sqrt{2\pi}\left[e^{-|a\alpha|} - \frac{1}{2}\cdot\right]\text{sgn}(\alpha)$.

30. Find the inverse Fourier sine transforms of the following

(i) $\frac{\pi}{2}$ (ii) $\frac{\alpha}{1+\alpha^2}$ (iii) $\sqrt{\frac{\pi}{2}}e^{-2\alpha}$ (iv) $\frac{e^{-a\alpha}}{\alpha}$, $a>0$ (v) $\tan^{-1}\left(\frac{\alpha}{2}\right)$

$\left[\textit{Ans.}\ \text{(i)}\sqrt{\frac{\pi}{2}}\cdot\frac{1}{x}\ \text{(ii)}\sqrt{\frac{\pi}{2}}e^{-x}\ \text{(iii)}\left(\frac{x}{x^2+4}\right)\ \text{(iv)}\sqrt{\frac{2}{\pi}}\tan^{-1}\left(\frac{x}{\alpha}\right)\ \text{(v)}\sqrt{\frac{\pi}{2}}\frac{1}{x}e^{-2x}\right]$

31. Evaluate $F_c\{f(x)\}$ of the following:

(i) $f(x)=e^{-ax^2}$, $a>0$ (ii) $f(x)=\cos\left(\frac{x^2}{2}\right)$ (iii) $f(x)=\sin\left(\frac{x^2}{2}\right)$

(iv) $f(x)=\frac{\cos x}{\sqrt{x}}$ (v) $f(x)=\begin{cases}\cos^2 x, & 0<x<1\\ 0, & x>1\end{cases}$

(vi) $f(x)=e^{-ax}\cos bx,\ a>0,\ b>0$ (vii) $f(x)=e^{-x}\sin^2 x$

(viii) $f(x)=\frac{\cos x-\cos^2 x}{x^2}$ (ix) $f(x)=\frac{1}{(a^2+x^2)^2}$, $a>0$

Ans. (i) $\frac{1}{\sqrt{2a}}e^{-\frac{\alpha^2}{4a}}$;

(ii) $\frac{1}{\sqrt{2}}\left(\cos\frac{\alpha^2}{2}+\sin\frac{\alpha^2}{2}\right)$;

(iii) $\frac{1}{\sqrt{2}}\left\{\cos\frac{\alpha^2}{2}-\sin\frac{\alpha^2}{2}\right\}$

(iv) $\frac{1}{2}\left\{\frac{1}{\sqrt{\alpha+1}}+\frac{1}{\sqrt{\alpha-1}}\right\}$; (v) $\frac{1}{2\sqrt{2\pi}}\left\{\frac{\sin(\alpha+2)}{\alpha+2}+2\frac{\sin\alpha}{\alpha}+\frac{\sin(\alpha-2)}{\alpha-2}\right\}$

(vi) $\frac{a}{\sqrt{2\pi}}\left\{\frac{1}{a^2+(\alpha+b)^2}+\frac{1}{a^2+(\alpha-b)^2}\right\}$;

(vii) $\frac{1}{\sqrt{2\pi}}\left[\frac{1}{\alpha^2+1}-\frac{1}{2}\left\{\frac{1}{(\alpha+1)^2+1}+\frac{1}{(\alpha-1)^2+1}\right\}\right]$

(viii) $\frac{\pi}{2}(2-\alpha)$, $0\le\alpha\le 1$; 0 for $\alpha>1$ (ix) $\sqrt{\frac{\pi}{2}}\frac{1}{2a^3}(1+a\alpha)e^{-a\alpha}$

32. If $a>b>0$, prove that

$$F_c\left\{\frac{1}{x}(e^{-bx}-e^{-ax})\right\}=\frac{1}{\sqrt{2\pi}}\log\frac{a^2+\alpha^2}{b^2+\alpha^2}$$

33. Show that $\cos\left(\frac{1}{2}x^2 - \frac{\pi}{8}\right)$ is self-reciprocal w.r.t. Fourier cosine transforms

34. Find the inverse Fourier cosine transforms of the following:

(i) $\frac{1}{1+\alpha^2}$ (ii) $\frac{\sqrt{\alpha+2}+\sqrt{\alpha-2}}{\sqrt{\alpha^2-4}}$ (iii) $\varphi(\alpha) = \begin{cases} \frac{1}{\sqrt{2\pi}}\left(a - \frac{\alpha}{2}\right), & \alpha < 2a \\ 0, & \alpha > 2a > 0 \end{cases}$

$$\left[Ans. \text{ (i)} \sqrt{\frac{\pi}{2}} e^{-x}; \text{ (ii)} \frac{2\cos 2x}{\sqrt{x}}; \text{ (iii)} \frac{\sin^2 ax}{\pi x^2} \right]$$

35. Find the Fourier cosine transforms of the following functions:

(i) $f(x) = \begin{cases} 1, & 0 < x < a \\ 0, & x > a \end{cases}$ $\left[Ans. \sqrt{\frac{2}{\pi}} \frac{\sin \alpha a}{\alpha} \right]$

(ii) $f(x) = \begin{cases} x, & 0 < x < 1 \\ 2-x, & 1 < x < 2 \\ 0, & x > 2 \end{cases}$ $\left[Ans. \sqrt{\frac{2}{\pi}} \left[\frac{2\cos\alpha - \cos 2\alpha - 1}{\alpha^2} \right] \right]$

(iii) $f(x) = \begin{cases} 4x, & 0 < x < 1 \\ 4-x, & 1 < x < 4 \\ 0, & x > 4 \end{cases}$ $\left[Ans. \frac{\sin\alpha}{\alpha} - \frac{1}{\alpha^2}(1 - 2\cos\alpha + \cosh\alpha \right]$

2.2 Hankel Transform

Introduction

In this section, we shall discuss another type of transform known as Hankel Transform. This transform is a branch of mathematics which has extensive applications in many disciplines of engineering.

We will first define the infinite Hankel Transform and derive a few properties. Next we will consider the Transforms of a few standard functions and associated forms. A list of useful results on Bessel functions and on infinite integrals involving Bessel functions is built up. The solutions of differential equations by method of Hankel Transforms is also touched upon.

It will be helpful to recapitulate the following few results (Table 2.4) concerning Bessel function and infinite integrals involving Bessel function.

Table 2.4

I Bessel function of first kind

$$J_n(x) = \sum_{r=0}^{\infty} \frac{(-1)^r}{r!\Gamma(n+r+1)} \cdot \left(\frac{x}{2}\right)^{n+2r}$$

II Recurrence formula for $J_n(x)$

(1) $xJ_n'(x) = nJ_n(x) - xJ_n + 1(x).$

(2) $xJ_n'(x) = -nJ_n(x) + xJ_n + 1(x).$

(3) $2J_n(x) = J_n - 1(x) - J_n + 1(x).$

(4) $\frac{d}{dx}[x^{-n} J_n(x)] = -x^{-n} J_n + 1(x).$

(5) $\frac{d}{dx}[x^n J_n(x)] = x^n J_n - 1(x).$

(6) $\frac{d}{dx}[x^n J_n(x)] = x^n J_n - 1(x).$

III Infinite integrals involving Bessel functions.

(1) $\int_0^{\infty} e^{-ax} J_0(sx)dx = (a^2 + s^2)^{-\frac{1}{2}}.$

(2) $\int_0^{\infty} e^{-ax} J_1(sx)dx = \frac{1}{s} - \frac{a}{s(a^2 + s^2)^{\frac{1}{2}}}.$

(3) $\int_0^{\infty} xe^{-ax} J_0(sx)\,dx = a(a^2 + s^2)^{-\frac{3}{2}}.$

(4) $\int_0^{\infty} xe^{-ax} J_1(sx)\,dx = s(a^2 + s^2)^{-\frac{3}{2}}.$

(5) $\int_0^{\infty} \frac{e^{-ax}}{x} J_1(sx)\,dx = \frac{(a^2 + s^2)^{\frac{1}{2}} - a}{s}.$

Hankel, Hermann (1839–1873), German analyst and geometer, proved that the system of complex numbers cannot be enlarged if field properties are to be preserved. Let $f(x)$ be a function of 'x' defined for all positive values of 'x'. Consider the definite integral.

Integral $= \int_0^\infty f(x) \cdot xJ_n(sx)\,dx$ [1]

Where $J_n(sx)$ is the Bessel function of the first kind of order *n.* If the integral [1] exists, evidently it must be a function of the variable '*s*'. Thus, the process (of $f(x)$ multiplying by $nJ_n(sx)$ and integrating the product between zero and infinity) transforms the function *f(x)* to a new functions of the new variable '*s*'. This process (or operation) is called the Hankel transform of order *n* of the function $f(x)$. The resulting function of '*s*' is called the Hankel transform of $f(x)$ and is denoted by $H\{f(x)\}$; or $\overline{\phi(s)}$ or $H\{f(x);n\}$

Note: If the order of the function (i.e., $xJ_n(sx)$ is specified, then, we denote Hankel transform by $H\{f(x), n = \text{order}\}$

Definition (Infinite Hankel transform of order *n*)

If $f(x)$ is a function of *x,* defined for all positive values of *x,* then the Hankel transform $f(x)$ of order *n*

$$\int_0^\infty f(x) \cdot xJ_n(sx)\,dx = H\{f(x);n\} = H\{f(x)\} = \overline{\phi(s)} \qquad [1]$$

Note: $J_n(sx)$ is called the kernel of the transformation

Inverse Transform

The function of *x* which has $\overline{\phi(s)}$ as its Hankel transform is known as the Inverse Hankel transform of the function $\overline{\phi(s)}$

Thus, if $\overline{\phi(s)} = H\{f(x)\}$, $f(x)$ then is said to be the Inverse transform of $\overline{\phi(s)}$ and is written as $H^{-1}\{\overline{\phi(s)}\}$

Thus, we get if $H\{f(x)\} = \overline{\phi(s)}$ then, $f(x) = H^{-1}\{\overline{\phi(s)}\}$.

The inversion formula for Hankel transform is

$$f(x) = H^{-1}\{\overline{\phi(s)}\} = \int_0^\infty \overline{\phi(s)} \cdot sJ_n(sx)\,ds. \qquad [2]$$

Thus, we have the following results:

(i) $\overline{\phi(s)} = H\{f(x)\} = \int_0^\infty f(x) \cdot xJ_n(sx)dx$

(ii) $f(x) = H^{-1}\{\overline{\phi(s)}\} = \int_0^\infty \overline{\phi(s)} \cdot sJ_n(sx)ds$

Properties of Hankel transform

P$_1$: (Linearity Property) If $f(x)\, g(x)$ and are two functions and $a, b,$ be two constants, then

$$H\{af(x)+bg(x)\}=aH\{f(x)\}+bH\{g(x)\}$$

Proof: LHS is $H\{af(x)+bg(x)\}$

$$=\int_0^\infty x[af(x)+bg(x)]J_n(sx)dx$$

$$=a\int_0^\infty xf(x)\cdot J_n(sx)dx+b\int_0^\infty xg(x)\cdot J_n(sx)dx$$

$$=aH\{f(x)\}+bH\{g(x)\}.$$

We also get,

$$H[cf(x)]=cH\{f(x)\},$$

$$H(f+g)=H(f)+H(g).$$

P$_2$: (Change of scale property)

If $\quad H\{f(x)\}=\overline{\phi(s)},$ then

$$H\{f(ax)\}=\frac{1}{a^2}\overline{\phi}\left(\frac{s}{a}\right)$$

Proof: Let $H\{f(x)\}=\overline{\phi(s)}$

$$H\{f(ax)\}=\int_0^\infty xf(ax)\cdot J_n(sx)dx$$

(put $ax=u$ to evaluate the above integral)

$$=\int_0^\infty \frac{u}{a}\cdot f(u)\cdot J_n\left(\frac{su}{a}\right)\frac{du}{a}$$

$$\because \quad H\{f(ax)\}=\frac{1}{a^2}\int_0^\infty u\cdot f(u)\cdot J_n\left[\left(\frac{s}{a}\right)\cdot u\right]du \qquad [2]$$

$$=\frac{1}{a^2}\cdot\overline{\phi}\left(\frac{s}{a}\right)$$

$\because$ by definition of Hankel transform. We have

$$H\{f(x)\}=\overline{\phi(s)}=\int_0^\infty f(x)\cdot xJ_n(sx)dx$$

if [1] is denoted by $\overline{\phi(s)}$, then [2] must be denoted by $\bar{\phi}\left(\frac{s}{a}\right)$.

P$_3$: (Hankel transform of the derivative of f i.e., $\left(\frac{df}{dx}\right)$

if $\quad H\{f(x),n\}=\overline{\phi_n(s)}$

and $\quad H\{f'(x),n\}=\overline{\phi_n'(s)}$

then, $\overline{\phi_n'(s)}=-\frac{s}{2n}\left[(n+1)\overline{\phi_{n-1}(s)}-(n-1)\overline{\phi_{n+1}(s)}\right]$.

Proof: By definition of Hankel transform, we have

$$\overline{\phi_n(s)}=\int_0^{\infty} xf(x)J_n(sx)dx$$

$$\overline{\phi_n'(s)}=\int_0^{\infty} x\frac{df}{dx}\cdot J_n(sx)dx$$

$$=\int_0^{\infty}\frac{df}{dx}\cdot[xJ_n(sx)]dx.$$

Integrating by parts and assuming that $xf(x)\to 0$ as $x\to 0$ and $xf(x)\to 0$ as $x\to\infty$, we obtain

$$\overline{\phi_n'(s)}=\left[f(x)\cdot xJ_n(sx)\right]_0^{\infty}-\int_0^{\infty} f(x)\cdot\frac{d}{dx}[xJ_n(sx)]\cdot dx$$

$$=0-\int_0^{\infty} f(x)\cdot[J_n(sx)+sxJ_n'(sx)]dx \qquad [1]$$

Now, we know that

$$xJ_n'(x)=-nJ_n(x)+xJ_{n-1}(x)$$

(Recurrence formula for $J_n(x)$)

replacing x by sx, we get

$$sxJ_n'(sx)=-nJ_n(sx)+sxJ_{n-1}(sx)$$

or $\quad sxJ_n'(sx)+J_n(sx)=(1-n)J_n(sx)+sxJ_{n-1}(sx)$

Using this in [1], we get

$$\overline{\phi_n'(s)}=-\int_0^{\infty} f(x)\cdot[(1-n)J_n(sx)+J_{n-1}(sx)]dx$$

$$=-(1-n)\int_0^{\infty} f(x)\cdot J_n(sx)dx-s\int_0^{\infty} xf(x)\cdot J_{n-1}(sx)dx$$

$$\overline{\phi_n'(s)} = -(1-n)\int_0^\infty f(x)\cdot J_n(sx)dx - s\overline{\phi_{n-1}(s)} \qquad [2]$$

Again from a recurrence formula for $J_n(x)$,

we have that $2nJ_n(x) = x[J_{n-1}(x) + J_{n+1}(x)]$

replacing x by sx,

$$2nJ_n(sx) = sx[J_{n+1}(sx) + J_{n-1}(sx)] \qquad [3]$$

(multiplying [3] both sides by $f(x)$ and then integrating between 0 and ∞ we find,

$$2n\int_0^\infty f(x)\cdot J_n(sx)dx$$

$$= s\left[\int_0^\infty xf(x)J_{n-1}(sx)dx + \int_0^\infty xf(x)\cdot J_{n+1}(sx)dx\right]$$

$$= s\left[\overline{\phi_{n-1}}(s) + \overline{\phi_{n+1}}(s)\right].$$

or $$\int_0^\infty f(x)\cdot J_n(sx)dx = \frac{s}{2n}\left[\overline{\phi_{n-1}(s)} + \overline{\phi_{n+1}(s)}\right]$$

In this event [2] becomes

$$\overline{\phi_n'(s)} = \frac{(1-n)s}{2n}\left[\overline{\phi_{n-1}}(s) + \overline{\phi_{n+1}}(s)\right] - s\overline{\phi_{n-1}}(s)$$

$$= \frac{s}{2n}\left[(n-1)\overline{\phi_{n-1}}(s) + (n-1)\overline{\phi_{n+1}}(s) - 2n\overline{\phi_{n-1}}(s)\right]$$

$$= \frac{s}{2n}\left[-(n+1)\overline{\phi_{n-1}}(s) + (n-1)\overline{\phi_{n+1}}(s)\right]$$

or $$\overline{\phi_n'(s)} = -\frac{s}{2n}\left[(n+1)\overline{\phi_{n-1}}(s) - (n-1)\overline{\phi_{n+1}}(s)\right] \qquad [4]$$

Corollary 1: When $n = 1$

$$\overline{\phi_1'(s)} = -s\overline{\phi_0(s)}$$

$\Rightarrow \quad H\{f'(x), n=1\} = -sH\{f(x), n=0\}$

(This result is obtained by putting $n = 1$ in [4])

Corollary 2: When $n = 2$

$$\overline{\phi_1'(s)} = -\frac{s}{4}\left[3\overline{\phi_1(s)} - \overline{\phi_3(s)}\right]$$

(This result is obtained by putting $n = 2$ in [4])

Corollary 3: When $n = 3$

$$\overline{\phi_n'(s)} = -\frac{s}{6}\left[4\overline{\phi_2(s)} - 2\overline{\phi_4(s)}\right]$$

(This result is obtained by putting $n = 3$ in [4])

Corollary 4: To prove that

$$\overline{\phi_n''(s)} = \frac{s^2}{4}\left[\left(\frac{n+1}{n-1}\right)\overline{\phi_{n-2}}(s) - 2\left(\frac{n^2-3}{n^2-1}\right)\overline{\phi_n}(s) + \left(\frac{n-1}{n+1}\right)\overline{\phi_{n+2}}(s)\right].$$

Proof: From equation [4], we have

$$\overline{\phi_n'(s)} = -s\left[\left(\frac{n+1}{2n}\right)\overline{\phi_{n-1}}(s) - \left(\frac{n-1}{2n}\right)\overline{\phi_{n+1}}(s)\right] \quad [1]$$

Replacing ϕ by ϕ', we get

$$\overline{\phi''(s)} = -s\left[\left(\frac{n+1}{2n}\right)\overline{\phi_{n-1}'}(s) - \left(\frac{n-1}{2n}\right)\overline{\phi_{n+1}'}(s)\right] \quad [2]$$

Replacing n by n–1 and n+1 respectively in Equation [1], we get

$$\overline{\phi_{n-1}'}(s) = -s\left[\frac{n}{2n-2}\overline{\phi_{n-2}}(s) - \frac{n-2}{2(n-1)}\overline{\phi_n}(s)\right]$$

$$\overline{\phi_{n+1}'}(s) = -s\left[\frac{(n+2)}{2(n+1)}\overline{\phi_n}(s) - \frac{n}{2(n+1)}\overline{\phi_{n+2}}(s)\right] \quad [3]$$

Substituting these values of $\overline{\phi_{n-1}'}(s)$ and $\overline{\phi_{n+1}'}(s)$ from [3], into Equation [2], We find

$$\overline{\phi_n''(s)} = \frac{s^2}{4}\left[\left(\frac{n+1}{4}\right)\left\{\frac{n}{n-1}\overline{\phi_{n-2}}(s) - \left(\frac{n-2}{n-1}\right)\overline{\phi_n}(s)\right\}\right.$$

$$\left. - \left(\frac{n-1}{2}\right)\left\{\left(\frac{n+2}{n+1}\right)\overline{\phi_n}(s) - \left(\frac{n}{n+1}\right)\overline{\phi_{n+2}}(s)\right\}\right]$$

$$= \frac{s^2}{4}\left[\left(\frac{n+1}{n-1}\right)\overline{\phi_{n-2}}(s) + \left(\frac{n-1}{n+1}\right)\overline{\phi_{n+2}}(s) - \frac{2(n^2-3)}{n^2-1}\overline{\phi_n}(s)\right]$$

P$_4$: To find the Hankel transform of

$$\frac{d^2 f}{dx^2} + \frac{1}{x}\cdot\frac{df}{dx} - \frac{n^2}{x^2}f.$$

Proof: By definition

$$H\left\{\frac{d^2 f}{dx^2}, n\right\} = \int_0^\infty \frac{d^2 f}{dx^2} \cdot xJ_n(sx)\,dx$$

$$= \left[\frac{df}{dx} \cdot xJ_n(sx)\right]_0^\infty - \int_0^\infty \frac{df}{dx}[J_n(sx) + sxJ_n'(sn)]dx$$

(Letting $xf(x) \to 0$ when $x \to 0$ or when $x \to \infty$)

$$H\left(\frac{d^2 f}{dx^2}\right) + \int_0^\infty \frac{f'(x)}{x} xJ_n(sx)dx = -s\int_0^\infty xf'(x)J_n'(sx)dx$$

or $$H\left(\frac{d^2 f}{dx^2} + \frac{f'(x)}{x}\right) = -s\left[xf(x)J_n'(sx)\right]_0^\infty + s\int_0^\infty f(x)\frac{d}{dx}[xJ_n'(sx)]dx$$

Again, letting $xf(x) \to 0$ as $x \to 0$ or when $x \to \infty$)

$$H\left(\frac{d^2 f}{dx^2} + \frac{f'(x)}{x}\right) = s\int_0^\infty f\frac{d}{dx}[xJ_n'(sx)]dx \qquad [1]$$

Now, $J_n(x)$ satisfies the differential equation,

$$\frac{d}{dx}\left\{x\frac{dy}{dx}\right\} + \left(1 - \frac{n^2}{x^2}\right)xy = 0.$$

This yields

$$\frac{d}{dx}\{xJ_n(x)\} = -\left(1 - \frac{n^2}{x^2}\right)xJ_n(x).$$

Replacing x by sx, we get

$$\frac{1}{s} \cdot \frac{d}{dx}\{sxJ_n'(sx)\} = \frac{n^2}{sx}J_n(sx) - sxJ_n(sx)$$

Using this in Equation [1], we get

$$H\left(\frac{d^2 f}{dx^2} + \frac{f'(x)}{x}\right) = s\int_0^\infty f\frac{n^2}{sx}J_n(sx)dx - s\int_0^\infty f \cdot sxJ_n(sx)dx$$

$$= \int_0^\infty \left(\frac{n^2 f}{x^2}\right) sJ_n(sx)dx - s^2\int_0^\infty fxJ_n(sx)dx$$

or $$H\left(\frac{d^2 f}{dx^2} + \frac{f'(x)}{x} - \frac{n^2}{x^2 f}\right) = -s^2\overline{\phi_n(s)}$$

Finally, $$\int_0^\infty \left[\frac{d^2 f}{dx^2} + \frac{1}{x}\cdot\frac{df}{dx} - \frac{n^2}{x^2} f\right] xJ_n(sx)dx = -s^2 \overline{\phi_n(s)} \qquad [2]$$

Corollary 1: When $n = 0$ is Equation [2], we get

$$\int_0^\infty \left(\frac{d^2 f}{dx^2} + \frac{1}{x}\cdot\frac{df}{dx}\right) xJ_n(sx)dx = -s^2 \overline{\phi_0(s)}.$$

where $\overline{\phi_0(s)}$ is the Hankel Transform of *f(x)* of order zero. i.e., $H\{f(x), n = 0\} = \overline{\phi_0(s)}$

if $H\{f(x), n = 0\} = \overline{\phi_0(s)}$, then

$$H\left\{\frac{d^2 f}{dx^2} + \frac{1}{x}\cdot\frac{df}{dx}\right\} = -s^2 \overline{\phi(s)}$$

Corollary 2: When $n = 1$ in Equation [2], we get

$$\int_0^\infty \left(\frac{d^2 f}{dx^2} + \frac{1}{x}\cdot\frac{df}{dx} - \frac{1}{x^2} f\right) xJ_1(sx)dx$$

$$= -s^2 \overline{\phi_1(s)}$$

P$_5$: (Parseval's Theorem)

Given $H\{f(x)\} = \phi(s)$, if $\overline{\phi_1(s)}$ and $\overline{\phi_2(s)}$ be the Hankel transforms of $f(x)$ and $g(x)$ respectively, then

$$\int_0^\infty xf(x)\cdot g(x)dx = \int_0^\infty s\overline{\phi_1(s)}\,\overline{\phi_2(s)}\,ds$$

Proof: By data,

$$\overline{\phi_1(s)} = H\{f(x)\} = \int_0^\infty f(x)\cdot xJ_n(sx)dx$$

$$\overline{\phi_2(s)} = H\{g(x)\} = \int_0^\infty g(x)\cdot xJ_n(sx)dx$$

$$\int_0^\infty s\overline{\phi_1(s)}\,\overline{\phi_2(s)}ds = \int_0^\infty s\overline{\phi_1(s)}\left[\int_0^\infty g(x)\cdot xJ_n(sx)dx\right]ds$$

$$= \int_0^\infty s\overline{\phi_1(s)}ds \int_0^\infty g(x)\cdot xJ_n(sx)dx$$

$$= \int_0^\infty xg(x)dx \int_0^\infty s\overline{\phi_1(s)} J_n(sx)ds$$

(on interchanging order of integration)

$$= \int_0^\infty xg(x)dx \cdot f(x).$$

$$\left(\text{For } f(x) = \int_0^\infty \overline{\phi_1(s)} \cdot sJ_n(sx)ds \right) \text{ (Inversion formula)}$$

$$\int_0^\infty xf(x) \cdot g(x)dx$$

$$\int_0^\infty xf(x) \cdot g(x)dx = \int_0^\infty s\overline{\phi_1(s)}\overline{\phi_2(s)}ds$$

(B) EXAMPLES

1. Find the Hankel transform of each of the following:

(i) e^{-x} (ii) e^{-ax} (iii) $\dfrac{e^{-x}}{x}$

(iv) $\dfrac{e^{-ax}}{x}$ of order zero (or taking $xJ_0(sx)$ as the kernel)

Solution: (i) By definition, of Hankel transform of order n, we have,

$$H\{f(x)\} = \int_0^\infty f(x) \cdot xJ_n(sx)dx \quad [1]$$

Here, $f(x) = e^{-x}$ and we have to evaluate Hankel transform of order zero.

Hence, we write $H\left\{e^{-x}, n=0\right\} = \int_0^\infty e^{-x} \cdot xJ_0(sx)dx$

$$= (1+s^2)^{-\frac{3}{2}}$$

$$\because \quad \int_0^\infty xe^{-ax} J_0(sx)dx = (1+s^2)^{-\frac{3}{2}}$$

(ii) Here, $f(x) = e^{-ax}$; and we have to evaluate Hankel transform of order zero.

Hence, we write

$$H\left\{e^{-ax}, n=0\right\} = \int_0^\infty e^{-ax} \cdot xJ_0(sx)dx$$

$$= a(a^2+s^2)^{-\frac{3}{2}}$$

(iii) Here, $f(x)=\frac{e^{-x}}{x}$; and we have find the Hankel transform of order zero.

Hence, we write

$$H\left\{\frac{e^{-x}}{x}, n=0\right\}=\int_0^{\infty}\frac{e^{-x}}{x}xJ_0(sx)dx$$

$$=\int_0^{\infty}e^{-x}\cdot J_0(sx)dx$$

$$=(1+s^2)^{-\frac{1}{2}}$$

(iv) Here, $f(x)=\frac{e^{-ax}}{x}; n=0;$

$$\therefore \quad H\left\{\frac{e^{-ax}}{x}, n=0\right\}$$

$$=\int_0^{\infty}\frac{e^{-ax}}{x}\cdot xJ_0(sx)dx$$

$$=\int_0^{\infty}e^{-ax}.J_0(sx)dx$$

$$=(a^2+s^2)^{-\frac{1}{2}}$$

2. Find the Hankel transform of $\frac{\sin ax}{x}$ taking $xJ_0(sx)$ as the kernel.

Solution: Here, $f(x)=\frac{\sin ax}{x}; \; n=0$

$$\therefore \quad H\left\{\frac{\sin ax}{x}, n=0\right\}=\int_0^{\infty}\frac{\sin ax}{x}\cdot x\cdot J_0(sx)dx$$

$$=\text{Imaginary part of }\int_0^{\infty}e^{-iax}J_0(sx)dx$$

$$=\text{Imaginary part of }(i^2a^2+s^2)^{-\frac{1}{2}} \qquad \left\{\begin{array}{r}\int_0^{\infty}e^{-ax}J_0(sx)dx\\ =(a^2+s^2)^{-\frac{1}{2}}\end{array}\right\}$$

$$= \text{Imaginary part of } (-a^2+s^2)^{-\frac{1}{2}}$$

$$= \begin{cases} 0, & \text{if } s > a \\ (a^2-s^2), & \text{if } 0 < s < a \end{cases}$$

3. Find the Hankel transform of $\dfrac{\cos ax}{x}$; $n=0$ taking $xJ_0(sx)$ as the kernel

***Solution*:** Here, $f(x) = \dfrac{\cos ax}{x}$; $n = 0$

$$\therefore \quad H\left\{\frac{\cos ax}{x}, n=0\right\} = \int_0^\infty \frac{\cos ax}{x} \cdot xJ_0(sx)dx$$

$$= \int_0^\infty \cos ax \cdot J_0(sx)dx$$

$$= \text{real part of } \int_0^\infty e^{-\mathrm{i}ax} J_0(sx)\,dx$$

$$= \text{real part of } (\mathrm{i}^2a^2+s^2)^{-\frac{1}{2}}$$

$$= \text{real part of } (s^2-a^2)^{-\frac{1}{2}} = \begin{cases} (s^2-a^2)^{-\frac{1}{2}} & \text{if } s > a \\ 0, & \text{if } 0 < s < a \end{cases}$$

4. Find the Hankel transform of

$$f(x) = \begin{cases} 1, & 0 < x < a, \quad n = 0 \\ 0, & x > a, \quad n = 0 \end{cases}$$

***Solution*:** By definition of Hankel transform, we have,

$$H\{f(x), n=0\} = \int_0^\infty f(x) \cdot xJ_0(sx)dx = \overline{\phi(s)}$$

$$= \int_0^a f(x) \cdot xJ_0(sx)dx + \int_a^\infty f(x) \cdot xJ_0(sx)dx$$

$$= \int_0^a 1 \cdot xJ_0(sx)dx + \int_a^\infty 0.xJ_0(sx)dx$$

$$= \int_0^a xJ_0(sx)dx + 0 \quad [1]$$

But by a recurrence formula for Bessel's function we have

$\frac{d}{dx}\left\{x^n J_n(x)\right\} = x^n J_{n-1}(x)$. Changing ($n$ and x) by (1 and sx) respectively,

$$\frac{1}{s}\cdot\frac{d}{dx}\{sxJ_1(sx)\} = sxJ_0(sx).$$

or $\quad \frac{d}{dx}\{xJ_1(sx)\} = sxJ_0(sx) \quad [2]$

Integrating [2] from $x = 0$ to $x = a$,

$$[xJ_1(sx)]_0^a = \int_0^a xJ_0(sx)dx$$

or $\quad \int_0^a xJ_0(sx)dx = \frac{a}{s}J_1(sa).$

Now [1] becomes

$$H\{f(x), n = 0\} = \frac{a}{s}J_1(sa)$$

5. Find the Hankel transform of

$$f(x) = \begin{cases} a^2 - x^2 & 0 < x < a, n = 0, \\ 0, & x > a, \quad n = 0 \end{cases}$$

Solution: By definition of Hankel transform. We have,

$$H\{f(x), n = 0\} = \int_0^\infty f(x)\cdot xJ_0(sx)dx$$

$$= \int_0^a f(x)\cdot xJ_0(sx)dx + \int_a^\infty f(x).xJ_0(sx)\,dx$$

$$= \int_0^a (a^2 - x^2)xJ_0(sx)dx + \int_a^\infty 0\cdot xJ_0(sx)dx$$

$$= \int_0^a (a^2 - x^2)xJ_0(sx)dx + 0 \quad [1]$$

$$= a^2\int_0^a xJ_0(sx)\,dx \cdot \int_0^a x^3 J_0(sx)dx$$

$$= a^2 I_1 - I_2. \quad [2]$$

where $\quad I_1 = \int_0^a aJ_0(sx)dx$

and $\quad I_2 = \int_0^a x^3 J_0(sx)dx$

By a recurrence formula for Bessel's function

We have, $\dfrac{d}{dx}\left\{x^n J_n(x)\right\} = x^n J_{n-1}(x)$

Changing x by sx, we get

$$\frac{1}{s}\cdot\frac{d}{dx}\left\{(sx)^n J_n(sx)\right\} = (sx)^n J_{n-1}(sx)$$

or, $$\frac{1}{s}\cdot\frac{d}{dx}\left\{x^n J_n(sx)\right\} = x^n J_{n-1}(sx) \qquad [3]$$

For $n = 1$, this gives

$$\frac{1}{s}\cdot\frac{d}{dx}\{xJ_1(sx)\} = aJ_0(sx) \qquad [4]$$

Integrating, we get

$$\int_0^a xJ_0(sx)dx = \frac{1}{s}\left[xJ_1(sx)\right]_0^a = \frac{a}{s}J_1(sa)$$

or $$I_1 = \int_0^a xJ_0(sx)\,dx = \frac{a}{s}J_1(sa) \qquad [5]$$

From [4], $I_2 = \int_0^a x^3 J_0(sx)dx$

$$= \int_0^a \frac{x^2}{s}\cdot\frac{d}{dx}\{xJ_1(sx)dx\}$$

$$= \frac{1}{s}\left[\left\{x^2\cdot xJ_1(sx)\right\}_0^a - \int_0^a 2x\cdot xJ_1(sx)dx\right]$$

$$= \frac{1}{s}\left[a^3 J_1(sa) - 2\int_0^a x^2\cdot J_1(sx)dx\right]$$

$$= \frac{1}{s}\left[a^3 J_1(sa) = 2\int_0^a \frac{1}{s}\cdot\frac{d}{dx}\left\{x^2 J_2(sx)dx\right\}\right]$$

$$= \frac{a^3}{s}J_1(sa) - \frac{2}{s^2}\left\{x^2 J_2(sx)\right\}_0^a$$

$$= \frac{a^2}{s}J_1(sa) - \frac{2a^2}{s^2}J_2(sa). \qquad [6]$$

Writing [2] with the help of [5] and [6],

$$H\{f(x), n=0\} = a^2 \cdot \frac{a}{s} J_1(sa) - \left[\frac{a^3}{s} J_1(sa) - \frac{2a^2}{s^2} J_2(sa)\right]$$

$$H\{f(x), n=0\} = \frac{2a^2}{s^2} J_2(sa) \qquad [7]$$

6. Show that the Hankel transform of order zero of

$$f(x) = \begin{cases} a^2 - x^2, & 0 < x < a \\ 0, & x > a \end{cases} \quad \text{is} \quad \frac{4a}{s^3} J_1(sa) - \frac{2a^2}{s^2} J_0(sa)$$

Solution: We have, from Equation [7], of the previously solved Example 5 that

$$H\{f(x), n=0\} = \frac{2a^2}{s^2} J_2(sa) \qquad [1]$$

Now, by a recurrence formula for Bessel's function,

$$2nJ_n(x) = x\left[J_{n-1}(x) + J_{n+1}(x)\right] \qquad [2]$$

$$\left[\begin{array}{l} \text{put } n = 1 \text{ in [2] to get} \\ 2J_1(x) = x[J_0(x) + J_2(x)] \end{array}\right]$$

Replacing x by sa, we get

$$2J_1(sa) = sa[J_0(sa) + J_2(sa)]$$

or $$\frac{2}{sa} J_1(sa) - J_0(sa) = J_2(sa).$$

Multiplying this by $\dfrac{2a^2}{s^2}$

$$\frac{4a}{s^3} J_1(sa) - \frac{2a^2}{s^2} J_0(sa) = \frac{2a^2}{s^2} J_2(sa)$$

Substituting from equation [3] into equation [1], we get

$$H\{f(x), n=0\} = \frac{4a}{s^3} J_1(sa) - \frac{2a^2}{s^2} J_0(sa).$$

7. Find that Hankel transform of order zero of the function

$$f(x) = \begin{cases} 4 - x^2, & \therefore 0 < x < 2 \\ 0, & x > 2 \end{cases}$$

Solution: We can prove as in the previous problem by taking $a = 2$, in

$$f(x)\begin{cases} a^2 - x^2, & 0 < x < a \\ 0, & x > a \end{cases}$$

$$\therefore \quad H\{f(x), n=0\} = \frac{4a}{s^3} j_1(sa) - \frac{2a^2}{s^2} J_0(sa)$$

$$= \frac{8}{s^3} J_1(2s) - \frac{8}{s^2} J_0(2s)$$

8. Find the Hankel transform of $\frac{e^{-5x}}{x}$, taking $xJ_0(sx)$ as the kernel.

Solution: By definition, we have

$$H\left\{\frac{e^{-5x}}{x}, n=0\right\} = \int_0^\infty \frac{e^{-5x}}{x} \cdot xJ_0(sx)dx$$

$$= \int_0^\infty e^{-5x} \cdot J_0(sx)dx$$

$$= (5^2 + s^2)^{-\frac{1}{2}} = (25 + s^2)^{-\frac{1}{2}}$$

$$\left(\because \quad \int_0^\infty e^{-ax} J_0(sx)\, dx = (a^2 + s^2)^{-\frac{1}{2}}\right)$$

9. Find the Hankel transform of $\frac{1}{x}$ of order zero.

Solution: Here, $f(x) = \frac{1}{x}; n = 0$

By definition, $H\{f(x), n=0\} = \int_0^\infty \frac{1}{x} . xJ_0(sx)dx$

$$= \int_0^\infty J_0(sx)dx$$

$$= \int_0^\infty e^{-ox} J_0(sx)\, dx$$

$$= (o^2 + s^2)^{-\frac{1}{2}}$$

$$= \frac{1}{s}$$

or, $$H\{f(x), n=0\} = \overline{\phi(s)} = \frac{1}{s}$$

10. Find the Hankel transform of $x^{-2}e^{-x}$ order 1.

or

Find the Hankel transform of $x^{-2}e^{-x}$, taking $xJ_1(sx)$ as the kernel.

Solution: By definition of Hankel transform we have

$$H\{f(x)\} = \int_0^\infty f(x)\cdot xJ_n(sx)dx \qquad [1]$$

$$= \overline{\phi(s)}$$

Here, $f(x) = x^2e^{-x}; n=1$

$$\therefore \quad H\left\{x^{-2}e^{-x}, n=1\right\} = \int_0^\infty x^{-2}e^{-x}\cdot xJ_1(sx)dx$$

$$= \int_0^\infty \frac{e^{-x}}{x}\cdot J_1(sx)dx$$

$$= \frac{(1+s^2)^{\frac{1}{2}}-1}{s}$$

$$\left(\because \int_0^\infty \frac{e^{-ax}}{x}J_1(sx)dx\right) = \frac{(a^2+s^2)^{\frac{1}{2}}}{s}.$$

11. Find the Hankel transform of $\frac{e^{-ax}}{x}$ of order one.

Solution: By definitions, we have,

$$H\{f(x)\} = \int_0^\infty f(x)\cdot xJ_1(sx)\,dx = \overline{\phi(s)}$$

Here, $f(x) = \frac{e^{-ax}}{x}; n=1$

$$\therefore \quad H\left\{\frac{e^{-ax}}{x}, n=1\right\} = \int_0^\infty \frac{e^{-ax}}{x} \cdot xJ_1(sx)dx$$

$$= \int_0^\infty e^{-ax} \cdot J_1(sx)dx$$

$$= \frac{1}{s} - \frac{a}{s(a^2+s^2)^{\frac{1}{2}}}$$

$$\left(\because \int_0^\infty e^{-ax} \cdot J_1(sx)dx = \frac{1}{s} - \frac{a}{s(a^2+s^2)^{\frac{1}{2}}}\right)$$

12. Find the Hankel transform of e^{-ax} of order one.

***Solution*:** By definition, we have

$$H\{f(x)\} = \int_0^\infty f(x).xJ_1(sx)dx = \overline{\phi(s)}$$

Here, $f(x) = e^{-ax}; n = 1$

$$\therefore \quad H\{e^{-ax}, n=1\} = \int_0^\infty e^{ax} \cdot xJ_1(sx)dx$$

$$= s(a^2+s^2)^{-\frac{3}{2}}$$

$$\left(\because \int_0^\infty xe^{-ax} J_1(sx)dx = s(a^2+s^2)^{-\frac{3}{2}}\right)$$

13. Find the Hankel transform of order *n* of the function $\frac{\partial^2 f}{\partial t^2}$ where *f* is a function of *x* and *t*.

***Solution*:** By definition, we have

$$H\left\{\frac{\partial^2 f}{\partial t^2}, n\right\} = \int_0^\infty x \cdot \frac{\partial^2 f}{\partial t^2} \cdot J_n(sx)dx$$

$$= \frac{\partial^2}{\partial t^2}\int_0^\infty f(x,t) \cdot xJ_n(sx)dx = \frac{d^2\overline{\phi_n(s)}}{dt^2}$$

where $\overline{\phi_n(s)}$ is the Hankel transform of order *n* of *f(x)*.

14. Show that, if $f = \dfrac{e^{-ax}}{x}$ then $H\left\{\dfrac{df}{dx}, n=1\right\} = -s\overline{\phi_0(s)} = -s(a^2+s^2)^{-\frac{1}{2}}$

Solution: $H\left\{\dfrac{df}{dx}, n=1\right\} = \int_0^\infty x\cdot\dfrac{df}{dx} J_1(sx)dx$

$$= -s\overline{\phi_0(s)}$$

$$= -s\cdot\int_0^\infty e^{-ax}\cdot J_0(sx)dx$$

$$= -s(a^2+s^2)^{-\frac{1}{2}}$$

15. Find the Hankel transform of order zero of the function $\dfrac{d^2 f}{dx^2} + \dfrac{1}{x}\cdot\dfrac{df}{dx}$, where $f(x) = e^{\frac{-ax}{x}}$.

Solution: We know that

$$\int_0^\infty \left(\frac{d^2 f}{dx^2} + \frac{1}{x}\cdot\frac{df}{dx} - \frac{n^2 f}{x^2}\right) xJ_n(sx)dx = -s^2\overline{\phi_n(s)} \qquad \text{(by property 4)}$$

For order zero, i.e., for $n = 0$, this gives

$$\int_0^\infty \left(\frac{d^2 f}{dx^2} + \frac{1}{x}\cdot\frac{df}{dx}\right) xJ_0(sx)dx = -s^2\overline{\phi_0(s)}$$

or $$H\left\{\frac{d^2 f}{dx^2} + \frac{1}{x}\cdot\frac{df}{dx}; n=0\right\} = -s^2\overline{\phi_0(s)}$$

$$= -s^2\int_0^\infty f(x)\cdot x\cdot J_0(sx)dx$$

$$= -s^2\int_0^\infty \frac{e^{-ax}}{x}\cdot xJ_0(sx)dx$$

$$= -s^2\int_0^\infty e^{-ax} J_0(sx)dx$$

$$= -s^2(a^2+s^2)^{-\frac{1}{2}}$$

$$\left(\because \int_0^\infty e^{-ax} J_0(sx)dx = (a^2+s^2)^{-\frac{1}{2}}\right)$$

16. Prove that $H\left[\sum_{i=1}^{3} f_i(x)\right] = \sum_{i=1}^{3} H\left[f_i(x)\right]$

***Solution*:** Applying definition to LHS, we get

$$H\left[\sum_{i=1}^{3} f_i(x)\right] = \int_0^\infty \left\{\sum_{i=1}^{3} f_i(x)\right\} xJ_n(sx)dx$$

$$= \int_0^\infty (f_1 + f_2 + f_3)xJ_n(sx)dx$$

$$= \int_0^\infty f_1 xJ_n(sx)dx + \int_0^\infty f_2 \cdot xJ_n(sx)dx$$

$$= H(f_1) + H(f_2) + H(f_3)$$

$$= \sum_{i=1}^{3} H\left[f_i(x)\right] = \text{RHS}$$

17. Find the Hankel transform of $f(x) = \begin{cases} x^n, & 0 < x < a \\ 0, & x > a, \end{cases}$ $n > -1$ taking $xJ_n(sx)$ as kernel

***Solution*:** By definition, we have

$$H\{f(x)\} = \int_0^\infty f(x) \cdot xJ_n(sx)dx$$

$$= \int_0^a f(x) \cdot xJ_n(sx)dx + \int_a^\infty f(x) \cdot xJ_n(sx)dx$$

$$= \int_0^a x^n \cdot xJ_n(sx)dx + \int_a^\infty 0 \cdot xJ_n(sx)dx$$

or $\quad \overline{\phi_n(s)} = \int_0^a x^{n+1} J_n(sx)dx$

Now, by a recurrence formula for $J_n(x)$, we have

$$\frac{d}{dx}\left\{x^n J_n(x)\right\} = x^n J_{n-1}(x)$$

Replacing x by sx,

$$\frac{1}{s} \cdot \frac{d}{dx}\left\{x^n J_n(sx)\right\} = x^n J_{n-1}(sx)$$

Replacing *n by n*+1, we get

$$\frac{1}{s}\cdot\frac{d}{dx}\left\{x^{n+1}J_{n+1}(sx)\right\}=x^{n+1}J_n(sx)$$

$$\overline{\phi_n(s)}=\int_0^a\frac{1}{s}\cdot\frac{d}{dx}\left\{x^{n+1}J_{n+1}(sx)\right\}dx$$

$$=\frac{1}{s}\left[x^{n+1}J_{n+1}(sx)\right]_0^a$$

$$\overline{\phi_n(s)}=\frac{a^{n+1}}{s}J_{n+1}(sa).$$

18. Show that $H^{-1}\left[\frac{1}{s}\right]=\frac{1}{x}$ for order $n = 0$.

***Solution*:** Applying inversion formula,

$$f(x)=H^{-1}\left\{\overline{\phi(s)}\right\}=\int_0^\infty\overline{\phi(s)}\cdot sJ_0(sx)ds.$$

Here, $\overline{\phi(s)}=\frac{1}{s};\ n=0$

$$=\int_0^\infty\frac{1}{s}\cdot s\cdot J_0(sx)ds$$

$$=\int_0^\infty J_0(sx)ds=(0^2+x^2)^{-\frac{1}{2}}$$

$$=\frac{1}{x}$$

$\therefore\quad H^{-1}\left\{\frac{1}{s}\right\}=\frac{1}{x}$

19. Show that $H^{-1}\left\{s^{-2}\cdot e^{-as}\right\}=\frac{(a^2+x^2)^{\frac{1}{2}}-a}{x}$ for order $n=1$.

***Solution*:** Applying inversion formula, we get

$$f(x)=H^{-1}\left\{s^{-2}e^{-as},\ n=1\right\}$$

$$=\int_0^\infty s^{-2}\cdot e^{-as}\cdot sJ_1(sx)ds=\int_0^\infty\frac{e^{-as}}{s}J_1(sx)ds$$

$$= \frac{(a^2 + x^2)^{\frac{1}{2}}}{x}$$

$$\left(\because \int_0^\infty \frac{e^{-ax}}{x} J_1(sx)\, dx = \frac{(a^2 + s^2)^{\frac{1}{2}} - a}{s} \right)$$

$$\int_0^\infty \frac{e^{-as}}{s} J_1(sx) ds = \frac{(a^2 + x^2)^{\frac{1}{2}} - a}{x}$$

20. Evaluate: $H^{-1}\left\{e^{-as}, n = 0\right\}$

***Solution*:** By the definition of inverse formula for Hankel transform, we have

$$f(x) = H^{-1}\left\{e^{-as}, n = 0\right\} = \int_0^\infty e^{-as} sJ_0(sx)\, ds$$

$$= a(a^2 + x^2)^{-\frac{3}{2}}$$

21. Evaluate: $H^{-1}\left\{\frac{e^{-as}}{s}, n = 0\right\}$

***Solution*:** By definition, we have

$$f(x) = H^{-1}\left\{\frac{e^{-as}}{s}, n = 0\right\} = H^{-1}\left\{\overline{\phi(s)}\right\}$$

$$= \int_0^\infty \frac{e^{-as}}{s} \cdot sJ_0(sx) ds$$

$$= \int_0^\infty e^{-as} J_0(sx) ds = (a^2 + x^2)^{-\frac{1}{2}}.$$

$$\left(\because \int_0^\infty e^{-ax} J_0(sx) dx = (a^2 + s^2)^{-\frac{1}{2}} \right)$$

22. Evaluate: $H^{-1}\left\{\frac{e^{-as}}{s}, n = 1\right\}$

Solution: Applying inversion formula, we have

$$H^{-1}\left\{\frac{e^{-as}}{s}, n=1\right\} = \int_0^\infty \frac{e^{-as}}{s}\cdot sJ_1(sx)ds$$

$$= \int_0^\infty e^{-as}\cdot J_1(sx)ds$$

$$= \frac{1}{x} - \frac{a}{x(a^2+x^2)^{\frac{1}{2}}}$$

$$\left(\therefore \int_0^\infty e^{-ax}\cdot J_1(sx)dx = \frac{1}{s} - \frac{a}{s(a^2+s^2)^{\frac{1}{2}}}\right)$$

23. Find the Hankel transform of $\frac{1}{r}$ and then apply the inversion formula to get original function.

Solution: Let $f(r) = \frac{1}{r}$.

$$\therefore \quad H\{f(x)\} = \overline{\phi(s)} = \int_0^\infty \frac{1}{r}\cdot rJ_0(sr)dr \quad (\text{for } n = 0)$$

$$= \int_0^\infty e^{-0s}J_0(sr)dr = (0^2+s^2)^{-\frac{1}{2}}$$

$$= \frac{1}{s}.$$

$$\therefore \quad \overline{\phi(s)} = \frac{1}{s}$$

Now applying the inversion formula, we have

$$f(r) = H^{-1}\{\overline{\phi(s)}\}$$

$$= \int_0^\infty \overline{\phi(s)}\, sJ_0(sr)\,ds$$

$$= \int_0^\infty J_0(sr)\,dr = \frac{1}{r}$$

24. Find the potential $V(v,z)$ of a field due to a flat circular disc with its centre at the origin and radius equal to one. The axis of the disc is along z-axis and it satisfies the differential equations $\frac{\partial^2 v}{\partial^2 r^2}+\frac{1}{r}\cdot\frac{\partial v}{\partial r}+\frac{\partial^2 v}{\partial z^2}=0,\ 0\le r\le\infty, z\ge 0$ and the boundary conditions $V=V_0$ when $Z=0,\ 0\le r\le 1$ and $\frac{\partial v}{\partial z}=0$, when $Z=0,\ r>1$.

Solution: Taking the Hankel transform of the differential equation for $n = 0$, we get

$$\int_0^\infty\left(\frac{\partial^2 v}{\partial v^2}+\frac{1}{v}\cdot\frac{\partial v}{\partial v}\right)rJ_0(sr)\,dr+\int_0^\infty\frac{\partial^2 v}{\partial z^2}rJ_0(sr)\,dr=0$$

or $$-s^2\overline{v}+\frac{d^2\overline{v}}{dz^2}=0,\ \text{where } \overline{v}=\int_0^\infty vrJ_0(sr)\,dr$$

As the solution $\overline{V}=Ae^{sz}+Be^{-sz}$, V is finite and so $\overline{V}$ is finite as $z\to\infty$, consequently, A = 0, otherwise $\overline{V}$ becomes infinite as $z\to\infty$, now we have $\overline{V}=Be^{-sz}$

Applying inversion formula,

$$V(v,z)=\int_0^\infty Be^{-sz}sJ_0(sr)\,ds \qquad [1]$$

Here, $B=B(s)\quad\Rightarrow\quad \frac{\partial V}{\partial z}=\int_0^\infty Be^{-sz}(-s^2)J_0(sr)\,ds \qquad [2]$

putting z= 0 in [1] and [2], we get

$$\int_0^\infty sBJ_0(sr)\,ds=(V)_{z=0}=0\le r<1$$

$$\int_0^\infty -s^2BJ_0(sr)\,ds=\left(\frac{\partial V}{\partial z}\right)_{z=0}=0, r>1$$

or $$\left.\begin{aligned}&\int_0^\infty s^2BJ_0(sv)ds=V_0,\ 0\le r<1\\&\int_0^\infty s^2BJ_0(sv)ds=0,\ r>1\end{aligned}\right\} \qquad [3]$$

Comparing [3] with the well established integrates,

$$\int_0^\infty\frac{\sin s}{s}J_0(sr)\,ds=\frac{\pi}{2},\ 0\le r<1:$$

$$\int_0^\infty \sin s\cdot J_0(sr)\,ds=0,\ r>1$$

We obtain $B = \frac{2}{\pi} V_0 \frac{\sin s}{s^2}$, putting this in Equation [1],

$$V(r,z) = \frac{2V_0}{\pi} \int_0^\infty e^{-sz} \cdot \frac{\sin s}{s} \cdot J_0(sr)\, ds.$$

25. The vibrations of very large membrane are governed by the equation

$$\frac{\partial^2 U}{\partial r^2} + \frac{1}{r} \cdot \frac{\partial U}{\partial r} = \frac{1}{c^2} \cdot \frac{\partial^2 U}{\partial t^2},\ r \ge 0,\ t \ge 0 \text{ with } U = f(x),\ \frac{\partial U}{\partial t} = g(r), \text{ when } t = 0.$$

show that for $t > 0$,

$$U(r,t) = \int_0^\infty \overline{s\phi_1(s)} \cos(sct) J_0(sr)\, ds$$

$$+ \frac{1}{c} \int_0^\infty \overline{\phi_2 \sin(sct) J_0(sr) ds},$$

where $\phi_1(s)$ and $\phi_2(s)$ are the zero order Hankel transform of $f(r), g(r)$ respectively.

***Solution*:** Taking Hankel transform of order zero of the given equation,

$$\int_0^\infty \left(\frac{\partial^2 U}{\partial r^2} + \frac{1}{r} \cdot \frac{\partial U}{\partial r} \right) r J_0(sr)\, dr = \frac{1}{c^2} \int_0^\infty \frac{\partial^2 U}{\partial t^2} r J_0(sr)\, dr$$

or $\quad -s^2 U = \frac{1}{c^2} \cdot \frac{d^2 U}{dt^2} \quad \Rightarrow \quad [D^2 + (sc)^2] \overline{U} = 0$

Its solution is:

$$\overline{U} = A \cos sct + B \sin sct \qquad [1]$$

$$\therefore \quad \frac{d\overline{U}}{dt} = cs[-A \sin sct + B \cos sct]$$

At $\quad t = 0,\ \frac{d\overline{U}}{dt} = Bcs \qquad [2]$

$$U = f(r),\ t = 0 \quad \Rightarrow \quad \overline{U} = \int_0^\infty f(r) r J_0(sr)\, dr$$

$$= \overline{\phi(s)}$$

$$\therefore \quad \overline{U} = \overline{\phi(s)} \text{ at} = 0$$

$H\left(\frac{\partial U}{\partial t}\right) = H\{g(r)\} \quad \Rightarrow \quad \frac{d\overline{U}}{dt} = \overline{\phi_2(s)} \sin sct$, applying inversion formula:

$$U(r,t)=\int_0^{\infty}\left[\overline{\phi_1(s)\cos(sct)}\right]sJ_0(sr)ds+\int_0^{\infty}\left[\frac{1}{cs}\overline{\phi_2(s)}\sin(sct)\right]sJ_0(sr)ds$$

thus, the result follows.

26. The magnetic potential V for a circular disc of radius a and strength w, magnetized parallel to its axis, satisfies Laplace's equation is equal to $2\pi w$ on the disc itself and vanishes at exterior points in the plane of the disc. Show that at the point (v, z), $z > 0$.

$$V = 2\pi\omega a\int_0^{\infty} e^{-sz} J_0(rs)J_1(as)\,ds$$

***Solution*:**. Here, the magnetic potential V satisfies the Laplace's equation

$$\frac{\partial^2 v}{\partial r^2}+\frac{1}{r}\cdot\frac{\partial v}{\partial r}+\frac{\partial^2 V}{\partial z^2}=0 \qquad [1]$$

where $0 < r < \infty$

with the boundary conditions

$$\begin{Bmatrix}\text{(i) } V = 2\pi\omega, & 0\le r\le a, & z=0\\ \text{(ii) } V=0, & r>a, & z=0\end{Bmatrix}$$

Taking the Hankel transform for $n = 0$ of Equation [1], and the boundary conditions we have,

$$\int_0^{\infty}\left(\frac{\partial^2 V}{\partial r^2}+\frac{1}{r}\cdot\frac{\partial V}{\partial r}\right)rJ_0(sr)dr+\int_0^{\infty}\frac{\partial^2 v}{\partial z^2}vJ_0(sr)dr=0$$

$$\Rightarrow \quad -s^2\overline{V}+\frac{d^2\overline{V}}{dz^2}=0 \qquad [2]$$

(Using property P_4)

where $\overline{V} = H\{V, n=0\}$

$$=\int_0^{\infty} V.r.J_0(sr)\,dr$$

The solution of Equation [2] is $\overline{V} = Ae^{sz} + Be^{-sz}$

(but V is bounded $\Rightarrow$ $\overline{V}$ is bounded)

$\Rightarrow$ $A = 0$; otherwise $\overline{V}\to\infty$ as $z\to\infty$

$$\therefore \quad \overline{V} = Be^{-sz} \qquad [3]$$

To find B:

$$\left[\text{put } z = 0, \text{then } B = [\overline{V}]_{z=0}\right] \qquad [4]$$

and $\overline{V} = \int_0^\infty V.rJ_0(sr)dr$

or $\quad B = \left(\overline{V}\right)_{z=0} = \int_a^\infty (V)_{z=0} rJ_0(sr)dr$

$$= \int_0^a (V)_{z=0} rJ_0(sr)dv + \int_a^\infty (V)_{z=0} rJ_0(sr)dv$$

$$= \int_0^a 2\pi\omega rJ_0(sr)dr + \int_a^\infty 0rJ_0(sr)\,dr$$

$$= 2\pi\omega\int_0^a rJ_0(sr)dr + 0$$

$$= 2\pi\omega\int_0^a \frac{1}{s}\cdot\frac{d}{dr}[rJ_1(sr)]dr$$

or $\quad B = \dfrac{2\pi\omega}{s} aJ_1(sa) \qquad [5]$

Now [3] becomes

$$\overline{V} = \frac{2\pi\omega a}{s} J_1(sa)\cdot e^{-sz}$$

Using inversion formula of Hankel transform for $n = 0$.

$$V = \int_0^\infty \overline{V}\cdot SJ_0(rs)ds$$

$$= \int_0^\infty \frac{2\pi\omega a}{s} J_1(sa)\cdot Se^{-sz} J_0(sr)ds$$

or $\quad V(r,z) = 2\pi\omega a\int_0^\infty e^{-sz}\cdot J_1(sa)\cdot J_0(sr)ds$

27. Heat is supplied at a constant rate Q per unit area per unit time over a circular area of radius a in the z-plane; $z = 0$ to an infinite solid of conductivity k. Show that the steady temperature at a distance r from the axis of circular area and distance z from the plate $z = 0$ is given by

$$\frac{Qa}{2k}\int_0^\infty e^{-sz} J_0(pr)J_1(sa)\cdot s^{-1}ds$$

***Solution*:** Let $V(r, z)$ denote temperature at a point (r, z) then it is governed by the equation:

$$\frac{\partial V}{\partial t} = k\left[\frac{\partial^2 V}{\partial r^2} + \frac{1}{r}\cdot\frac{\partial V}{\partial r} + \frac{\partial^2 V}{\partial z^2}\right]$$

Since the surface is circular and therefore we have taken cylindrical coordinates (r, θ, z). Here temperature is steady and so $\frac{\partial V}{\partial t} = 0$. Hence

$$\frac{\partial^2 V}{\partial r^2} + \frac{1}{r} + \frac{\partial v}{\partial r} + \frac{\partial^2 v}{\partial z^2} = 0$$

Taking Hankel transform of order zero i.e., $n = 0$

$$\int_0^{\infty}\left(\frac{\partial^2 V}{\partial r^2} + \frac{1}{r}\cdot\frac{\partial V}{\partial r}\right) rJ_0(sr)dr + \int_0^{\infty}\frac{\partial^2 V}{\partial z^2} rJ_0(sr)dr = 0$$

writing $\quad H(V, n = 0) = \int_0^{\infty} VrJ_0(sr)dr = \overline{V},$

We get $\quad -s^2\overline{V} + \frac{d^2\overline{V}}{dz^2} = 0$

or $\quad (D^2 - s^2)\overline{V} = 0$

Its solution is $\overline{V} = Ae^{sz} + Be^{-sz}$

But $\overline{V}$ must tend to zero as $z \to \infty$ so that

$$A = 0; \quad \therefore \quad \overline{V} = Be^{-sz} \qquad [1]$$

This $\Rightarrow \frac{d\overline{V}}{dz} = -Be^{-sz} \Rightarrow \frac{d\overline{V}}{dz} = -Bs$ at $z = 0$ $\qquad$ [2]

$\therefore$ the boundary conditions are

(i) $2\left(-k\frac{\partial V}{\partial z}\right) = Q,\ 0 \le r \le a,\ z = 0$

(ii) $2\left(-k\frac{\partial V}{\partial z}\right) = 0,\ r > a,\ z = 0$

(i) $\Rightarrow \quad \frac{\partial V}{\partial z} = -\frac{Q}{2k},\ 0 \le r < a,\ z = 0$

Also $\quad \frac{\partial V}{\partial z} = 0,\ r > a$ $\qquad$ [3]

Consider the case when $z = 0$

$$\frac{d\overline{V}}{dz} = H\left\{\frac{\partial V}{\partial z}\right\} = \int_a^{\infty}\frac{\partial v}{\partial z} vJ_0(sr)dr$$

$$\int_0^a \frac{\partial V}{\partial Z} rJ_0(sv)dr + \int_a^{\infty} rJ_0(sv)dr$$

Using [3], $\dfrac{d\overline{V}}{dz} = \int_0^a \left(-\dfrac{Q}{2k}\right) vJ_0(sr) + \int_a^\infty \dfrac{\partial V}{\partial z} vJ_0(sr)dr$

or $$\frac{d\overline{V}}{dz} = -\frac{Q}{2k}\int_0^a vJ_0(sr)dr \qquad [4]$$

Since $\dfrac{d}{dx}[x^n J_n(x)] = x^n J_{n-1}(x)$

put $n = 1$ and replace x by sr

$$\frac{d}{Sdr}[(sr)J_1(sr)] = (srJ_0(sr)$$

or $\dfrac{d}{dr}[rJ_1(sr)] = srJ_0(sr)$

Using this in [4],

$$\frac{d\overline{V}}{dz} = -\frac{Q}{2k}\int_0^a \frac{1}{s}\cdot\frac{d}{dr}[rJ_0(sr)]dr$$

or $$\frac{d\overline{V}}{dz} = -\frac{Q}{2pk}\left[rJ_1(sr)\right]_{r=0}^a$$

$$= -\frac{Qa}{2sk}J_1(sa)$$

Putting this in [1],

$$\overline{V} = \frac{Qa}{2ks^2}J_1(sa)e^{-sz}$$

Applying inversion formula of Hankel transform

$$V(r,z) = \int_0^\infty \left[\frac{Qa}{2ks^2}J_1(sa)e^{-sz}\right]sJ_0(sr)ds$$

or $$V(r,z) = \frac{Qa}{2k}\int_0^\infty s^{-1}J_1(sa)J_0(sr)e^{-sz}ds.$$

28. Applying Hankel transform to solve the differential equation $\dfrac{\partial^2 V}{\partial r^2} + \dfrac{1}{r}\dfrac{\partial V}{\partial r} + \dfrac{\partial^2 V}{\partial z^2} = 0$ for the region $r \geq 0$, $z \geq 0$ satisfying the conditions:

$V \to 0$ as $z \to \infty$ and $r \to \infty$

$V = f(r)$ on $z = 0$, $r \geq 0$

Solution: We have already proved the result earlier,

$$\overline{V} = Be^{-sz} \qquad [1]$$

At $z = 0, V = f(r) \Rightarrow \overline{V} = \overline{\phi(s)} = \int_0^\infty f(r) r J_0(sr) dr$

$\therefore \quad \overline{V}(s,0) = \overline{\phi}(s).$

Using this in [1], $\overline{\phi}(s) = B.$

or $\quad \overline{V} = \overline{\phi}(s)e^{-sz}$

Using inversion formula,

$$V = \int_0^\infty \overline{V} s J_0(sr) ds = \int_0^\infty \overline{\phi}(s) e^{-sz} J_0(sr) ds.$$

EXERCISES

1. Find the Hankel transform of $f(x) = \begin{cases} 1, & 0 < x < a, \\ 0, & x > a, \end{cases}$ taking $xJ_0(sx)$ as the kernel.

$\left[Ans.\ \frac{a}{s} J_1(as) \right]$

2. Find $H\left\{ \frac{e^{-7x}}{x}, n = 0 \right\}$.

$\left[Ans.\ (49 + s^2)^{-\frac{1}{2}} \right]$

3. Find the Hankel transform of order zero of $f(x) = \begin{cases} 1 - x^2, & 0 < x < 1, \\ 0, & x > a. \end{cases}$

Hint. Refer to Example 5 by putting $a = 1$.

$\left[Ans.\ \frac{2}{s^2} J_2(s) \right]$

4. Find the Hankel transform of $\frac{\partial^2 f(x,t)}{\partial t^3}$.

$\left[Ans.\ \frac{d^3 \overline{f}(s,t)}{dt^3} \right]$

5. Find Hankel transform of $\frac{\sin ax}{x}$ taking $xJ_0(px)$ as the kernel.

$\left[Ans.\ \begin{cases} 0, & \text{if } p > a \\ -(p^2 - a^2)^{-\frac{1}{2}}, & \text{if } 0 < p < a \end{cases} \right]$

6. Find Hankel transform of order zero of function $f(x)=\begin{cases}4-x^2, & 0<x<2\\ 0, & x>2\end{cases}$

$$\left[\textit{Ans.}\ \frac{4}{s^3}J_1(s)-\frac{2}{s^2}J_0(s)\right]$$

7. Find Hankel transform of order zero of $\frac{1}{x}$ and then apply the inversion formula toget the original function.

$$\left[\textit{Ans.}\ \frac{1}{s};\ \frac{1}{x}\right]$$

8. Find Hankel transform of $x^{-2}e^{-x}$ of order one.

$$\left[\textit{Ans.}\ \frac{\left(1+s^2\right)^{-1}-1}{s}\right]$$

The finite Hankel transform

So far in this chapter, use has only been made of Hankel transform in which the range of integration has been infinite. In our solutions of boundary-value problems we have thus only been able to include a variable with range $(0,\infty)$ or $(-\infty,\infty)$.

Definition: (Finite Hankel transform) If $f(x)$ be a function defined and satisfies Dirichlet's conditions in the closed interval $[0,a]$, then its finite Hankel transform of order n is denoted by $\overline{\phi}(s_i)$ and is defined as

$$H_n\{f(x)\}=\overline{\phi}(s_i)=\int_0^a f(x)xJ_n(xs_i)dx \qquad [1]$$

where a is positive root of the transcendental equation

$$J_n(as_i)=0 \qquad [2]$$

If $f(x)$ is continuous at any point of the interval $[0,a]$, then the inversion formula

for $\overline{\phi}(s_i)$ is $f(x)=\dfrac{2}{a^2}\sum_i\overline{\phi}(s_i)\dfrac{J_n(xs_i)}{[J_n'(as_i)]^2}$ [3]

Where the sum is taken over all the positive roots of the Equation [2].

Note: (i) if we denote $f(x)$ by generalised Fourier Bessel Series

$$f(x)=\sum C_iJ_n(xs_i),\ 0\le x\le a,$$

then the coefficient C_i is given by

$$C_i = \frac{2}{a^2 J_{n+1}^2(as_i)} \int_0^a f(x).xJ_n(xs_i)dx$$

$$= \frac{2\bar{\phi}(s_i)}{a^2[J_{n+1}(as_i)]^2}$$

$$= \frac{2\bar{\phi}(s_i)}{a^2[J_n')as_i)]^2}.$$

(ii) Since $xJ_n'(x) = nJ_n(x) - xJ_{n+1}(x)$. replacing x, by as_i,

$$as_iJ_n'(as_i) = nJ_n(as_i) - as_iJ_{n+1}(as_i)$$

Using [2], we get

$$as_iJ_n'(as_i) = -as_iJ_{n+1}(as_i)$$

or $J_n'(as_i) = -J_{n+1}(as_i)$ [4]

Consequently, $f(x) = \frac{2}{a^2} \sum_i \bar{\phi}(s_i) \frac{J_n(xs_i)}{[J_n'(as_i)]^2}$.

(iii) **Another form of finite Hankel transform: Let the field of variation of the variable does not include the origin i.e., when the variable lies in the interval** $0 < b \le r \le a$.

For such cases we define the transform by

$$\bar{\phi}(s_i) = \int_a^b rf(r)[J_n(s_i r)y_n(as_i) - y_n(s_i r)J_n(as_i)dr \qquad [1]$$

Where y_n is the Bessel function of order of second kind, and s_i is a root of the equation

$$J_n(s_i a)y_n(s_i b) - J_n(s_i b)y_n(s_i a) = 0 \qquad [2]$$

Inversion formula

If $f(x)$ satisfies Dirichlet's conditions in the range $a \le r \le b$ and if its finite transform in that range is defined by [i] in which s_i is a root of the transcendental equation (ii) then at each point in the interval (*a*, *b*) at which the function $f(r)$ is continuous

$$f(r) = \sum_i \frac{2s_i^2 J_n^2(s_i a)\phi(s_i)}{J_n^2(bs_i) - J_n^2(as_i)}.$$

$$[J_n(as_i)y_n(bs_i) - J_n(bs_i)y_n(s_i a)].$$

The summation extending over all the positive roots of Equation [2].

Note: We can take the definition of finite Hankel transform as

$$\bar{\phi}(s_i) = \int_0^1 f(x).xJ_n(xs_i)dx. \qquad [1]$$

where s_i is a root of the transcandenal equation $sJ_n'(s) \pm hJ_n(s) = 0$ properties of finite Hankel transform.

P$_1$: To find the finite Hankel transform of $\frac{df}{dx}$ i.e., $H_n\left(\frac{df}{dx}\right) = \int_0^a \frac{df}{dx}.xJ_n(sx)dx$ where s is any root of $J_n(sa) = 0$.

To prove that

$$\overline{\phi_n'(s)} = \frac{s}{2n}\left[(n-1)\overline{\phi_{n+1}}(s) - (n+1)\overline{\phi_{n-1}}\right] \qquad \text{i.e.,}$$

to prove

$$H_n\left\{\frac{df}{dx}\right\} = \frac{s}{2n}\left[(n-1)H_{n+1}\{f(x) - (n+1)\}H_{n-1}\{f(x)\}\right],$$

where s is a root of the equation $J_n(sa) = 0$.

Proof: By definition of finite Hankel transform we have

$$H_n\left\{\frac{df}{dx}\right\} = \int_0^a \frac{df}{dx} \cdot xJ_n(sx)dx,$$

where s is a root of $J_n(sa) = 0$

$$= \left[f(x) \cdot xJ_n(sx)\right]_{x=0}^a - \int_0^a f(x) \cdot \frac{d}{dx}\{xJ_n(sx)\}dx \quad \text{(by parts)}$$

$$= -\int_0^a f(x)\frac{d}{dx}\{xJ_n(sx)\}dx. \qquad [1]$$

(for $J_n(sa) = 0$)

But by a recurrence formula,

$$2J_n'(x) = J_{n-1}(x) - J_{n+1}(x)$$

and $2nJ_n(x) = x[J_{n-1}(x) + J_{n+1}(sx)$

Replacing x by sx in both equations

$$2J_n'(sx) = J_{n-1}(sx) - J_{n+1}(sx) \quad [2]$$

and $2nJ_n(sx) = sx[J_{n-1}(sx) + J_{n+1}(sx)]$ [3]

Now, $\frac{d}{dx}\{xJ_n(sx)\} = J_n(sx) + sxJ_n'(sx)$

$$= \frac{sx}{2n}[J_{n-1}(sx) + J_{n+1}(sx)] + \frac{sx}{2}[J_{n-1}(sx) - J_{n+1}(sx)]$$

$$= \frac{sx}{2n}[J_{n-1}(sx)\cdot(1+n) + (1-n)J_{n+1}(sx)]$$

(follows from [2] and [3])

Now [1] becomes

$$H_n\left\{\frac{df}{dx}\right\} = -\int_0^a f(x)\cdot\frac{sx}{2n}[J_{n-1}(sx)\cdot(1+n) + (1-n)J_{n+1}(sx)]dx$$

$$= -\frac{s}{2n}\int_0^a [f(x)\cdot xJ_{n-1}(sx)\cdot(1+n) - (n-1)f(x)xJ_{n+1}(sx)]dx$$

$$= -\frac{s}{2n}[(1+n)H_{n-1}\{f\} - (n-1)H_{n+1}\{f\}]$$

$$= \frac{s}{2n}[(n-1)H_{n+1}\{f(x)\} - (1+n)H_{n-1}\{f\}] \quad [4]$$

Corollary 1: If we put $n=1$, then the Equation [4] gives

$$H_1\left\{\frac{df}{dx}\right\} = \frac{s}{2}\left[0 - 2H_0\{f(x)\}\right] = -sH_0\{f(x)\}.$$

P_2: Hankel transform of $\frac{d^2f}{dx^2} + \frac{1}{x}\frac{df}{dx}$ where s is a root of the equation $J_n(sa) = 0$.

Proof: $H_n\left\{\frac{d^2f}{dx^2} + \frac{1}{x}\frac{df}{dx}\right\} = \int_0^a x\left\{\frac{d^2f}{dx^2} + \frac{1}{x}\frac{df}{dx}\right\}J_n(sx)dx$

$$= \int_0^a x\frac{d^2f}{dx^2}J_n(sx)\,dx + \int_0^a \frac{df}{dx}J_n(sx)dx$$

$$= H_n\left\{\frac{d^2f}{dx^2}\right\} + \int_0^a \frac{df}{dx}J_n(sx)dx \quad [1]$$

replacing f by $\dfrac{df}{dx}$ in the result proved under P_1: (i.e., in the Equation [4])

$$H_n\left\{\frac{d^2 f}{dx^2}\right\} - \frac{s}{2n}\left[-(1+n)H_{n-1}\left(\frac{df}{dx}\right)+(n-1)H_{n+1}\left\{\frac{df}{dx}\right\}\right] \qquad [2]$$

Now, from recurrence formula for Bessel's function, we have

$$2nJ_n(x) = x[J_{n-1}(x)+J_{n+1}(x)].$$

replacing x by sx, we have

$$J_n(sx) = \frac{sx}{2n}[J_{n-1}(sx)+J_{n+1}(sx)]$$

$$\therefore \quad \int_0^a \frac{df}{dx} J_n(sx)\,dx = \int_0^a \frac{df}{dx}\cdot\frac{sx}{2n}[J_{n-1}(sx)+J_{n+1}(sx)]dx$$

$$= \frac{s}{2n}\int_0^a x\frac{df}{dx}J_{n-1}(sx) + \frac{s}{2n}\int_0^a x\frac{df}{dx}J_{n+1}(sx)$$

$$= \frac{s}{2n}\left[H_{n-1}\left\{\frac{df}{dx}\right\}+H_{n+1}\left\{\frac{df}{dx}\right\}\right] \qquad [3]$$

Substituting from Equation [2] and [3] into Equation [1], we have

$$H_n\left\{\frac{d^2 f}{dx^2}+\frac{1}{x}\frac{df}{dx}\right\} = \frac{s}{2n}\left[-nH_{n-1}\left\{\frac{df}{dx}\right\}+nH_{n+1}\left\{\frac{df}{dx}\right\}\right]$$

or $$H_n\left\{\frac{d^2 f}{dx^2}+\frac{1}{x}\frac{df}{dx}\right\} = \frac{s}{2}\left[-H_{n-1}\left\{\frac{df}{dx}\right\}+H_{n+1}\left\{\frac{df}{dx}\right\}\right]$$

P_3: Hankel transform of

$$\frac{d^2 f}{dx^2}+\frac{1}{x}\cdot\frac{df}{dx}-\frac{n^2}{x^2}f(x)$$

where s in a root of the equation $J_n(sa)=0$

Proof: $$H_n\left\{\frac{d^2 f}{dx^2}+\frac{1}{x}\frac{df}{dx}-\frac{n^2}{x^2}f\right\}x\cdot J_n(sx)dx$$

$$= \int_0^a\left\{\frac{d^2 f}{dx^2}+\frac{1}{x}\cdot\frac{df}{dx}-\frac{n^2}{x^2}f\right\}x\cdot J_n(sx)dx$$

$$= \int_0^a \frac{d^2 f}{dx^2} \cdot xJ_n(sx)dx + \int_0^a \frac{df}{dx} \cdot J_n(sx)dx - n^2 \int_0^a \frac{f}{x} \cdot J_n(sx)dx$$

Now, (integrating first integral by parts), we get

$$= \left[\frac{df}{dx} \cdot xJ_n(sx)\right]_0^a - \int_0^a \frac{df}{dx}\frac{d}{dx} \cdot \{xJ_n(sx)\}\,dx$$

$$+ \int_0^a \frac{df}{dx} \cdot J_n(sx)dx - n^2 \int_0^a \frac{f}{x} \cdot J_n(sx)dx$$

$$= 0 - \int_0^a \frac{df}{dx}[J_n(sx) + sxJ_n'(sx)]dx + \int_0^a \frac{df}{dx} \cdot J_n(sx)dx \; - n^2 \int_0^a \frac{f}{x} \cdot J_n(sx)dx$$

(For $J_n(sa) = 0$)

$$= -s\int_0^a \frac{df}{dx} \cdot xJ_n'(sx)dx - n^2 \int_0^a \frac{f}{x} . J_n(sx)dx \qquad \text{(by parts)}$$

$$= -s\left[f(x) \cdot xJ_n'(sx)\right]_0^a + s\int_0^a f(x) \cdot \frac{d}{dx}[xJ_n'(sx)]dx \; - n^2 \int_0^a \frac{f}{x} \cdot J_n(sx)dx$$

$$= -sf(a) \cdot aJ_n'(sa) + s\int_0^a f(x) \cdot [J_n'(sx) + sxJ''(sx)]dx \; - n^2 \int_0^a \frac{f}{x} \cdot J_n(sx)dx \quad [1]$$

$J_n(x)$ is a solution of Bessel's equation:

$$x^2 \frac{d^2 y}{dx^2} + x\frac{dy}{dx} + (x^2 - n^2)y = 0$$

$$x^2 \frac{d^2 J_n(x)}{dx^2} + x\frac{d}{dx} J_n(x) + (x^2 - n^2)J_n(x) = 0$$

or $\quad x^2 J_n''(x) + xJ_n'(x) + (x^2 - n^2)J_n(x) = 0$

replacing x by sx, we get

or $\quad s^2x^2 J_n''(sx) + sxJ_n'(sx) + (s^2x^2 - n^2)J_n(sx) = 0$

$$s^2x^2 J_n''(sx) + J_n'(sx) = \left(-sx + \frac{n^2}{sx}\right) J_n(sx)$$

Using this in Equation [1],

$$H_n\left\{\frac{d^2 f}{dx^2} + \frac{1}{x}\frac{df}{dx} - \frac{n^2}{x^2} f\right\}$$

$$= -saf(a) \cdot J_n'(sa) + s\int_0^a f(x) \cdot \left[-sx + \frac{n^2}{sx}\right] J_n(sx)dx \; - n^2 \int_0^a \frac{f}{x} \cdot J_n(sx)dx$$

$$= -saf(a)\cdot J_n'(sa) + s\int_0^a -sx\cdot f(x)J_n(sx)dx$$

$$= -saf(a)\cdot J_n'(sa) - s^2 H_n\{f\}.$$

Finally,

$$H_n\left\{\frac{d^2 f}{dx^2} + \frac{1}{x}\frac{df}{dx} - \frac{n^2}{x^2 f}\right\}$$

$$= -saf(a)J_n'(sa) - s^2 H_n\{f\}. \qquad [2]$$

Corollary 2: If $a = 1$, [2] gives

$$H_n\left\{\frac{d^2 f}{dx^2} + \frac{1}{x}\frac{df}{dx} - \frac{n^2}{x^2} f\right\}$$

$$= -sf(1)J_n'(s) - s^2 H_n\{f\} \qquad [3]$$

Again if $n = 0$, [3] gives

$$H_n\left\{\frac{d^2 f}{dx^2} + \frac{1}{x}\frac{df}{dx}\right\}$$

$$= -sf(1)\cdot J_0'(s) - s^2 H_0\{f\}$$

$$= -sf(1)J_1(s) - s^2 H_0\{f\}$$

$$(\because \ J_0' = -J_1)$$

$$\frac{d}{dx}\left\{x^{-n}J_n(x)\right\} = -x^{-n}J_{n+1}(x);$$

For $n = 0,\ J_0'(x) = -J_1(x)$

(C) EXAMPLES

1. Show that $\int_0^a rJ_0(sr)\,dr = \frac{a}{s}J_1(as).$

Solution: By *a* recurrence formula for $J_n(x)$, we have

$$\frac{d}{dr}[r^n J_n(r)] = r^n J_{n-1}(r).$$

[put $n = 1$ and replace r by sr], we get

$$\frac{1}{s}\cdot\frac{d}{dr}[srJ_1(sr)] = rJ_0(sr)$$

or $$\frac{1}{s}\cdot\frac{d}{dr}[rJ_1(sr)] = rJ_0(sr)$$

$\therefore$ $$\int_0^a rJ_0(sr)dr = \frac{1}{s}\left[rJ_1(sr)\right]_0^a$$

$$= \frac{a}{s}J_1(as)$$

or $$\int_0^a rJ_0(sr)dr = \frac{a}{s}J_1(as)$$

2. Show that $H_0(c) = \dfrac{ca}{s}J_1(as)$ c being a constant $(c \neq 0)$

Solution: If $f(r) = c$, then

$$\bar{\phi}(s_i) = H_0\{f(r)\} = H_0\{c\} = \int_0^a cxJ_0(sx)dx$$

$$= c\int_0^a xJ_0(sa)dx \qquad [1]$$

but by a recurrence formula for $J_n(x)$, $\dfrac{d}{dx}\left[x^n J_n(x)\right] = x^2 J_{n-1}(x)$

[put n = 1 and replace x by sx], we get

$$\frac{1}{s}\cdot\frac{d}{dx}[sxJ_1(sx)] = sxJ_0(sx)$$

or $$\frac{d}{dx}\{xJ_1(sx)\} = sxJ_0(sx)$$

Using this in [1], we get

$$H_0\{c\} = \frac{c}{s}\int_0^a \frac{d}{dx}\{xJ_1(sx)\}dx = \frac{c}{s}[xJ_1(sx)]_0^a$$

$$= \frac{ca}{s}J_1(sa).$$

3. Find the Hankel transform of $x^{n-1} (n > 1)$, if $xJ_{n-1}(s_i x)$ is the kernal of the transform.

Solution: If $f(x) = x^{n-1}$, then

$$\bar{\phi}(s_i) = \int_0^a x^{n-1} \cdot xJ_{n-1}(s_i x)dx$$

$$= \int_0^a x^n J_{n-1}(s_i x)\,dx.$$

Now, replacing x by $s_i x$, in recurrence formula, for

$$J_n(x); \frac{d}{dx}\left\{x^n J_n(x)\right\} = x^2 J_{n-1}(x),$$

We get, $\frac{1}{s_i} \cdot \frac{d}{dx}\left\{x^n J_n(xs_i)\right\} = x^n J_{n-1}(xs_i)$

Using this is [1],

$$\bar{\phi}(s_i) = \int_0^a x^n J_{n-1}(s_i x)dx = \frac{1}{s_i}\int_0^a \frac{d}{dx}[x^n J_n(xs_i)]dx$$

$$= \frac{1}{s_i}\left[x^n J_n(s_i x)\right]_0^a$$

$$= \frac{a^n}{s_i} J_n(s_i a).$$

4. Find the Hankel transform of $x^n, (n > -1)$, if $xJ_n(s_i x)$ is the kernal of the transform

Solution: If $f(x) = x^n$,

Then $\bar{\phi}(s_i) = \int_0^a x^n \cdot xJ_n(s_i x)dx$

$$= \int_0^a x^{n+1} J_n(s_i x)dx \qquad [1]$$

Now, replacing x by $s_i x$ and n by $n + 1$ in recurrence formula given by

$\frac{d}{dx}\left\{x^n J_n(x)\right\} = x^n J_{n-1}(sx)$, we have

$$\frac{1}{s_i}\frac{d}{dx}[x^{n+1} J_{n+1}(s_i x)] = x^{n+1} J_n(s_i x)$$

$$\therefore \quad \bar{\phi}(s_i) = \frac{1}{s_i}\int_0^a \frac{d}{dx}[x^{n+1} J_{n+1}(s_i x)]dx$$

$$= \frac{a^{n+1}}{s_i} J_{n+1}(as_i)$$

5. Find the Hankel transform of x^2 if $xJ_0(sx)$ is the kernel of the transform.

or

Show that $\int_0^a r^3 J_0(sr)dr = \frac{a^2}{s^2}[2J_0(sa)] + \left(as - \frac{4}{as}\right)J_1(sa)$

***Solution*:** We have the following recurrence formula for

$$J_n(x)\begin{cases} \frac{2n}{x}J_n(x) = J_{n-1}(x) + J_{n+1}(x) & [1] \\ \frac{d}{dx}[x^n J_n(x)] = x^n J_{n-1}(x) & [2] \end{cases}$$

[put $n = 1$ in [2] and replace x by sr] we have

$$\frac{1}{s}\frac{d}{dr}[rJ_1(sr)] = rJ_0(sr)$$

$$\therefore \quad \int_0^a r^3 J_0(sr)dr = \int_0^a r^2 \cdot \frac{1}{s}\frac{d}{dr}\{rJ_1(sr)\}dr \quad \text{(by parts)}$$

$$= \left[\frac{r^2}{s} \cdot rJ_1(sr)\right]_0^a - \int_0^a \frac{2r}{s} \cdot rJ_1(sr)dr$$

$$= \frac{a^3}{s}J_1(sa) - \frac{2}{s}\int_0^a \frac{1}{s}\frac{d}{dr}\left\{r^2 \cdot J_2(sr)\right\}dr$$

[writing $n = 2$ and $x = sr$ in [2]]

$$= \frac{a^3}{s}J_1(sa) - \frac{2}{s^2}\left[r^2 J_2(sr)\right]_0^a$$

$$= \frac{a^3}{s} \cdot J_1(sa) - \frac{2a^2}{s^2}J_2(sa) \quad [3]$$

From recurrence formula (given in Equation [1]) for Bessel function; put $n = 1$ and $x = sa$, we have

$$J_2(sa) = \frac{2n}{sa}J_1(sa) - J_0(sa)$$

$\therefore$ From [3], we get

$$\int_0^a r^3 J_0(pr)dr = \frac{a^3}{s}J_1(sa) - \frac{2a^2}{s^2}\left[\frac{2}{sa}J_1(sa) - J_0(sa)\right]$$

or

$$\int_0^a r^3 J_0(sr)dr = \frac{a^2}{s^2}\left[2J_0(sa) + \left(as - \frac{4}{as}\right)J_1(sa)\right]$$

6. Find the Hankel transform of $(a^2 - x^2)$, if $xJ_0(sx)$ is the kernel of the transform.

Solution: By definition, we have

$$H_0\left\{(a^2 - x^2)\right\} = \int_0^a (a^2 - x^2)\cdot xJ_0(sa) = 0$$

where s is a root $J_0(sa) = 0$

$$= a^2 \int_0^a xJ_0(sx)\,dx - \int_0^a x^2 \cdot xJ_0(sx)dx$$

$$= a^2 \cdot \frac{a}{s} J_1(sa) - \frac{a^2}{s^2}\left[\left(as - \frac{4}{as}\right) J_1(sa) + 2J_0(sa)\right] \quad [1]$$

$$\left(\frac{a^3}{s} - \frac{a^3}{s} + \frac{4a}{s^3}\right) J_1(sa) - \frac{2a^2}{s^3} J_0(sa),$$

$$H_0\left\{(a^2 - x^2)\right\} = \frac{4a}{s^3} J_1(sa) - \frac{2a^2}{s^2} J_0(sa)$$

Note: In Equation [1], we have used the result of solved examples (2) and (5), here you can reproduce those solutions.

7. Find the finite Hankel transform of $(1 - x^2)$, taking $xJ_0(sx)$ as the kernel

Solution: Here s is a root of $J_0(as) = 0$

$$H_0\left\{(1 - x^2)\right\} = \int_0^a (1 - x^2) xJ_0(sx)\,dx$$

$$= \int_0^a [xJ_0(sx) - x^3 J_0(sx)]dx$$

$$= I_1 - I_2 \quad [1]$$

where $\quad I_1 = \int_0^a xJ_0(sx)dx = \int_0^a \frac{1}{s}\cdot\frac{d}{dx}[xJ_1(sx)]\,dx$

$$\left[\text{For } \frac{d}{dx}[x^n J_n(x)] = x^n J_{n-1}(x)\right]$$

$$= \frac{1}{s}\left[xJ_n(sx)\right]_0^a = \frac{a}{s} J_1(as) \quad [2]$$

$$I_2 - \int_0^a x^3 J_0(sx)dx = H_0\left\{x^2\right\}$$

$$= \frac{a^2}{s^2}\left[\left(as -^4 as\right)J_1(sa) + 2 \cdot J_0(sa)\right] \quad [3]$$

Substituting from equation [2] and [3] into equation [1], we get

$$H_0\left\{(1-x^2)\right\} = \frac{a}{s}J_1(as) - \frac{a^2}{s^2}\left[\left(as - \frac{4}{as}\right)J_1(sa) + 2J_0(sa)\right]$$

Note: If the limit are from $n = 0$ to $x = 1$ in place of $x = 0$ and $x = a$, then

$$H_0\left\{(1-x^2)\right\} = \frac{4}{s^3}J_1(s) - \frac{2}{s^2}J_0(s)$$

which is obtained by putting $a = 1$

8. Show that $\int_0^1 \frac{J_n(ax)}{J_n(a)} \cdot xJ_n(sx)dx = \frac{s}{a^2 - s^2}J_n'(s)$.

Show that the Hankel transform of $\frac{J_n(ax}{J_n(a)}$ of order n is $\frac{s}{a^2 - s^2}J_n'(s)$, where $J_n(s) = 0$.

Solution: $J_n(x)$ is the solution of Bessel differential equation,

$$x^2\frac{d^2y}{dx^2} + x\frac{dy}{dx} + (x^2 - n^2)y = 0.$$

$$\therefore \quad x^2\frac{d^2}{dx^2}J_n(x) + x\frac{dJ_n(x)}{dx} + (x^2 - n^2)J_n(x) = 0$$

replacing x by sx, we have

$$x^2\frac{d^2J_n}{dx^2}(sx) + x \cdot \frac{dJ_n}{dx}(sx) + (s^2x^2 - n^2)J_n(sx)dx = 0 \quad [1]$$

Similarly $x^2\frac{d^2}{dx^2}J_n(ax) + x\frac{d}{dx}J_n(ax) + (a^2x^2 - n^2)J_n(ax)dx = 0$ [2]

Multiplying [1] by $J_n(ax)$ and [2] by $J_n(sx)$ and then subtracting, we have

$$x^2\left[J_n(ax)\cdot\frac{d^2J_n}{dx^2}(sx) - J_n(sx)\cdot\frac{d^2J_n}{dx^2}(\alpha x)\right]$$

$$+x\left[J_n(ax)\cdot\frac{dJ_n}{dx}(\alpha x) - J_n(sx)\frac{dJ_n}{dx}(\alpha x)\right]$$

$$+x^2(s^2 - a^2)J_n(sx)\cdot J_n(\alpha x) = 0$$

or $(a^2 - s^2)xJ_n(sx)J_n(ax)$

$$= \frac{d}{dx}\left[x\left\{J_n(ax)\cdot\frac{d}{dx}J_n(sx) - J_n(sx)\frac{d}{dx}J_n(\alpha x)\right\}\right]$$

or $xJ_n(sx)J_n(\alpha x)$

$$= \frac{1}{(a^2 - s^2)}\cdot\frac{d}{dx}\left[x\cdot\left\{J_n(ax)\cdot\frac{d}{dx}J_n(sx) - J_n(sx)\cdot\frac{d}{dx})ax)\right\}\right]$$

Hence $\int_0^1 x\cdot\frac{J_n(ax)}{J_n(a)}J_n(sx)\cdot dx$

$$= H_n\left\{\frac{J_n(ax)}{J_n(a)}\right\} = \frac{1}{(\alpha^2 - s^2)J_n(a)}.$$

$$= \frac{1}{(a^2 - s^2)J_n(a)}\left[x\left\{J_n(ax)\cdot\frac{d}{dx}J_n(sx) - J_n(sx)\cdot\frac{d}{dx}J_n(ax)\right\}\right]_0^1$$

$$= \frac{1}{(a^2 - s^2)J_n(\alpha)}[sJ_n(a)\cdot J_n'(s) - aJ_n(s)\cdot J_n'(a)]$$

$$= \frac{s}{(a^2 - s^2)}J_n'(s), \text{ If } J_n(s) = 0.$$

9. Find the finite Hankel transform of $\frac{1}{r}\cdot\frac{\partial}{\partial r}\left(r\frac{\partial V}{\partial r}\right) - \frac{n^2 V}{r^2}$,

Where $V = \begin{cases} 0, & \text{when } r = 0 \\ V_1 & \text{when } r = 1 \end{cases}$

Solution: Let $f(r) = \frac{1}{r}\cdot\frac{\partial}{\partial r}\left(r\frac{\partial V}{\partial r}\right) - \frac{n^2}{r^2}V.$

Then $f(r)$ is expressible as

$$f(r) = \frac{1}{r}\left(\frac{\partial V}{\partial r} + r\frac{\partial^2 V}{\partial r^2}\right) - \frac{n^2 V}{r^2}$$

$$= \frac{\partial^2 V}{\partial r^2} + \frac{1}{r}\frac{\partial V}{\partial r} - \frac{n^2}{r^2}V$$

$$H_n\{f(r)\} = H\left\{\frac{\partial^2 V}{\partial r^2} + \frac{1}{r}\frac{\partial V}{\partial r} - \frac{n^2}{r^2}V\right\}$$

$$= -sf(1)J_n'(s) - s^2 H_n\{V\}$$

$$= -sV_1 J_n'(s) - s^2 H_n\{V\}$$

[Using the result of P_3]

10. Prove that the finite Hankel transform of $f(x),\quad 0 \le x \le 1$ is $s^{n-m}J_m(s)$, where

$$f(x) = \frac{2^{1+n-m}}{\Gamma(m-n)} \cdot x^n (1-x^2)^{m-n-1}.$$

Solution: We have $H_n\{f(x)\} = \int_0^1 f(x) \cdot xJ_n(sx)\,dx$

$$= \int_0^1 \frac{2^{1+n-m}}{\Gamma(m-n)} \cdot x^n (1-x^2)^{m-n-1} \cdot x \cdot \sum_{r=0}^{\infty} \frac{(-1)^r}{r!\Gamma(n+r+1)}\left(\frac{sx}{2}\right)^{n+2r}_{dx}$$

$$= \frac{1}{\Gamma(m-n)} \cdot \sum_{r=0}^{\infty} \frac{(-1)^r \cdot s^{n+2r}}{r!\Gamma(n+r+1)\cdot 2^{-1+m+2r}} \int_0^1 x^{2(n+v)}(1-x^2)^{m-n-1}\,xdx$$

(Integrating term by term)

$$= \frac{1}{\Gamma(m-n)} \cdot \sum_{r=0}^{\infty} \frac{(-1)^r s^{n+2r}}{r!\Gamma(n+r+1)\cdot 2^{n+2r}} \int_0^1 y^{(n+r+1)} \cdot (1-y)^{(m-n)-1}\,dy$$

$$\left[\left(\text{Put } x^2 = y;\text{ so that } xdx = \frac{dy}{2}\right)\right]$$

$$= \frac{1}{\Gamma(m-n)} \sum_{r=0}^{\infty} \frac{(-1)^r \cdot s^{n+2r}}{r!\Gamma(n+r+1)\cdot 2^{m+2r}} \int_0^1 y^{(n+r+1)-1}(1-y)^{(m-n)-1}\,dy$$

$$= \frac{1}{\Gamma(m-n)} \sum_{r=0}^{\infty} \frac{(-1)^r s^{n+2r}}{r!\Gamma(n+r+1)\cdot 2^{m+2r}} \cdot \frac{\Gamma(n+r+1)\cdot\Gamma(m-n)}{\Gamma(n+r+1+m-n)}$$

$$= \sum_{r=0}^{\infty} \frac{(-1)^r s^{n+2r}}{r!2^{m+2r}\cdot\Gamma(m+r+1)}$$

$$= s^{n-m} \sum_{r=0}^{\infty} \frac{(-1)^r}{r!\Gamma^{(m+r+1)}} \cdot \left(\frac{s}{2}\right)^{m+2r}$$

$$= s^{n-m} J_m(s).$$

11. Find the Hankel transform of $\dfrac{d^2 f}{dx^2}+\dfrac{1}{x}\dfrac{df}{dx}$ where s is a root of the equation

$sJ_0'(s)+hJ_0(s)=0$

Solution: $H_0\left\{\dfrac{d^2 f}{dx^2}+\dfrac{1}{x}\dfrac{df}{dx}\right\}$

$$=\int_0^1\left(\frac{d^2 f}{dx^2}+\frac{1}{x}\frac{df}{dx}\right)\cdot xJ_0(sx)dx$$

$$=\int_0^1\frac{d^2 f}{dx^2}\cdot xJ_n(sx)dx+\int_0^1\frac{df}{dx}J_0(sx)dx$$

$$=\left[\frac{df}{dx}\cdot xJ_0(sx)\right]_0^1-\int_0^1\frac{df}{dx}\frac{d}{dx}\{xJ_0(sx)\}dx+\int_0^1\frac{df}{dx}J_0(sx)dx$$

(Integrate 1st integral by parts)

$$=\left(\frac{df}{dx}\right)_{x=1}J_0(s)-\int_0^1\frac{df}{dx}\{xsJ_0'(sx)+J_0(sx)\}dx+\int_0^1\frac{df}{dx}J_0(sx)dx$$

$$=\left(\frac{df}{dx}\right)_{x=1}J_0(s)-s\int_0^1\frac{df}{dx}\cdot xJ_0'(sx)dx$$

$$=\left(\frac{df}{dx}\right)_{x=1}J_0(s)-s\cdot\left[f(x)\cdot xJ_0'(sx)\right]_0^1+s\int_0^1 f(x)\cdot\frac{d}{dx}\{xJ_0'(sx)\}dx$$

$$=\left(\frac{df}{dx}\right)_{x=1}J_0(s)-sf(1)J_0'(s)+s\int_0^1 f(x)\{xsJ_0''(sx)+J_0'(sx)\}dx$$

Now $J_0(x)$ is the solution of Bessel's differential equation $x\dfrac{d^2y}{dx^2}+\dfrac{dy}{dx}+xy=0$

$$\therefore\quad xJ_0''(x)+J_0'(x)+xJ_0(x)=0$$

replacing x by sx, we have

$$sxJ_0''(sx)+J_0'(sx)=-sxJ_0(sx)$$

Hence

$$H_n\left\{\frac{d^2 f}{dx^2}+\frac{1}{x}\frac{df}{dx}\right\}=\left(\frac{df}{dx}\right)_{x=1}J_0(s)-sf(1)\left\{-\frac{hJ_0(s)}{s}\right\}$$

$$-s^2\int_0^1 xf(x)J_0(sx)dx$$

(Since $sJ_0'(s)+hJ_0(s)=0$)

$$=\left\{\left(\frac{df}{dx}\right)_{x=1}+hf(1)\right\}J_0(s)-s^2H_0\{f(x)\}$$

$$=\left\{\frac{df}{dx}+hf(x)\right\}_{n=1}J_0(s)-s^2H_0\{f(x)\}$$

12. Using Hankel transform technique, solve the differential equation:

$$\frac{\partial^2 V}{\partial r^2}+\frac{1}{r}\frac{\partial V}{\partial r}=\frac{1}{k}\cdot\frac{\partial V}{\partial t},$$

$(0\le r\le 1,\, t>0)$

where $\frac{\partial V}{\partial r}+hV=0$ when $r=1, t>0$

$V=1$ when $t=0,\ 0\le r<1$

Solution: Let $\overline{V}=\int_0^a V_r J_0(sr)dr$, choosing s as a positive root of the equation

$$sJ_0'(s)+hJ_0(s)=0$$

Take finite Hankel transform of

$$\frac{\partial^2 V}{\partial r^2}+\frac{1}{r}\frac{\partial V}{\partial r}=\frac{1}{k}\cdot\frac{\partial V}{\partial r},$$

We get, $\int_0^1\frac{\partial^2 V}{\partial r^2}\cdot rJ_0(sr)dr+\int_0^1\frac{1}{r}\frac{\partial V}{\partial r}rJ_0(sr)dr$

$$=\int_0^1\frac{1}{k}\frac{\partial V}{\partial t}r\cdot J_0(sr)dr \qquad [1]$$

LHS of [1]

$$=\left[\frac{\partial V}{\partial t}rJ_0(sr)\right]_{r=0}^1-\int_0^1\frac{\partial V}{\partial r}\cdot[J_0(sr)+rsJ_0'(sr)]dr+\int_0^1\frac{\partial V}{\partial r}J_0(sr)dr$$

$$=\left[r\frac{\partial V}{\partial r}J_0(s)\right]_{r=0}^1-\int_0^1\frac{\partial V}{\partial r}\cdot rsJ_0'(sr)dr$$

$$\left[\because\ \frac{\partial V}{\partial r}+hV=0 \text{ when } r=1\right]$$

$$= -hJ_0(s)(V)_{r=1} - s\left[\left\{V_r J_0'(sr)\right\}_0^1\right] - \int_0^1 V\left[J_0'(sr) + srJ_0''(sr)\right]dr$$

$$= sJ_0'(s)(V)_{r=1} - sJ_0'(s)(V)_{r=1} + s\int_0^1 V\left[J_0'(sv) + srJ_0''(sr)\right]dr$$

$$= 0 + s\int_0^1 V\left[J_0'(sr) + SrJ_0''(sr)\right]dr$$

Finally, LHS of [1]

$$= s\int_0^1 V\left[srJ_0''(sr) + J_0'(sr)\right]dr \qquad [2]$$

$J_0(x)$ is the solution of Bessel's equations

$$x^2 \cdot \frac{d^2y}{dx^2} + x\frac{dy}{dx} + (x^2 - n^2)y = 0$$

$\therefore$ $J_0(x)$ is the solution

or $x^2\dfrac{d^2y}{dx^2} + x\dfrac{dy}{dx} + x^2y = 0$

i.e., of $x\dfrac{d^2y}{dx^2} + \dfrac{dy}{dx} + xy = 0$

$$\Rightarrow \quad x\frac{d^2J_0(x)}{dx^2} + \frac{d}{dx}J_0(x) + xJ_0(x) = 0$$

$$\Rightarrow \quad xJ_0''(x) + J_0'(sx) + xJ_0(x) = 0$$

replacing x by sx

$$sxJ_0''(sx) + J_0'(sx) + sxJ_0(sx) = 0$$

or $\quad sxJ_0''(sr) + J_0'(sr) + srJ_0(sr) = 0$

or $\quad sxJ_0''(sr) + J_0'(sr) = -srJ_0'(sr)$

In this event [2] becomes

LHS of [1] $= s\int_0^1 V[-srJ_0(sr)]dr$

$$= -s^2\int_0^1 VrJ_0(sr)\,dr = -s^2\overline{v}$$

Now [1] becomes, $-s^2\overline{v} = \int_0^1 \dfrac{1}{k}\dfrac{\partial V}{\partial t} \cdot r \cdot J_0(sr)dr$

$$= \frac{1}{k}\frac{d\overline{v}}{dt}$$

or $\quad \dfrac{d\overline{v}}{dt} = -ks^2\overline{v}$

The solution of this is $\overline{V} = Ae^{-ks^2t}$

or $\quad \overline{V}(s,t) = Ae^{-ks^2t} \qquad [3]$

$\Rightarrow \quad \left(\overline{V}\right)_{t=0} = A$

$\Rightarrow \quad \int_0^1 (V)_{t=0}\, rJ_0(sr)dr = A$

$\Rightarrow \quad \int_0^1 rJ_0(sr)dr = A$

$\Rightarrow \quad A = \dfrac{1}{s}J_1(s)$

[Prove it as in example (1)]

Hue $a = 1$

Now [3] becomes

$$\overline{V} = \frac{1}{s}J_1(s)e^{-ks^2t}.$$

Applying inversion formula

$$V = 2\sum_s e^{-ks^2t}\cdot\frac{s^2}{h^2+s^2}\cdot\frac{J_1(s)}{s}\cdot\frac{J_0(sx)}{J_0^2(s)}$$

or $$V = 2\sum_s e^{-ks^2t}\cdot\frac{sJ_1(s)}{h^2+s^2}\cdot\frac{J_0(sr)}{J_0^2(s)}$$

Where the summation is taken over all the positive roots of $sJ_0'(s) + hJ_0(s) = 0$.

EXERCISES

1. Evaluate $H_n\{f(x)\}$, where $0 \le r \le a$

(i) $f(r) = r$, Kernel of transform being $rJ_1(s_i r)$

(ii) $f(r) = r^2$, kernel of transform being $rJ_0(s_i r)$

$$\left[\begin{array}{l} \text{Ans. (i) } \dfrac{ca}{s} J_1(sa) \\ \text{(ii) } \dfrac{a^2}{s^2}[2J_0(sa)] + \left(as - \dfrac{4}{as}\right) J_1(sa) \end{array}\right]$$

2. Find the Hankel transform of $1 - r^2$, taking $rJ_0(sr)$ as the kernel, where $0 \le r \le 1$.

$$\left[\text{Ans. } \left\{\frac{a}{s} J_1(as) - \frac{a^2}{s^2}\left[\left(as - \frac{4}{as}\right) J_1(sa) + 2J_0(sa)\right]\right\}\right]$$

3. Show that $\int_0^a r^2 J_0(sr)\, dr = \dfrac{a^2}{s^2}\left[2J_0(sa) + \left(as - \dfrac{4}{as}\right) J_1(sa)\right]\cdot$

4. Find the Hankel transform of $\dfrac{x^2 - b^2}{x}$, the kernel of the transform being $J_n(xs)y_n(bs) - J_n(bs)y_n(xs)\cdot$

$$\left[\text{Ans. } \frac{b^2 - a^2}{as^2} \frac{J_1(sb)}{J_1(sa)}\right]$$

5. Given: $H\{f(x)\} = \dfrac{c}{s} J_1(as)$; Prove that $f(x) = \dfrac{2}{a^3} \sum_i \dfrac{J_0(sx)}{J_1(as)}$, where the sum is taken over all the positive roots of the equation $J_0(s_i a) = 0$.

6. Find the Hankel transform of $\dfrac{d^2 f}{dx^2} + \dfrac{1}{x}\dfrac{df}{dx}$ if s is a root of the equation: $J_n(sa)y_n(sb) - J_n(sb)y_n(sa) = 0$.

$$\left[\text{Ans. } -s^2 \bar{f}_0(s)\right]$$

7. If $f(a) = 0 = f(b)$, then deduce that

$$H_0\left\{\frac{d^2 f}{dx^2} + \frac{1}{x}\frac{df}{dx}\right\} = -s^2 H_0\{f(x)\}.$$

2.3 Mellin Transform

Let $f(x)$ be a function of x defined for all positive values of x, i.e., in the interval $(0,\infty)$. Consider the definite integral: $\int_0^\infty x^{p-1} f(x)\,dx,\ p>0$. If this integral exists, evidently it must be a function of the variable 'p'. Thus, the process of multiplying $f(x)$ by x^{p-1} and integrating the product between 0 and ∞ transforms the function $f(x)$ to a new function of the new variable 'p'. This operation (or process) is called the Mellin transformation of $f(x)$. The resulting function of 'p' is called the Mellin transform of $f(x)$ and is denoted by $M\{f(x)\}$; or $\overline{\phi(p)}$. [Mellin, Robert Hjalmar (1854-1933). Finnish Analyst and mathematical physicist.]

Definition: (Mellin transform). If $f(x)$ is a function of x defined for all positive values of x in $(0,\infty)$, then the Mellin transform of $f(x)$

$$= \int_0^\infty x^{p-1} f(x)dx = M\{f(x)\}$$

$$= \overline{\phi(p)}$$

Properties of Mellin transform

P$_1$: (Linearity property). If $f(x)$ and $g(x)$ are two functions of x and a,b are two constants, then $M[af(x)+bg(x);p] = aM[f(x),p]+bM[g(x),p]$

Proof: We have

$$M[af(x)+bg(x);p] = \int_0^\infty x^{p-1}\{af(x)+bg(x)\}dx$$

$$= a\int_0^\infty x^{p-1} f(x)\,dx + b\int_0^\infty x^{p-1} g(x)dx$$

$$= aM[f(x);p]+bM[g(x);p]$$

P$_2$: $M[f(ax)] = \dfrac{1}{a^p}\overline{\phi(p)}$

Proof: By definition of Mellin transform, we have

$$M[f(ax)] = \int_0^\infty x^{p-1} f(ax)\,dx\,\cdot$$

$$= \int_0^\infty \left(\frac{t}{a}\right)^{p-1} f(t)\frac{dt}{a}; \qquad (\because\ ax = t)$$

$$= \frac{1}{a^p} \int_0^\infty t^{p-1} f(t) dt$$

$$= \frac{1}{a^p} \int_0^\infty x^{p-1} f(x) dx = \frac{1}{a^p} \overline{\phi(p)}$$

P_3: $M\left[f(x^a)\right] = \frac{1}{a} \overline{\phi\left(\frac{p}{a}\right)} \quad a > 0$

Proof: By definition of Mellin transform, we have

$$M[f(x^a)] = \int_0^\infty x^{p-1} f(x^a) dx \qquad [1]$$

$$\left(\text{put } x^a = t' \ x = t^{\frac{1}{a}}\right)$$

$$= \int_0^\infty t^{\frac{p-1}{a}} \cdot f(t) \left\{ \frac{1}{a} t^{\frac{1}{a}-1} dt \right\}$$

$$= \frac{1}{a} \int_0^\infty t^{\frac{p}{a}-1} f(t) dt$$

$$= \frac{1}{a} \int_0^\infty x^{\frac{p}{a}-1} f(x) dx$$

$$M[f(x^a)] = \frac{1}{a} \overline{\phi\left(\frac{p}{a}\right)}$$

P_4: $M[x^a f(x)] = \overline{\phi}(p+a)$

Proof: By definition,

$$M[x^a f(x)] = \int_0^\infty x^{p-1} [x^a f(x)] dx$$

$$= \int_0^\infty x^{(p+a)-1} f(x) dx = \int_0^\infty x^{s-1} f(x) dx$$

$$= \overline{\phi(s)} = \overline{\phi}(p+a); (p+a=s)$$

P₅: $M\left[\frac{1}{x}f\left(\frac{1}{x}\right)\right]=\overline{\phi(1-p)}$

Proof: By definition,

$$M\left[\frac{1}{x}f\left(\frac{1}{x}\right)\right]=\int_0^{\infty}x^{p-1}\left[\frac{1}{x}f\left(\frac{1}{x}\right)\right]dx$$

$$=\int_0^{\infty}x^{p-2}f\left(\frac{1}{x}\right)dx\cdot \qquad \left(\begin{aligned} put\frac{1}{x}&=t;\\ x&=\frac{1}{t}\end{aligned}\right)$$

$$=\int_{\infty}^{0}(t^{-1})^{p-2}f(t)\left(-\frac{dt}{t^2}\right)$$

$$=\int_0^{\infty}t^{-p}f(t)$$

$$=\int_0^{\infty}x^{-p}f(x)\,dx$$

$$=\int_0^{\infty}x^{(1-p)-1}f(x)\,dx$$

$$=\overline{\phi(1-p)}.$$

P₆: $M[(\log x)f(x)]=\frac{d}{ds}\overline{\phi(p)}=\overline{\phi'(p)}$

Proof: By definition,

$$M[(\log x)f(x)]=\int_0^{\infty}x^{p-1}[(\log x)f(x)]\,dx \qquad [1]$$

and $\quad M[f(x)]=\overline{\phi(p)}=\int_0^{\infty}x^{p-1}f(x)\,dx \qquad [2]$

Differentiating [2] w.r.t. p, we get

$$\frac{d}{dp}\overline{\phi(p)}=\overline{\phi'(p)}=\int_0^{\infty}x^{p-1}(\log x)f(x)\,dx$$

$$=M[(\log x)f(x)],\ \text{Using [1]}$$

P$_7$: (Mellin transform of derivatives)

If $M[f(x)] = \bar{\phi}(p)$, then

(a) $M[f^{(n)}(x)] = (-1)^n \dfrac{\Gamma(p)}{\Gamma(p-n)}\bar{\phi}(p-n)$

(b) $M\left[\left(x\dfrac{d}{dx}\right)^n f(x)\right] = (-p)^n \bar{\phi}(p)$

Proof: (a) We have,

$$M[f'(x):p] = \int_0^\infty x^{p-1} f'(x)\,dx$$

$$= \left[x^{p-1} f(x)\right]_0^\infty - (p-1)\int_0^\infty x^{p-2} f(x)\,dx$$

(by parts)

$$= 0 - (p-1)\int_0^\infty x^{(p-1)-1} f(x)\,dx$$

$$= -(p-1)M[f(x):p-1]$$

or, $\quad M[f'(x):p] = -(p-1)\bar{\phi}(p-1)$ [1]

Where $\quad \bar{\phi}(p) = \int_0^\infty x^{p-1} f(x)\,dx$

Applying [1], we have

$$M[f''(x):p] = -(p-1)M[f'(x);p-1]$$

$$= -(p-1)\{-(p-2)\}\cdot M[f(x):p-2]$$

or $\quad M[f''(x),p] = (-1)^2(p-1)(p-2)\bar{\phi}(p-2)$ [2]

Continuing the process again until n times, we get

$$M[f^{(n)}(x);p] = (-1)^n(p-1)(p-2)\ldots(p-n)\bar{\phi}(p-n)$$

or $\quad M[f^{(n)}(x),p] = (-1)^n \dfrac{\Gamma(p)}{\Gamma(p-n)}\bar{\phi}(p-n)$ [3]

(b) By the property (P$_4$), we have

$$M[x^a f(x):p] = \bar{\phi}(p+a) = M[f(x);p+a]$$

$\therefore \quad M[xf'(x);p = M[f'(x):p+1]$

$$= -\{(p+1)-1\}\bar{\phi}\{(p+1)-1\},$$

or $$M\left[\left(x\frac{d}{dx}\right)f(x);p\right]=-p\bar{\phi}(p) \qquad [4]$$

Now supposed

$$x\frac{d}{dx}f(x)=g(x)$$

$$\therefore \quad M[xg'(x):p]=M[g'(x);p+1]$$
$$=-\{(p+1)-1\}M[g(x);(p+1)-1]$$
$$=(-p)\cdot M[xf'(x);p]$$
$$=(-p)(-p\bar{\phi}(p) \qquad \text{Using [4]}$$

or $$M\left[x\frac{d}{dx}\left\{x\frac{d}{dx}f(x)\right\}:p\right]=(-p)^2\bar{\phi}(p)$$

or $$M\left[\left(x\frac{d}{dx}\right)^2 f(x);p\right]=(-p)^2\bar{\phi}(p) \qquad [5]$$

Proceeding in this way, we get

$$M\left[\left(x\frac{d}{dx}\right)^n f(x):p\right]=(-p)^n\bar{\phi}(p) \qquad [6]$$

P_8: (Mellin transform (integrals)

Proof: Let $\phi(x)=\int_0^x f(u)\,du$

Now $\mathrm{M}[\phi'(x);p]=-(p-1)\bar{\phi}(p-1)$

$$=-(p-1)M[\phi(x);p-1]$$

or $$M[f(x);p]=-(p-1)M\left[\int_0^x f(x)\,dx;p-1\right]$$

or, $$M\left[\int_0^x f(u)\,du;p-1\right]=-\frac{1}{p-1}M[f(x);p]$$

Replacing p by $p+1$, we get

$$M\left[\int_0^x f(u)\,du;p\right]=-\frac{1}{p}M[f(x);p+1]$$

or $$M\left[\int_0^x f(u)du;p\right]=-\frac{1}{p}\bar{\phi}(p+1).$$

P_9: (Convolution or felting theorem)

If $\overline{\phi_1}(p)$ and $\overline{\phi_2}(p)$ are the Mellin transforms of the functions $f(x)$ and $g(x)$ respectively, then

$$M[f(x)g(x);p]=\frac{1}{2\pi i}\int_{c-i\infty}^{c+i\infty}\overline{\phi_1}(u)\overline{\phi_2}(p-u)\,du$$

Proof: By definition,

$$M[f(x)g(x);p]=\int_0^\infty x^{p-1}f(x)g(x)\,dx$$

$$=\int_0^\infty x^{p-1}g(x)\left\{\frac{1}{2\pi i}\cdot\int_{c-i\infty}^{c+i\infty}\overline{\phi_1}(u)\cdot x^{-u}du\right\}dx$$

$$=\frac{1}{2\pi i}\int_{c-i\infty}^{c+i\infty}\overline{\phi_1}(u)\left\{\int_0^\infty x^{(p-u)-1}\cdot g(x)dx\right\}du$$

(Changing the order of integration)

$$\therefore\quad M[f(x)g(x);p]=\frac{1}{2\pi i}\int_{c-i\infty}^{c+i\infty}\overline{\phi_1}(u)\overline{\phi_2}(p-u)\,du$$

(D) EXAMPLES

1. Find Mellin transform of e^{-x}

***Solution*:** We have $M\left\{e^{-x}\right\}=\int_0^\infty x^{p-1}e^{-x}dx$

$$=\Gamma p,$$

$(r.p.$ of $(p)>0)$

2. Find the Mellin transform of e^{-x}.

***Solution*:** We have, $M\left\{e^{-x}\right\}=\int_0^\infty x^{p-1}e^{-x}dx$

$$=\int_0^\infty\left\{(Z)^{\frac{1}{2}}\right\}^{p-1}\cdot e^{-z}\cdot\frac{dz}{2\cdot\sqrt{z}}$$

$$=\frac{1}{2}\int_0^\infty z^{\frac{p-1}{2}-\frac{1}{2}}\cdot e^{-z}dz$$

$$=\frac{1}{2}\int_0^\infty Z^{\frac{p-2}{2}}\cdot e^{-z}dz$$

$$=\frac{1}{2}\int_0^\infty z^{\left(\frac{p}{2}\right)-1}\cdot e^{-z}dz$$

$$=\frac{1}{2}\Gamma\left(\frac{p}{2}\right)$$

3. Prove that $M[(1+x)^{-\infty}]=\dfrac{\Gamma(\alpha-p)\Gamma(p)}{\Gamma(\alpha)}$

Solution: We know that $B(m,n)=\int_0^\infty \dfrac{x^{n-1}}{(1+x)^{m+n}}dx,$

$$M[(1+x)^{-\infty}]=\int_0^\infty \frac{x^{p-1}}{(1+x)^\infty}dx$$

[put $\alpha = p+m$ and so $\alpha > p$]

$$=\int_0^\infty \frac{x^{p-1}}{(1+x)^{p+m}}dx$$

$$=B(p,m).$$

$$=\frac{\Gamma(p)\Gamma(m)}{\Gamma(p+m)}$$

$$=\frac{\Gamma(p)\Gamma(\alpha-p)}{\Gamma(\alpha)}.$$

4. Prove that $M(\sin x)=\Gamma(p)\sin\dfrac{p\pi}{2}.$

Solution: By definition, we have

$$M(\sin x)=\int_0^\infty x^{p-1}(\sin x)dx$$

$$=(\text{I.P.of})\left[-\int_0^\infty x^{p-1}\cdot e^{ix}dx\right]$$

(put $ix = u$)

$$= (\text{I.P.of})\left[-\int_0^{\infty}\left(\frac{u}{\mathrm{i}}\right)^{p-1} e^{-u}\left(\frac{du}{\mathrm{i}}\right)\right]$$

$$= (\text{I.P.of})\left[-\int_0^{\infty}\frac{u^{p-1}}{\mathrm{i}^p}e^{-u}du\right]$$

$$\left[\text{But } \mathrm{i}^p = \left(e^{\mathrm{i}\frac{\pi}{2}}\right)^p = e^{\mathrm{i}\frac{\pi p}{2}}\right]$$

$$= (\text{I.P.of})\left[-e^{\mathrm{i}\frac{\pi p}{2}}\int_0^{\infty}u^{p-1}e^{-u}du\right]$$

$$= (\text{I.P.of})\left[-e^{-\mathrm{i}\frac{\pi p}{2}}\Gamma(p)\right]$$

$$= \Gamma(p)\sin\left(\frac{\pi p}{2}\right).$$

5. Prove that $M\left\{x^n f^n(x); p\right\} = (-1)^n \dfrac{\Gamma(p+n)}{\Gamma(p)}\bar{\phi}(p)$

Solution: By property (P_4); we have

$$M\left\{x^n f(x); p\right\} = \bar{\phi}(p+n) = M[f(x) : p+n]$$

$$\therefore \quad M[x^n f^n(x); p] = M[f^n(x); p+n]$$

$$= (-1)^n \frac{\Gamma(p+n)}{\Gamma(p+n-n)}\overline{\phi(p+n-n)}$$

(replacing p by $p+n$ in the equation [3] of (p_7),

$$= (-1)^n \frac{\Gamma(p+n)}{\Gamma(p)}\bar{\phi}(p)$$

6. Prove that $M\left[x^2\dfrac{d^2f}{dx^2}+x\dfrac{df}{dx};p\right]=p^2\bar{\phi}(p)$

***Solution*:** We know that by Mellin transform of derivatives (property P_7,(b), Equation [5])

$$M\left[\left(x\frac{d}{dx}\right)^2 f(x);p\right]=(-p)^2\bar{\phi}(p)$$

or $$M\left[x\frac{d}{dx}\left(x\frac{d}{dx}f(x)\right);p\right]=p^2\bar{\phi}(p)$$

or $$M\left[x\frac{d^2f}{dx^2}+x\frac{df}{dx};p\right]=p^2\bar{\phi}(p)$$

7. Prove that $M\left\{(1+x^a)^{-\alpha}\right\}=\dfrac{\Gamma\left(\frac{p}{a}\right)\Gamma\left(\alpha-\frac{p}{a}\right)}{a\Gamma(\alpha)}$,

$[0<\text{Re}(p)<\text{Re}(a\alpha)]$

***Solution*:** We have proved that in Example 3

$$M\left\{(1+x)^{-\alpha}\right\}=\frac{\Gamma(\alpha-p)\Gamma(p)}{\Gamma(\alpha)}=\bar{\phi}(p).$$

and by (P_3);

$$M\left\{f(x^a)\right\}=a^{-1}\bar{\phi}\left(\frac{p}{a}\right),\ a>0$$

Where $\bar{\phi}(p)=M[f(x)]$

$$\therefore\quad M\left\{(1+x^a)^{-\alpha}\right\}=a^{-1}\bar{\phi}\left(\frac{p}{a}\right)$$

$$=\frac{1}{a}\cdot\frac{\Gamma\left(\alpha-\frac{p}{a}\right)\Gamma\left(\frac{p}{a}\right)}{\Gamma(\alpha)}$$

Where $0<\text{Re}(p)<\text{Re}(a\alpha)$

EXERCISES

1. Find the Mellin transforms of

(a) $(1+n)-1$ $\left[Ans.\ \dfrac{\pi}{\sin(\pi s)}\right]$

(b) $\cos x$ $\left[Ans.\ \Gamma(s)\cos\left(\dfrac{s\pi}{2}\right)\right]$

(c) $(1+x^a)^{-1}$ $\left[Ans.\ \left(\dfrac{\pi}{a}\right)\cos\left(\dfrac{\pi s}{a}\right)\quad 0<\text{Re}(s)<\text{Re}(a)\right]$

2. Prove that $M[f(x)g(x):1]=\dfrac{1}{2\pi i}\int_{c-i\infty}^{c+i\infty}\phi_1(u)\phi_2(1-u)\,du$

[Hint: put $s=1$]

CHAPTER 3

Partial Differential Equations

Introduction

A partial differential equation (PDE) occurs in any mathematical or physical problem with a function of two or more independent variables involving its partial derivatives. This partial differential equation arises in many diverse subjects, e.g., meteorology, elasticity, heat transfer, electromagnetic theory etc. The order of a partial differential equation is the order of the highest differential coefficient occurring in it.

For example,

$$x^2 \frac{\partial z}{\partial x} + y^2 \frac{\partial z}{\partial y} = z^2, \qquad [1]$$

$$\frac{\partial^2 z}{\partial t^2} = a^2 \frac{\partial^2 z}{\partial x^2} \qquad [2]$$

$$\frac{\partial^2 V}{\partial x^2} + \frac{\partial^2 V}{\partial y^2} + \frac{\partial^2 V}{\partial z^2} = 0 \qquad [3]$$

are all partial differential equations, [1] is a PDE of the first order in which Z is the dependent variable considered as a function of two independent variables x and y. [2] is a PDE of second order involving two independent variables x and t. [3] is a PDE of the second order involving three independent variables.

3.1 Linear partial differential equations with constant coefficients of second order and their classification

The linear PDE with constant coefficients of second order is given by

$$a_0 \frac{\partial^2 z}{\partial x^2} + a_1 \frac{\partial^2 z}{\partial x \partial y} + a_2 \frac{\partial^2 z}{\partial y^2} = 0 \qquad [1]$$

where the coefficient (a_0, a_1, a_2) are constants.

Denoting $\dfrac{\partial}{\partial x}$ by D, $\dfrac{\partial}{\partial y}$ by D' and $\dfrac{\partial^2}{\partial x \partial y}$ by DD'.

Equation [1] can also be written as

$$F(x,y) = (a_0 D^2 + a_1 DD' + a_2 D'^2)$$

i.e., $$F(x,y) = f(D,D')z \qquad [2]$$

the complete solution of [2] consists of two parts, the complementary function and the particular integral where the complementary function is the solution of

$$F(D,D') = 0 \qquad [3]$$

Linear PDE with constant coefficient are of two types viz. homogenous and nonhomogenous.

When all the derivatives appearing in [1] or [2] are of the same order, then the equation is known as linear homogenous PDE with constant coefficients otherwise it is called nonhomogenous.

Working rule for finding C.F.

Step 1. Write the given equation in the symbolic form $F(D,D')z \equiv (a_0 D^2 + a_1 DD' + a_2 D'^2)$ [1]
Step 2. Replace D by m and D' by 1 in $F(D,D')$ and obtain the auxiliary equation. $a_0 m^2 + a_1 m + a_2 = 0$ [2]
Step 3. Find the roots of the auxiliary equation [2] the cases which arise: Case (i) If the roots of the auxiliary equation are real and distinct; say m_1, m_2 then complementary function $= F_1(y + m_1 x) + F_2(y + m_2 x)$ Case (ii) If the auxiliary equation has repeated roots (or equal roots), say (m_1, m_1) then the corresponding part of complementary function is given by complementary function $= F_1(y + m_1 x) + xF_2(y + m_1 x)$ where F_1 and F_2 are arbitrary functions.
Suppose the equation: $(a_0 D^n + a_1 D^{n-1} D' + a_2 D^{n-2} D'^2 + \cdots + a_n D'^n)z$ has $m_1, m_2, \cdots, m_n$ distinct roots of auxiliary equation, then the complementary function is $= F_1(y + m_1 x) + F_2(y + m_2 x) + \cdots + F_n(y + m_n x)$ [A] and if $m_1 = m_2 = \cdots = m_r$. Then, complementary function $= F_1(y + m_1 x) + xF_2(y + m_2 x) + \cdots + z^{r-1} F_r(y + m_r)$

Particular integral: To find particular integral in various cases, we consider the following standard formulae:

1. $\dfrac{1}{f(D,D')}e^{ax+by} = \dfrac{1}{f(a,b)}e^{ax+by}$ if $f(a,b) \neq 0$

2. $\dfrac{1}{f(D,D')}x^r y^s = [f(D,D')]^{-1} x^r y^s$ which is found by the expansion of $[f(D,D')]^{-1}$ in powers of D and D'.

3. $\dfrac{1}{f(D^2,DD',D'^2)}\sin(ax+by) = \dfrac{1}{f(-a^2,-ab,-b^2)}\sin(ax+by)$ if $f(-a^2,-ab,-b^2) \neq 0$. Similar formulae hold for $\cos(ax+by)$.

4. $\dfrac{1}{f(D,D')}e^{ax+by}F(x,y) = e^{ax+by}\dfrac{1}{f(D+a,D'+b)}F(x,y)$.

5. $\dfrac{1}{f(D^2,D'^2)}\sin ax \sin by = \dfrac{1}{f(-a^2,-b^2)}\sin ax \sin by$ if $f(-a^2,-b^2) \neq 0$. Similar formulae hold for $\cos ax \cos by$.

6. To find $\dfrac{1}{D-mD'}f(x,y)$. Change y to $y - mx$ in $f(x,y)$ and integrate w.r.t. x, treating y as a constant. In the resulting integral change y to $y + mx$

(A) EXAMPLES

$$\left(p = \frac{\partial z}{\partial x};\ q = \frac{\partial z}{\partial y};\ r = \frac{\partial^2 z}{\partial x^2};\ s = \frac{\partial^2 z}{\partial x \partial y} \text{ and } t = \frac{\partial^2 z}{\partial y^2}\right)$$

Solve the following partial differential equations.

1. $\dfrac{\partial^2 z}{\partial x^2} - \dfrac{\partial^2 z}{\partial y^2} = 0$

***Solution*:** The complementary function of z is found such that $(D^2 - D'^2)z = 0$

The auxiliary equation is: $(D = m;\ D' = 1)$

$$\left\{\begin{matrix}(m^2-1)=0 \\ \Rightarrow \quad m=\pm 1\end{matrix}\right\} \quad \text{different roots} \quad \left(\begin{matrix}\text{since RHS is 0,} \\ \text{particular integral} = 0\end{matrix}\right)$$

Hence the complementary function is given by

$$z = F_1[y+(1)x] + F_2[y+(-1)x]$$

or $\quad z = F_1(y+x) + F_2(y-x)$

where F_1 and F_2 are arbitrary functions.

2. $\dfrac{\partial^2 z}{\partial x^2} - 3\dfrac{\partial^2 z}{\partial x \partial y} + 2\dfrac{\partial^2 z}{\partial y^2} = 0$

Solution: The complete function of z is found such that

$$(D^2 - 3DD' + 2D')z = 0 \qquad \left(\begin{matrix}\text{since RHS is 0,} \\ \text{particular integral} = 0\end{matrix}\right)$$

The auxiliary equation is $(D = m;\ D' = 1)$

$$\left\{\begin{matrix} & (m^2 - 3m + 2) = 0 \\ \Rightarrow & (m-2)(m-1) = 0 \\ \Rightarrow & m = 1, 2\end{matrix}\right\}$$

Hence the complete function is given by

$$z = F_1(y+x) + F_2(y+2x)$$

3. $r = a^2 t$

Solution: Given equation is $\dfrac{\partial^2 z}{\partial x^2} = a^2 \dfrac{\partial^2 z}{\partial y^2}$ or $(D^2 - a^2 D'^2)z = 0$

The auxiliary equation is

$$\left\{\begin{matrix} & (m^2 - a^2) = 0 \\ \Rightarrow & m = \pm a; \text{(roots are different)} \\ \text{Hence the complementary function is given by} & \\ z = F_1(y+ax) + F_2(y-ax) & \end{matrix}\right\}$$

$(D = m;\ D' = 1)$

4. $(D^2 + DD' - 2D'^2)z = 0.$

Solution: The auxiliary equation is $(D = m;\ D' = 1)$

$$\left.\begin{array}{ll} & (m^2 + m - 2) = 0 \\ \Rightarrow & (m+2)(m-1) = 0 \\ \Rightarrow & m = 1, -2 \end{array}\right\}$$

real and distinct

$\therefore$ Complementary function $z = F_1(y - 2x) + F_2(y + x)$.

5. $(D^2 - 2DD' + D'^2)z = 0.$

Solution: The auxiliary equation is

$$\left.\begin{array}{ll} & (m^2 - 2m + 1) = 0 \qquad (D = m;\ D' = 1) \\ \Rightarrow & (m-1)^2 = 0 \\ \Rightarrow & m = 1, 1. \text{ Hence the complementary function is given by} \\ & z = F_1(y + x) + xF_2(y + x) \end{array}\right\}$$

6. $(D^4 - D'^4)z = 0.$

Solution: The auxiliary equation is

$$\left.\begin{array}{ll} & (m^4 - 1) = 0; \qquad (D = m;\ D' = 1) \\ \Rightarrow & (m^2 + 1)(m^2 - 1) = 0 \\ \Rightarrow & m = 1, -1, \mathrm{i}, -\mathrm{i} \end{array}\right\}$$

Hence the complementary function is given by

$$z = F_1(y + x) + F_2(y - x) + F_3(y + \mathrm{i}x) + F_4(y - \mathrm{i}x)$$

7. $r + t + 2s = 0$

Solution: The given equation is $\dfrac{\partial^2 z}{\partial x^2} + \dfrac{\partial^2 z}{\partial y^2} + 2\dfrac{\partial^2 z}{\partial x \partial y} = 0.$

or $\quad (D^2 + D'^2 + 2DD')z = 0$

or $\quad (D+D')^2 z = 0$

The auxiliary equation is

$$\left\{\begin{aligned} & (m+1)^2 = 0; \quad (D=m;\ D'=1) \\ \Rightarrow \quad & m = -1, -1 \text{(equal)} \end{aligned}\right\}$$

Hence the complementary function is given by $z = F_1(y-x) + xF_2(y-x)$.

8. $(D^3 - 6D^3D' + 11DD'^2 - 6D'^3)z = 0$

Solution: The auxiliary equation is

$$\left\{\begin{aligned} & (m^3 - 6m^2 + 11m - 6) = 0; \quad (D=m;\ D'=1) \\ \Rightarrow \quad & (m-1)(m-2)(m-3) = 0; \\ \Rightarrow \quad & m = 1,\ 2,\ 3 \qquad \text{(real and distinct roots)} \end{aligned}\right\}$$

Hence the complementary function is given by

$$z = F_1(y+x) + F_2(y+2x) + F_3(y+3x).$$

9. $25r - 40s + 16t = 0$

Solution: The given equation is $25\dfrac{\partial^2 z}{\partial x^2} - 40\dfrac{\partial^2 z}{\partial x \partial y} + 16\dfrac{\partial^2 z}{\partial y^2} = 0.$

or $\quad (25D^2 - 40DD' + 16D'^2)z = 0$

The auxiliary equation is

$$\left\{\begin{aligned} & (25m^2 - 40m + 16) = 0; \quad (D=m;\ D'=1) \\ \Rightarrow \quad & (5m-4)^2 = 0; \\ \Rightarrow \quad & m = \left(\frac{4}{5}, \frac{4}{5}\right) \qquad \text{(real and distinct roots)} \end{aligned}\right\}$$

Hence the complementary function is given by $z = F_1\left(y + \dfrac{4x}{5}\right) + xF_2\left(y + \dfrac{4x}{5}\right)$

10. $2r + 5s + 2t = 0$

Solution: The given equation is $(2D^2 + 5DD' + 2D'^2)z = 0$

the auxiliary equation is

$$\left\{\begin{array}{lll} & (2m^2 + 5m + 2) = 0; & (D = m;\ D' = 1) \\ \Rightarrow & (2m+1)(m+1) = 0 & \\ \Rightarrow & m = \left(\dfrac{-1}{2}, -2\right) & \text{(real and distinct roots)} \end{array}\right\}$$

Hence the complementary function is given by $z = F_1\left(y - \dfrac{1}{2}x\right) + F_2(y - 2x)$.

11. $(D^4 + D'^4 - 2D^2D'^2)z = 0$

Solution: The auxiliary equation is

$$\left\{\begin{array}{lll} & (m^4 + 1 - 2m^2) = 0; & (D = m;\ D' = 1) \\ \Rightarrow & (m^2 - 1)^2 = 0 & \\ \Rightarrow & [(m-1)(m+1)]^2 = 0 & \end{array}\right\}$$

Hence the complementary function is given by

$$z = [F_1(y + x) + xF_2(y + x)] + [F_3(y - x) + xF_4(y - x)]$$

where F_i's are arbitrary.

12. $\dfrac{\partial^4 z}{\partial x^4} - 2\dfrac{\partial^4 z}{\partial x^3 \partial y} + 2\dfrac{\partial^4 z}{\partial x \partial y^2} - \dfrac{\partial^4 z}{\partial y^4} = 0.$

Solution: The given equation is $(D^4 - 2D^3D' + 2DD'^3 - D'^4)z = 0$

The auxiliary equation is

$$\left\{\begin{array}{lll} & (m^4 - 2m^3 + 2m - 1) = 0; & (D = m;\ D' = 1) \\ \text{or} & (m^2 + 1)(m^2 - 1) - 2m(m^2 - 1) = 0 & \\ \Rightarrow & (m-1)(m+1)(m^2 + 1 - 2m) = 0 & \\ \Rightarrow & (m-1)^3(m+1) = 0 & \\ \Rightarrow & m = 1, 1, 1, -1 & \end{array}\right\}$$

$\therefore$ the complementary function is given by

$$z = F_1(y+x) + xF_2(y+x) + x^2F_3(y+x) + F_4(y-x)$$

13. Write the particular integral of $(D^2 + 3DD' + 2D'^2)z = x + y$

Solution: The auxiliary equation is

$$\left\{\begin{array}{lll} & (m^2+3m+2)=0; & (D=m;\ D'=1) \\ \Rightarrow & (m+1)(m+2)=0 & \\ \Rightarrow & m=-1,\ -2; & \text{(real and distinct)} \end{array}\right\}$$

the complementary function is given by $z = F_1(y-x) + F_2(y-2x)$.

To find particular integral:

$$\text{P.I} = \frac{1}{(D^2+3DD'+2D'^2)}(x+y)$$

$$= \frac{1}{D^2}\left(1 + \frac{3D'}{D} + \frac{2D'^2}{D^2}\right)^{-1}(x+y)$$

$$= \frac{1}{D^2}\left[1 - \left(\frac{3D'}{D} + \frac{2D'^2}{D^2}\right) + \left(\frac{3D'}{D} + \frac{2D'^2}{D^2}\right)^2 - \cdots\right](x+y)$$

(expanding by Binomial theorem)

$$= \frac{1}{D^2}\left[1 - \frac{3D'}{D} + \cdots\right](x+y)$$

$$= \frac{1}{D^2}\left[(x+y) - \frac{3}{D}(1)\right] = \frac{1}{D^2}[(x+y) - 3x]$$

$$= \frac{1}{D^2}(y-2x) = \frac{1}{D}\left[yx - x^2\right] = \frac{1}{2}yx^2 - \frac{1}{3}x^3$$

$\therefore$ required solution is z = complementary function + particular integral

i.e., $z = F_1(y-x) + F_2(y-2x) + \frac{1}{2}yx^2 - \frac{1}{3}x^3.$

14. Find the complementary function of $(D^2 - 4DD' + 4D'^2)z = x + y.$

Solution: The auxiliary equation is given by

$$\left\{\begin{aligned} & (m^2 - 4m + 4) = 0; \qquad (D = m;\ D' = 1) \\ \Rightarrow\quad & (m-2)^2 \\ \Rightarrow\quad & m = 2, 2; \qquad \text{(repeated roots)} \end{aligned}\right\}$$

Hence, the complementary function is given by $z = F_1(y+2x) + xF_2(y+2x).$

15. $(D^2 - 2DD' + D'^2)z = e^{x+2y} + x^3$

Solution: The auxiliary equation is

$$\left\{\begin{aligned} & (m^2 - 2m + 1) = 0; \\ \Rightarrow\quad & (m-1)^2 = 0; \\ \Rightarrow\quad & m = 1, 1 \end{aligned}\right\}$$

Hence, complementary function: $z = F_1(y+x) + xF_2(x+y).$

Particular integral (P.I) $= \dfrac{1}{D^2 + 2DD' + D'^2}\left(e^{x+2y} + x^3\right)$

$$= \frac{1}{D^2 - 2DD' + D'^2}\cdot e^{x+2y} + \frac{1}{D^2 - 2DD' + D'^2}x^3$$

$$= \frac{1}{1^2 - 2(2) + 2^2}\cdot e^{x+2y} + \frac{1}{D^2}\left(1 - \frac{D'}{D}\right)^{-1} x^3$$

$$= \frac{1}{5-4}\cdot e^{x+2y} + \frac{1}{D^2}\left(1 + \frac{2D'}{D} + \cdots\right)x^3$$

$$= e^{x+2y} + \frac{1}{20}x^5$$

Hence the complete solution is $z = F_1(y+x) + xF_2(y+x) + e^{x+2y} + \dfrac{1}{20}x^5.$

16. $(D+D')^2 z = e^{x-y}$

Solution: The complementary function of z is found such that

$$(D+D')^2 z = 0$$

The auxiliary equation is

$$\left\{\begin{array}{ll} & (m+1)^2 = 0; \\ \Rightarrow & (m+1)(m+1) = 0 \\ \Rightarrow & m=-1,\ m=-1 \end{array}\right\} \qquad (D=m;\ D'=1)$$

Hence the complementary function is $z = F_1(y-x) + xF_2(y-x)$

$$\text{P.I} = \frac{1}{(D+D')^2} e^{x-y} \times 1$$

$$= e^{x-y} \times \frac{1}{(D+1+D'\cdot 1)^2}(1)$$

$$= e^{x-y} \times \frac{1}{(D^2+2DD'+D'^2}(1)$$

$$= e^{x-y} \times \frac{1}{D^2\left(1+\frac{2D'}{D}+\frac{D'^2}{D^2}\right)}(1)$$

$$= e^{x-y} \cdot \frac{1}{D^2}\left[1+\left(\frac{2D'}{D}+\frac{D'^2}{D^2}\right)\right]^{-1}(1)$$

$$= e^{x-y} \cdot \frac{1}{D^2}\left[1+\left(\frac{2D'}{D}+\frac{D'^2}{D^2}\right)+\left(\frac{2D'}{D}+\frac{D'^2}{D^2}\right)^2 - \cdots\right](1)$$

$$= e^{x-y} \cdot \frac{1}{D^2}(1)$$

Since $D'(1) = 0 = D'^2(1)$ [1]

Hence the complete solution is $z = F_1(y-x) + xF_2(y-x) + \frac{x^2 e^{x-y}}{2}$.

17. $(D^2 - D'^2)z = e^{x+y}$

Solution: The auxiliary equation is

$$\left.\begin{aligned} & m^2 - 1 = 0 \\ \Rightarrow \quad & m = 1, -1 \end{aligned}\right\} \quad (D = m;\ D' = 1)$$

$\therefore$ Complementary function $= F_1(y+x) + F_2(y-x)$

$$\text{P.I} = \left(\frac{1}{(D^2 - D'^2)}\right) e^{x+y}$$

Here, $f(D, D') = D^2 - D'^2$ and $f(1, -1) = 0$

$$\text{Hence P.I} = \frac{1}{(D+D')(D-D')} e^{x+y}$$

$$= \left(\frac{1}{(1+1)(D-D')}\right) e^{x+y} = \frac{1}{2}\left(\frac{1}{D-D'}\right) e^{x+y}.$$

Changing y to $y–x$ in e^{x+y} to get e^y. Integrate w.r.t y, we get xe^y. Again changing y to $y + x$, we get

$$\text{P.I} = \frac{1}{2} x e^{x+y}$$

$\therefore$ the complete solution is $z = F_1(y+x) + F_2(y-x) + \dfrac{1}{2} x e^{x+y}$

18. $(D^2 + DD')z = \sin(x+y)$

Solution: The auxiliary equation is

$$\left.\begin{aligned} & (m^2 + m) = 0 \\ \Rightarrow \quad & m(m+1) = 0 \\ \Rightarrow \quad & m = 0, -1 \end{aligned}\right\} \quad (D = m;\ D' = 1)$$

$\therefore$ the complementary function $= F_1(y - 0x) + F_2(y-x)$

$$= F_1(y) + F_2(y-x)$$

$$\text{P.I} = \left(\frac{1}{D^2 + DD'}\right) \sin(x+y)$$

$$= \left(\frac{\sin(x+y)}{-1-1} \right) \sin(x+y)$$

$$= -\frac{1}{2}\sin(x+y)$$

$\therefore$ the complete solution is

z = complementary function + particular integral

$$= F_1(y) + F_2(y-x) - \frac{1}{2}\sin(x+y).$$

19. $(D^2 - 2DD' + 2D'^2)z = \sin(x-y)$

***Solution*:** The auxiliary equation is $m^2 - 2m + 2 = 0$

$$\therefore \quad m = \frac{2 \pm \sqrt{4-8}}{2} = 1 \pm \mathrm{i}$$

Complementary function $= F_1(y + (1+\mathrm{i})x) + F_2(y + (1-\mathrm{i})x)$

$$\text{P.I} = \left(\frac{1}{D^2 - 3DD' + 2D'^2)} \right) \sin(x-y)$$

$$= \frac{\sin(x-y)}{-1-2-2}$$

$$= -\frac{1}{5}\sin(x-y)$$

$\therefore$ the complete integral is z = complementary function + particular integral

$$\therefore \quad z = F_1(y + (1+\mathrm{i})x) + F_2(y + (1-\mathrm{i})x) - \frac{1}{5}\sin(x-y)$$

20. $(2D^2 - 5DD' + 2D'^2)z = 5\sin(2x-y)$

or $(2D - D')(D - 2D')z = 5\sin(2x-y)$

***Solution*:** The auxiliary equation is $(2m+1)(m-2) = 0$

$$\Rightarrow \quad m = -\frac{1}{2};\ m = 2$$

$\Rightarrow$ complementary function $z = F_1\left(y - \frac{1}{2}x \right) + F_2(y + 2x)$

To find $\text{P.I} = \dfrac{1}{2D^2 - 5DD' + 2D'^2} 5\sin(2x-y)$

$$= \frac{1}{2.4 - 5(-2) + 2(-1)^2}\{-5\sin(2x-y)\}$$

$$\left(\text{since} \int \sin t\, dt\, dt = -\sin t\right)$$

$$= -\frac{5}{20}\sin(2x-y)$$

Hence $z = F_1(2y+x) + F_2(y+2x) = \frac{1}{4}\sin(2x-y)$.

21. Solve $(D^2 - 4D'^2)z = \cos 4x \cos 3y$.

***Solution*:** The auxiliary equation is $m^2 - 4 = 0$

$\Rightarrow \quad m - 2$ or -2

$\therefore$ complementary function $= F_1(y+2x) + F_2(y-2x)$

$$\text{P.I} = \left(\frac{1}{D^2 - 4D'^2}\right)\cos 4x \cos 3y$$

$$= \frac{1}{-4^2 - 4(-3)^2}\cos 4x \cos 3y$$

$$= \frac{1}{20}\cos 4x \cos 3y$$

$\therefore$ the complete solution is $z = F_1(y+2x) + F_2(y-2x) + \frac{1}{20}\cos 4x \cos 3y$.

22. $(D^2 - DD')z = \sin x \sin 2y$

***Solution*:** The auxiliary equation is

$$\left\{\begin{aligned} & m^2 - m = 0 \\ \Rightarrow\quad & m(m-1) = 0 \\ \Rightarrow\quad & m = 0, 1 \end{aligned}\right\} \qquad (D = m;\ D' = 1)$$

Complementary function $= F_1(y) + F_2(y+x)$

$$\text{P.I} = \left(\frac{1}{(D^2 - DD')}\right)\sin x \sin 2y$$

$$= \left(\frac{1}{(D^2 - DD')}\right)\left[\frac{\cos(x-2y) - \cos(x+2y)}{2}\right]$$

$$= \frac{1}{2}\left(\frac{1}{D^2 - DD'}\right)\cos(x-2y) - \frac{1}{2}\left(\frac{1}{(D^2 - DD')}\right)\cos(x+2y)$$

$$= \frac{1}{2}\left[\frac{\cos(x-2y)}{-1+1(-2)}\right] - \frac{1}{2}\left[\frac{\cos(x+2y)}{-1^2+1\times 2}\right]$$

$$= \frac{\cos(x-2y)}{-6} - \frac{\cos(x+2y)}{2}$$

$\therefore$ the complete solution is z = complementary function + particular integral

$$\therefore \qquad z = F_1(y) + F_2(y+x) - \frac{\cos(x-2y)}{-6} - \frac{\cos(x+2y)}{2}.$$

23. $(D^2 - DD' - 2D'^2)z = (y-1)e^x$.

Solution: The auxiliary equation is

$$\left\{\begin{array}{lll} & (m^2 - m - 2) = 0; & (D = m;\ D' = 1) \\ \Rightarrow & (m-2)(m+1) = 0 & \\ \Rightarrow & m = -1, 2 & \text{(distinct roots)} \end{array}\right\}$$

Complementary function $= F_1(y-x) + F_2(y+2x)$

$$\text{P.I} = \left(\frac{1}{D^2 - DD' - 2D'^2}\right)(y-1)e^x$$

$$= \left(\frac{1}{(D-2D')(D+D')}\right)(y-1)e^x$$

$$= \left(\frac{e^x}{(D-2D')(D+D')}\right)(y-1)e^x$$

$$= \left(\frac{e^x}{[(D+1)-2D'][(D+1)+D']}\right)(y-1)$$

$$= \left(\frac{e^x}{[1-(2D'-D)][1+(D+D')]} \right)(y-1)$$

$$= e^x [1-(2D'-D)]^{-1} [1+(D+D')]^{-1} (y-1)$$

$$= e^x [1+(2D'-D)+\cdots][1-(D+D')+\cdots](y-1)$$

$$= e^x [1+(2D'-D)][1-(D+D')](y-1)$$

$$= e^x [1-(D+D')+(2D'-D)](y-1)$$

$$= e^x (1+D'-2D)(y-1)$$

$$= e^x (y-1+1)$$

$$= ye^x$$

Hence the complete solution is $z =$ complementary function + particular integral

i.e., $z = F_1(y-x) + F_2(y+2x) + ye^x$.

24. $(D^3 - 7DD'^2 - 6D'^3)z = \sin(x+2y) + e^{2x+y}$.

Solution: The auxiliary equation is $(m^3 - 7m - 6) = 0$

$$\left\{ \begin{array}{ll} \therefore & (m+1)(m+2)(m-3) = 0 \\ \Rightarrow & m = -1, -2, 3 \end{array} \right\}$$

$\therefore$ the complementary function $= F_1(y-x) + F_2(y-2x) + F_3(y+3x)$

$$\text{P.I}_1 = \left(\frac{1}{(D^3 - 7DD'^2 - 6D'^3)} \right) \sin(x+2y)$$

$$= \left(\frac{1}{-D + 7D \times 2^2 + 6 \times 2^2 D')} \right) \sin(x+2y)$$

$$= \left(\frac{1}{27D + 24D')} \right) \sin(x+2y)$$

$$= \left(\frac{27D - 24D'}{27^2 D^2 - 24D'^2)} \right) \sin(x+2y)$$

$$= \left(\frac{27D - 24D'}{-729 + 576 \times 2^2)} \right) \sin(x + 2y)$$

$$= \frac{27\cos(x+2y) - 48\cos(x+2y)}{1575}$$

$$= \frac{-21\cos(x+2y)}{1575}$$

$$= -\frac{1}{75}\cos(x+2y)$$

$$\text{P.I}_2 = \left(\frac{1}{(D^3 - 7DD'^2 - 6D'^3)} \right) e^{2x+y}$$

$$= \frac{e^{2x+y}}{8 - 14 - 6} = \frac{e^{2x+y}}{-12}$$

Hence, the complete solution is z = complementary function + P.I_1+ P.I_2

25. $\dfrac{\partial^2 y}{\partial t^2} - a^2 \dfrac{\partial^2 y}{\partial x^2} = E \sin pt$

Solution: Denoting $\dfrac{\partial y}{\partial t}$ by Dy and $\dfrac{\partial y}{\partial x}$ by $D'y$

we get $\dfrac{\partial^2 y}{\partial t^2} = D^2 y$ and $\dfrac{\partial^2 y}{\partial x^2} = D'^2 y$

The given equation reduces to $(D^2 - a^2 D'^2)y = E \sin pt$

$\therefore$ the auxiliary equation is $(m^2 - a^2) = 0$

$\Rightarrow \quad (m + a)(m - a) = 0$

$\Rightarrow \quad m = a, \ -a$

$\therefore$ the complementary function $= F_1(x + at) + F_2(x - at)$.

$$\text{P.I} = \left(\frac{1}{D^2 - a^2 D'^2} \right) (E \sin pt)$$

$$= \frac{E \sin pt}{-p^2 - a^2 \times 0}$$

$$= -\frac{E\sin pt}{p^2}$$

$\therefore$ the complete integral is z = complementary function + particular integral

or $\quad z = F_1(x+at) + F_2(x-at) - \dfrac{E\sin pt}{p^2}.$

26. $(D^3 - 7DD'^2 - 6D'^3)z = \sin(x+2y) + e^{2x+y}$

Solution: The auxiliary equation is $m^3 - 7m - 6 = 0$

$\Rightarrow \quad (m+1)(m^2 - m - 6) = 0$

$\Rightarrow \quad (m+1)(m-3)(m+2) = 0$

$\Rightarrow \quad m = -1, m = 3, m = -2$

Hence complementary function is; $z = F_1(y-x) + F_2(y+3x) + F_3(y-2x)$

To find particular integral:

$$\text{P.I} = \frac{1}{D^3 - 7DD'^2 - 6D'^3}\left[\sin(x+2y) + e^{2x+y}\right]$$

Now, $u_1 = \dfrac{1}{-D - 7D(-4) - 6(-4)} \cdot \sin(x+2y)$

$$(D^2 = -1,\ D'^2 = -4)$$

$$= \frac{1}{27D + 24D'}\sin(x+2y)$$

$$= \frac{1}{3(9D + 8D'}\sin(x+2y)$$

Nonhomogenous linear equation

Let us consider the equation

$$F(D, D')z = F(x, y) \qquad [1]$$

If $F(D,D')$ is not homogenous then Equation [1] is called nonhomogenous linear PDE.

Its complete solution is complementary function + particular integral.

To find complementary function:

Factorising $F(D,D')$ into the form $D-aD'-b$

then $(D-aD'-b)z=0$

$\Rightarrow \quad p-aq=bz$ [2]

the subsidiary equations are

$$\frac{dx}{1}=\frac{dy}{-a}=\frac{dz}{bz}$$

$\therefore$ from $\dfrac{dx}{1}=\dfrac{dy}{-a} \Rightarrow dy=-adx$

$\Rightarrow y=-ax+c$

$\Rightarrow y+ax=c$

Also from $\dfrac{dx}{1}=\dfrac{dz}{bz} \Rightarrow \dfrac{dz}{z}=bdx$

$\Rightarrow \log z=bx+\log d$

$\Rightarrow z=de^{bx}$

$\therefore$ solution of [2] is $z=e^{bx}\cdot F(y+ax)$

Note: The method to find particular integral are the same as those for homogenous linear equations.

27. $(D^2+DD'+D'-1)z=e^{-x}$

Solution: Let $z=ce^{hx+ky}$ be a trial solution of the equation

$$(D^2+DD'+D'-1)z=0,$$

then $h^2+hk+k-1=0$

$\Rightarrow \quad k(h+1)+h^2-1=0$

$\Rightarrow \quad k(h+1)+(h+1)(h-1)=0$

$\Rightarrow \quad (h+1)\cdot(h+k-1)=0$

$\therefore \quad h=-1$ and $h=1-k$

The complementary function $=\sum c_1e^{-x+ky}+\sum c_2e^{(1-k)x+ky}$

$$=\sum c_1e^{-x}e^{ky}+\sum c_2e^{x}e^{k(y-x)}$$

$$= e^{-x}F_1(y) + e^x F_2(y-x)$$

$$\text{P.I} = \left(\frac{1}{D^2 + DD' + D' - 1}\right)e^{-x}$$

$$= \left(\frac{1}{(D+1)(D+D'-1)}\right)e^{-x}$$

$$= \frac{1}{(-1+0-1}\left(\frac{e^{-x}}{D+1}\right) = -\frac{1}{2}xe^{-x}$$

$\therefore$ **the complete integral is z = complementary function + particular integral**

$$z = e^{-x}F_1(y) + e^x F_2(y-x) - \frac{1}{2}xe^{-x}.$$

28. $(2DD' + D'^2 - 3D')z = 3\cos(3x - 2y)$.

***Solution*:** Let $z = ce^{hx+ky}$ be a trial solution of $(2DD' + D'^2 - 3D')z = 0$

then $2hk + k^2 - 3k = 0$

$\Rightarrow$ $k(2h + k - 3) = 0$

$\therefore$ $k = 0$ or $k = 3 - 2h$

$\therefore$ the complementary function $= \sum c_1 e^{hx+0y} + \sum c_2 e^{hx+(3-2h)y}$

$$= \sum c_1 e^{hx} + \sum c_2 e^{3y} e^{h(x-2y)}$$

$$= F_1(x) + e^{3y}F_2(x-2y)$$

$$\text{P.I} = \left(\frac{1}{2DD' + D'^2 - 3D'}\right)3\cos(3x - 2y)$$

$$= \frac{3\cos(3x-2y)}{-2\times 3(-2) - (-2)^2 - 3D'}$$

$$= \frac{3\cos(3x-2y)}{8 - 3D'}$$

$$= \frac{3(8+3D')}{64 - 9D'^2}\cos(3x - 2y)$$

$$= \frac{3(8+3D')}{64 - 9(-2)^2}\cos(3x - 2y)$$

$$= \frac{3(8+3D')}{100}\cos(3x-2y)$$

$$= \frac{3}{100}[8\cos(3x-2y)+6\sin(3x-2y)]$$

$$= \frac{3}{50}[4\cos(3x-2y)+3\sin(3x-2y)]$$

$\therefore$ the complete integral is z = complementary function + particular integral

$$z = F_1(x)+e^{3y}F_2(x-2y)+\frac{3}{50}[4\cos(3x-2y)+3\sin(3x-2y)]$$

3.2 Partial Differential Equations of parabolic and hyperbolic type with illustrative examples

The classification of second order linear PDEs is very important, as it governs the number and nature of the conditions which must be associated. With the differential equation in order a unique solution may exist, in the sense that the solution changes continuously with the change in the prescribed conditions.

Consider the second order linear homogenous PDE.

$$A\frac{\partial^2 u}{\partial x^2}+B\frac{\partial^2 u}{\partial x\partial y}+C\frac{\partial^2 u}{\partial y^2}+F(x,y,u,p,q)=0 \qquad [1]$$

where A,B,C may be constants or functions of x and y and $p=\frac{\partial u}{\partial x}; q=\frac{\partial u}{\partial y}$.

Equation [1] is said to be

Parabolic if $B^2-4AC=0$

Hyperbolic equation if $B^2-4AC>0$

Elliptic equation if $B^2-4AC<0$

It is to be noted that the first order terms do not play any role in the classification of the equation.

The classifications of some of the well known equations are:

$\frac{\partial u}{\partial t}=C^2\cdot\frac{\partial^2 u}{\partial x^2}$ one dimensional heat flow—parabolic

$\frac{\partial u}{\partial t}=C^2\cdot\left(\frac{\partial^2 u}{\partial x^2}+\frac{\partial^2 u}{\partial y^2}\right)$ two dimensional heat flow—parabolic

$$\frac{\partial^2 u}{\partial t^2} = C^2 \cdot \frac{\partial^2 y}{\partial x^2} \text{ one dimensional wave equation—hyperbolic.}$$

$$\frac{\partial^2 u}{\partial x^2} + \frac{\partial^2 u}{\partial y^2} = 0 \text{ Laplace equation two dimensional—elliptic.}$$

(B) EXAMPLES

1. Determine which of the following equations are hyperbolic and elliptic.

(i) $3u_{xx} + 2u_{yy} = u_x + u$ (ii) $yu_{xx} + 2xu_{xy} + yu_{yy} = 0$

(iii) $f_{xx} + 2f_{xy} + f_{yy} = 0$ (iv) $f_{xy} - f_x = 0$

(v) $x^2 f_{xx} + (1 - y^2) f_{yy} = 0, \; (-\infty < x < \infty), \; (-1 < y < 1)$

(vi) $u_{tt} + 4u_{tx} + 4u_{xx} + 2u_x - u_t = 0$

***Solution*:** (i) Here $A = 3;\ B = 0;\ C = 2$

$\therefore \quad B^2 - 4AC = -24 < 0$

So, the equation is elliptic.

(ii) Here, $A = y;\ B = 2x;\ C = y$

$\therefore \quad B^2 - 4AC = 4x^2 - 4y^2 = 4(x^2 - y^2)$

which is > 0, if $x > y \Rightarrow$ Hyperbolic

< 0, if $x < y \Rightarrow$ Elliptic

$= 0$, if $x = y \Rightarrow$ Parabolic

(iii) Here, $A = 1;\ B = 2;\ C = 1$

$\therefore \quad B^2 - 4AC = 4 - 4 \times 1 \times 1 = 0$

So the equation is parabolic.

(iv) Here, $A = 0;\ B = 1,\ C = 0$

$B^2 - 4AC = 1$ and this is > 0

So the equation is hyperbolic.

(v) Here, $A = x^2,\ B = 0,\ C = (1 - y^2)$

$B^2 - 4AC = 0 - 4x^2(1 - y^2)$

$$= 4x^2(y^2 - 1)$$

For all values of x between $-\infty$ and ∞, x^2 is positive. As y lies between -1 and 1, y^2 is < 1. Hence $B^2 - 4AC$ is –ve in the given region, so the equation is elliptic.

(vi) Here, $A = 1;\ B = 2;\ C = 4;$

$$B^2 - 4AC = 4 - 4 \times 1 \times 4 = 0$$

So the equation is parabolic.

2. Classify the following partial differential equations:

(i) $u_{xx} + 2xu_{xy} + (1 - y^2)u_{yy} = 0$

(ii) $u_{tt} + (5 + 2x^2)u_{xt} + (1 + x^2)(4 + x^2)u_{xx} = 0$

(iii) $u_{xx} + 4u_{xy} + (x^2 + 4y^2)u_{yy} = \sin(xy)$

Solution: (i) Here, $A = 1;\ B = 2x;\ C = (1 - y^2)$

$$B^2 - 4AC = 4x^2 - 4 \times 1 \times (1 - y^2)$$

$$= 4x^2 - 4 + 4y^2$$

$$= 4\left[(x^2 + y^2) - 1\right]$$

If $x^2 + y^2 - 1 > 0$, hyperbolic

If $x^2 + y^2 - 1 = 0$, parabolic

If $x^2 + y^2 - 1 < 0$, elliptic

(ii) Here, $A = 1;\ B = (5 + 2x^2);\ C = (1 + x^2)(4 + x^2)$

$$\therefore \quad B^2 - 4AC = \left[\left(5 + 2x^2\right)^2 - 4 \times 1 \times \left(1 + x^2\right)\left(4 + x^2\right)\right]$$

$$= 25 + \cancel{20x^2} + 4x^4 - 16 - \cancel{4x^2} - \cancel{16x^2} - 4x^4$$

$$= 25 > 0, \text{ hyperbolic}$$

(iii) Here, $A = 1;\quad B = 2;\quad C = x^2 + 4y^2$

$$\therefore \quad B^2 - 4AC = 4 - 4 \times 1 \times (x^2 + 4y^2)$$

elliptic if $4 - 4(x^2 + 4y^2) < 0$

or, $\quad x^2 + 4y^2 > 1$

$$\Rightarrow \quad \frac{x^2}{4} + y^2 > 1$$

$\Rightarrow$ elliptic outside the ellipse $\left(\frac{x^2}{4} + y^2 = 1\right)$.

EXERCISES

1. Classify the following PDEs.

(i) $xf_{xx} + yf_{yy} = 0; \quad x > 0, \ y > 0$

(ii) $f_{xx} - 2f_{xy} = 0$

(iii) $f_{xx} + 2f_{xy} + 4f_{yy} = 0$

(iv) $f_{xx} - 2f_{xy} + f_{yy} = 0$

2. (i) $u_{xx} + u_{xy} = f(x)$

(ii) $u_{xx} - u_{xy} - u_{yy} = 0$

(iii) $(1-x)u_{xx} + 2yu_{xy} + (1+x)u_{yy} = 0$

(iv) $u_t = c^2 u_{xx} + u_x$

3. For what values of the constant c is

$$u_{xx} + cu_{xy} + u_{yy} = 0$$

elliptic; parabolic; hyperbolic?

[*Ans.* Elliptic for $|c| < 2$, Parabolic for $|c| = 2$, Hyperbolic for $|c| > 2$,]

3.3 Separation of variables method for solving partial differential equations

A method suitable for solving such equations is the method of separation of variables, also known as the product method.

In this method, we assume the solution in the form of a product of two functions, one of which is a function of x only, and the other is a function of y only. On separating the two variables, two ordinary differential equations are obtained and these are solved. Their product gives the solution of given equation.

(C) EXAMPLES

Solve the following PDEs by the method of separation of variables.

1. $x\dfrac{\partial z}{\partial x} - y\dfrac{\partial z}{\partial y} = 0$

2. $\dfrac{\partial z}{\partial x} + \dfrac{\partial z}{\partial y} = 2(x+y)z$

3. $\dfrac{\partial^2 z}{\partial x^2} - 2\dfrac{\partial z}{\partial x} + \dfrac{\partial z}{\partial y} = 0$

4. $\dfrac{\partial^2 z}{\partial x^2} - 2\dfrac{\partial z}{\partial u} - \dfrac{\partial z}{\partial y} = 0$

5. $\dfrac{\partial^2 z}{\partial x^2} = \dfrac{\partial z}{\partial y} + 2z$

6. $\dfrac{\partial z}{\partial x} = 2\dfrac{\partial z}{\partial y} + z$ where $z(x,0) = 6e^{-3x}$

7. $u_x + u_y = 0$, where $u(0,y) = 5e^{2y}$

8. $x^2 u_{yy} + u_x = 0$

1. ***Solution***: Given $x\dfrac{\partial z}{\partial x} - y\dfrac{\partial z}{\partial y} = 0$ [1]

We seek a solution of this equation in the form $Z = XY$

where $X = X(x),\ Y = Y(y)$

From this we find: $\dfrac{\partial z}{\partial x} = \dfrac{dX}{dx}Y,\ \dfrac{\partial z}{\partial y} = X\dfrac{dY}{dy}.$

Putting these into Equation [1] we get

$$x \cdot Y \cdot \frac{dX}{dx} - y \cdot X\frac{dY}{dy} = 0\ (\div XY)$$

$$\Rightarrow \quad \frac{xY}{XY}\cdot\frac{dX}{dx} - \frac{yX}{XY}\cdot\frac{dY}{dy} = 0$$

i.e.,
$$\frac{x}{X}\cdot\frac{dX}{dx} - \frac{y}{Y}\cdot\frac{dY}{dy} = 0$$

$$\Rightarrow \quad \frac{x}{X}\cdot\frac{dX}{dx} = \frac{y}{Y}\cdot\frac{dY}{dy}$$

LHS is a function of x only and RHS is a function of y only. Since x and y are independent variables. This expression can hold only if each side is a constant.

Therefore, we have: $\frac{x}{X} \cdot \frac{dX}{dx} = \frac{y}{Y} \cdot \frac{dY}{dy} = k$

Hence, we have two differential equations:

$$\frac{x}{X} \cdot \frac{dX}{dx} = k \quad \text{and} \quad \frac{y}{Y} \cdot \frac{dY}{dy} = k$$

or $\quad \frac{dX}{X} = \frac{k}{x} \cdot dx \Rightarrow \log(X) = k.\log x + \text{constant}$

or $\quad \log X = k \cdot \log x + \log c_1$

$$\log X - \log c_1 = k \cdot \log x$$

$\Rightarrow \quad \log\left(\frac{X}{c_1}\right) = \log\left(x^k\right)$

Hence $\frac{X}{c_1} = x^k$

$\Rightarrow \quad X = c_1 \cdot x^k$ [3]

Again, $\frac{y}{Y} \cdot \frac{dY}{dy} = k \Rightarrow \frac{dY}{Y} = \frac{k}{y} dy$

Its solution is $Y = c_2 y^k$ [4]

Hence, the complete solution is

$Z = X \cdot Y$ is represented as:

$$Z = XY = c_1 x^k \cdot c_2 y^k$$

$$= (c_1 \cdot c_2) x^k y^k$$

denoting $(c_1 \cdot c_2)$ as some other constant say c,

we find $\boxed{Z = c\, x^k \cdot y^k}$

2. ***Solution***: Given $\frac{\partial z}{\partial x} + \frac{\partial z}{\partial y} = 2(x+y)z$ [1]

We assume $Z = X(x)Y(y) = XY$ and substituting in [1]

$$\frac{\partial}{\partial x}(XY) + \frac{\partial}{\partial y}(XY) = 2(x+y)XY$$

$$\Rightarrow \quad Y\frac{dY}{dx} + X\frac{dY}{dy} = 2(x+y)XY \quad (\div XY)$$

we, have

$$\frac{1}{X}\frac{dX}{dx} + \frac{1}{Y}\cdot\frac{dY}{dy} = 2(x+y)$$

i.e., $\frac{1}{X}\frac{dX}{dx} - 2x = -\frac{1}{Y}\frac{dY}{dy} + 2y$

Equating both side to the same constant k, we find

$$\frac{1}{X}\frac{dX}{dx} - 2x = k \text{ and } -\frac{1}{Y}\frac{dY}{dy} + 2y = k$$

(a) Consider: $\frac{1}{X}\frac{dX}{dx} - 2x = k \Rightarrow \frac{dX}{dx} - 2xX = kX$

$$\frac{dX}{dx} - (2x+k)X = 0 \Rightarrow \frac{dX}{dx} = (2x+k)X$$

$$\Rightarrow \quad \log(X) = \left(2\frac{x^2}{2} + kx\right) + \text{Constant}$$

$$= (x^2 + kx) + \log(c_1)$$

$$\Rightarrow \quad \log\left(\frac{X}{c_1}\right) = (x^2 + kx)$$

$$\Rightarrow \quad X = c_1 \cdot e^{x^2+kx}$$

(b) Consider: $-\frac{1}{Y}\frac{dY}{dy} + 2y = k \Rightarrow \frac{dY}{dy} = (2y-k)Y$

It is a variable separable type, and hence the solution is

$$Y = c_2 e^{(y^2-ky)}$$

Hence, the complete solution is given by $Z = X \cdot Y$

$$= c_1 e^{(x^2+kx)} \cdot c_2 e^{(y^2-ky)}$$

$$= (c_1 c_2) e^{(x^2+kx)} e^{(y^2-ky)}$$

Let $c_1 \cdot c_2 = c$, we find

$$\boxed{Z = ce^{(x^2+kx)}e^{(y^2-ky)}}$$

3. ***Solution***: Given $\frac{\partial^2 z}{\partial x^2} - 2\frac{\partial z}{\partial x} + \frac{\partial z}{\partial y} = 0$

Assume a solution of the form $Z = XY$

where X is a function of x alone and Y is a function of y alone.

$$\therefore \quad \frac{\partial z}{\partial x} = X'Y; \quad \frac{\partial^2 z}{\partial x^2} = X''Y; \quad \frac{\partial z}{\partial y} = XY';$$

Hence the given equation reduces to

$$X''Y - 2X'Y + XY' = 0$$

$$\Rightarrow \quad (X'' - 2X')y = -XY' \quad (\div XY)$$

$$\frac{X'' - 2X'}{X} = -\frac{Y'}{Y} \qquad [2]$$

LHS is a function of x alone, while the RHS is a function of y alone and also x, y are independent.

$$\therefore \quad \frac{X'' - 2X'}{X} = -\frac{Y'}{Y} = k, \text{ (Constant)}$$

This reduces to two ordinary differential equations

$$X'' = 2X' - kX = 0 \qquad [3]$$

and $Y' + kY = 0$ [4]

If $D = \frac{d}{dx}$, then the first of these two becomes $(D^2 - 2D - k)X = 0$.

Its auxiliary equation is $m^2 - 2m - k = 0$

$$\Rightarrow \quad m = \frac{2 \pm \sqrt{4 + 4k}}{2} = 1 \pm \sqrt{1 + k}$$

$$\therefore \quad X = Ae^{(1+\sqrt{1+k})x} + Be^{(1-\sqrt{1+k})x}$$

Solving: $Y' + ky = 0$

$$\Rightarrow \quad \frac{Y'}{Y} = -k$$

$$\Rightarrow \quad \log Y = -ky + c$$

$$\Rightarrow \quad Y = e^{-ky+c} = e^c \cdot e^{-ky} = De^{-ky}$$

Hence $Z = XY$

$$= e^x\left[Ae^{\sqrt{1+k}x} + Be^{-\sqrt{1+k}y}\right]$$

$$\boxed{z = e^x \cdot e^{-ky}\left[ae^{\sqrt{1+k}x} + be^{-\sqrt{1+k}y}\right]}$$

4. ***Solution***: Given: $\dfrac{\partial^2 z}{\partial x^2} - 2\dfrac{\partial z}{\partial u} - \dfrac{\partial z}{\partial y} = 0$

Assume $Z = XY$ and we have

$$\frac{\partial^2}{\partial x^2}(XY) - 2\frac{\partial}{\partial x}(XY) - \frac{\partial}{\partial y}(XY) = 0$$

$$\Rightarrow \quad Y\frac{\partial^2 X}{\partial x^2} - 2Y\frac{dX}{dx} - X\frac{dY}{dy} = 0 \quad (\div XY)$$

$$\frac{1}{X}\frac{d^2X}{dx^2} - \frac{2}{X}\frac{dX}{dx} - \frac{1}{Y}\frac{dY}{dy} = 0$$

$$\Rightarrow \quad \frac{1}{X}\frac{d^2X}{dx^2} - \frac{2}{X}\frac{dX}{dx} = \frac{1}{Y}\frac{dY}{dy}$$

Equating both side to the same constant say k,

$$\frac{1}{X}\frac{d^2X}{dx^2} - \frac{2}{X}\frac{dX}{dx} = k$$

and $\dfrac{1}{Y}\dfrac{dY}{dy} = k$

(a) Consider: $\dfrac{1}{X}\dfrac{d^2X}{dx^2} - \dfrac{2}{X}\dfrac{dX}{dx} = k$

$$\frac{d^2X}{dx^2} - 2\frac{dX}{dx} = kX$$

$$\Rightarrow \quad \frac{d^2X}{dx^2} - 2\frac{dX}{dx} - kX = 0 \qquad [1]$$

This reduces to two ordinary differential equations

The auxiliary equation is $m^2 - 2m - k = 0$

$$\Rightarrow \quad m = \frac{\left[-(-2) \pm \sqrt{(-2)^2 - 4\cdot 1\cdot(-k)}\right]}{2\cdot 1}$$

$$= 1 \pm \sqrt{1+k} = \alpha, \beta$$

where $\alpha = 1 + \sqrt{1+k}$

$\beta = 1 - \sqrt{1+k}$

Hence, complementary function: $X = c_1 e^{\alpha x} + c_2 e^{\beta x}$ [2]

(b) Consider: $\dfrac{1}{Y}\dfrac{dY}{dy} = k \Rightarrow \dfrac{dY}{Y} = kdy$

$$\Rightarrow \quad \log(Y) = ky + \log(c_3)$$

$$\Rightarrow \quad Y = c_3 \cdot e^{(ky)} \qquad [3]$$

$$Z = XY = (c_1 e^{\alpha x} + c_2 e^{\beta x}) \cdot c_3 \cdot e^{(ky)}$$

$$Z = (c_1 \cdot c_3 e^{\alpha x} + c_2 \cdot c_3 e^{\beta x}) \cdot e^{(ky)}$$

$$= (A \cdot e^{\alpha x} + B e^{\beta x}) e^{ky} \qquad [4]$$

5. ***Solution***: Given $\dfrac{\partial^2 z}{\partial x^2} = \dfrac{\partial z}{\partial y} + 2Z$

Let $Z = XY$ be a solution

Hence, we have: $\dfrac{\partial^2}{\partial x^2}(XY) = \dfrac{\partial}{\partial y}(XY) + 2XY$

$$\Rightarrow \quad Y\frac{\partial^2 X}{\partial x^2} = X\frac{dY}{dy} + 2XY$$

Dividing throughout by XY:

$$\frac{1}{X}\frac{d^2X}{dx^2} = \frac{1}{Y}\frac{dY}{dy} + 2$$

$$\frac{1}{X}\frac{d^2X}{dx^2} - 2 = \frac{1}{Y}\frac{dY}{dy} = k\,.$$

Equate both sides to the same constant 'k'.

$$\frac{1}{X}\frac{d^2X}{dx^2} - 2 = k \text{ and } \frac{-1}{Y}\frac{dY}{dy} = k$$

(a) Consider: $\frac{1}{X}\frac{d^2X}{dx^2}-2=k$

$\Rightarrow \quad \frac{d^2X}{dx^2}-(2+k)X=0$

The auxiliary equation of the equation is $m^2-(2+k)=0$

$\Rightarrow \quad m=\pm\sqrt{2+k}$

Hence complementary function: $X=c_1e^{(\sqrt{2+k})x}+c_2^{-(\sqrt{2+k})x}$

(b) Consider: $-\frac{1}{Y}\frac{dY}{dy}=k \Rightarrow \frac{dY}{Y}=kdy$

$\Rightarrow \quad \log(Y)=ky+\log(c_3),$ (on integration)

$\Rightarrow \quad Y=c_3\cdot e^{(ky)}$

Hence the solution is $Z=XY$

$$=\left(c_1e^{\sqrt{2+k}x}+c_2e^{-\sqrt{2+k}x}\right)c_3\cdot e^{(ky)}$$

$$=\left(A\cdot e^{\sqrt{2+k}x}+B\cdot e^{-\sqrt{2+k}x}\right)c_3\cdot e^{(ky)}$$

6. ***Solution*:** Given: $\frac{\partial z}{\partial x}=2\frac{\partial z}{\partial y}+z$ [1]

with $z(x,0)=6e^{-3x}$

let $Z=XY$ be a solution of [1]

Hence, $\frac{dX}{dx}Y=2X\frac{dY}{dy}+XY$

$\Rightarrow \quad \frac{1}{X}\left(\frac{dX}{dx}-X\right)=\frac{2}{Y}\frac{dY}{dy}$

LHS is a function of x only and RHS is a function of y only. Therefore each side must be a constant, say 'k'

$$\frac{1}{X}\left(\frac{dX}{dx}-X\right)=k, \quad \frac{2}{Y}\frac{dY}{dy}=k,$$

$\Rightarrow \quad \frac{1}{X}\frac{dX}{dx}=k+1, \quad \frac{1}{Y}\frac{dY}{dy}=\frac{k}{2}$

or $\quad \frac{d}{dx}(\log X)=k+1, \quad \frac{d}{dy}(\log Y)=\frac{k}{2}$

Integration yields

$$\left.\begin{aligned}\log X &= (k+1)x + \log c_1 \\ \log Y &= \left(\frac{k}{2}\right)y + \log c_2\end{aligned}\right]$$

$$X = c_1 e^{(k+1)x}, \; Y = c_2 e^{\left(\frac{x}{2}\right)y}$$

$\therefore$ [1] becomes $Z = Ae^{(k+1)x} e^{\left(\frac{x}{2}\right)y}$ $\quad (A = c_1 c_2)$ [2]

It is given that $z(x,0) = 6e^{-3x}$ [3]

$\Rightarrow \quad z(x,0) = Ae^{(k+1)x}$ [4]

taking $y = 0$ in [2]

$\therefore \quad 6e^{-3x} = Ae^{(k+1)x}$

$\Rightarrow \quad A = 6, \; k = -4,$

On comparing the RHS of [3] and [4], [2] becomes

$$z = 6e^{-3x}e^{-2y} = 6e^{-(3x+2y)}$$

7. ***Solution***: Given: $u_x + u_y = 0$ with $u(0,y) = 5e^{2y}$ [1]

Let $u = XY$

$$\therefore \quad \frac{\partial u}{\partial x} = X' \cdot Y, \; \frac{\partial u}{\partial y} = XY'$$

$\therefore \quad$ from [1], we obtain

$$X'Y + XY' = 0$$

$$\Rightarrow \quad \frac{X'}{X} = -\frac{Y'}{Y}$$

LHS is a function of x only, RHS is a function of y only. Hence the equality is possible only when both the sides are equal to a constant

Let $\dfrac{X'}{X} = -\dfrac{Y'}{Y} = c$ (constant)

$\therefore$ we obtain two ordinary differential equations (ODEs) as follows:

$$\frac{X'}{X} = c \Rightarrow X' - cX = 0$$

and $\quad -\dfrac{Y'}{Y} = c \Rightarrow Y' + cY = 0$

whose solutions are $X = D_1 2^{cx}$ and $Y = D_2 e^{-cy}$

$\therefore$ the complete solution is

$$u(x, y) = D_1 D_2 \cdot e^{c(x-y)} \qquad [3]$$

Now, using $u(0, y) = D_1 D_2 e^{-cy} = 5e^{2y}$

$\Rightarrow \quad D_1 D_2 = 5$ and $c = -2$

$\therefore$ particular solution is $u(x, y) = 5e^{-2(x-y)}$

8. ***Solution***: Given $x^2 u_{yy} + u_x = 0$

Let $\quad u(x, y) = XY$ be a solution [1]

$$\frac{\partial^2 u}{\partial y^2} = XY'' = X \cdot \frac{d^2 Y}{dy^2}, \frac{\partial u}{\partial x} = X'Y = \frac{dX}{dx} \cdot y$$

$\therefore$ the given equation reduces to $x^2 XY'' + X'Y = 0$

$$\Rightarrow \quad \frac{Y'}{Y} = -\frac{X'}{x^2 \cdot X} = \text{constant} \qquad [2]$$

Let the constant $= c^2 > 0$, (positive)

we obtain two ODEs as follows

$$Y'' = c^2 y = 0$$

$$X' + c^2 y^2 X = 0$$

whose solutions are $Y = c_1 e^{cy} + c_2 e^{-cy}$ and $X = c_3 e^{\left(c^2 \cdot \frac{x^3}{3}\right)}$

$\therefore$ the general solution is $u(x, y) = c_3 e^{c^2 \cdot \frac{x^3}{3}} (c_1 e^{cy} + c_2 e^{-cy})$ [3]

Let constant $= -c^2 < 0$, (negative)

we obtain two ODEs as follows $Y'' + c^2 Y = 0$

$$X' - c^2 x^2 X = 0$$

whose solutions are $Y = c_1 \cos cy + c_2 \sin cy$ and $X = c_3 e^{-c^2 \cdot \frac{x^3}{3}}$

$\therefore$ the general solution is

$$u(x,y) = c_3 e^{-c^2 \cdot \frac{x^3}{3}} (c_1 \cos cy + c_2 \sin cy) \qquad [4]$$

Let constant = 0, then, we have

$$Y'' = 0, X' = 0$$

∴ the general solution is

$$u(x,y) = c_1 \cdot (c_2 y + c_3) \qquad [5]$$

3.4 Two dimensional heat equations, wave equations and Laplace equations

(i) Solution of two dimensional heat equation: $\frac{\partial^2 u}{\partial x^2} + \frac{\partial^2 u}{\partial y^2} = \left(\frac{1}{k}\right) \cdot \left(\frac{\partial u}{\partial t}\right)$, by the method of separation of variables.

Given: $\frac{\partial^2 u}{\partial x^2} + \frac{\partial^2 u}{\partial y^2} = \frac{1}{k} \cdot \frac{\partial u}{\partial t}$ [1]

Suppose that [1] has solutions of the form

$$\left.\begin{aligned} u(x,y,t) &= X(x)Y(y)T(t) \\ &= XYT \end{aligned}\right] \qquad [2]$$

Substituting this value of u in [1], we have $X''YT + XY''T = \frac{1}{k} XYT'$

$$\Rightarrow \quad \frac{X''}{X} + \frac{Y''}{Y} = \frac{1}{k} \cdot \frac{T'}{T} \qquad [3]$$

Since x, y and t are independent variables, [3] is true if each terms on each side is a constant, such that

$$\frac{X''}{X} = -n^2, \ \frac{Y''}{Y} = -m^2 \text{ and } \frac{T'}{kT} = -p^2 \qquad [4]$$

with $n^2 + m^2 = p^2$

We have chosen constants in [4] in such a way so that the solution u has the property that $u \to 0$ *as* $t \to \infty$. This is generally satisfied due to physical conditions of a real problem.

Solving differential equation given by [4], we have

$$X_n(x) = A_n \cos nx + B_n \sin nx,$$

$$Y_m(y) = C_m \cos my + D_m \sin my$$

and $T_p(t) = E_p e^{-p^2 kt} = F_{nm} e^{-(n^2+m^2)kt}$

Hence a suitable solution of [1] is

$$u_{nm}(x,y,t) = F_{nm}(A_n \cos nx + B_n \sin nx)(C_m \cos my + D_m \sin my)e^{-(n^2+m^2)kt}$$

(ii) Solution of wave equation: $\frac{\partial^2 y}{\partial t^2} = a^2 \cdot \frac{\partial^2 y}{\partial x^2}$ by the method of separation of variables

Given: $\frac{\partial^2 y}{\partial t^2} = a^2 \cdot \frac{\partial^2 y}{\partial x^2}$ [1]

Assume that a solution of [1] is of the form

$$y = X(x) \cdot T(t) = XT$$

where X is a function of x alone and T is a function of t only.

Then $\frac{\partial^2 y}{\partial t^2} = X \cdot T''$ and $\frac{\partial^2 y}{\partial x^2} X'' T$

Substituting these in [1], we get

$$XT'' = a^2 TX''$$

$$\Rightarrow \quad \frac{X''}{X} = \frac{T''}{a^2 T} \qquad [2]$$

The LHS of [2] is a function of x alone, while the RHS is a function of t alone. Since x and t are independent variables, [2] can be true only if each of the expressions is equal to a constant.

Let $\frac{X''}{X} = \frac{T''}{a^2 T} = k$

k being an arbitrary constant.

Hence, we get two ordinary linear differential equations

$$X'' - kX = 0 \qquad [3]$$

and $T'' - ka^2 T = 0$ [4]

The nature of the solution of [3] and [4] depends on the value of k. The following cases have to be considered.

Case I: Let $k = p^2 > 0$, (positive)

The solution of [3] and [4] are

$$X = A_1 e^{px} + A_2 e^{-px} \qquad [5]$$

and $T = A_3 e^{pat} + A_4 e^{-pat}$ [6]

Case II : Let $k = -p^2 < 0$, (negative)

then $X = A_5 \cos px + A_6 \sin px$ [7]

and $T = A_7 \cos pat + A_8 \sin pat$ [8]

Case III : Let $k = 0$

then $X = A_9 x + A_{10}$ [9]

and $T = A_{11} t + A_{12}$ [10]

The possible solutions of the wave equations are:

$$y = (A_1 e^{px} + A_2 e^{-px})(A_3 e^{pat} + A_4 e^{-pat}) \quad [11]$$

$$y = (A_5 \cos px + A_6 \sin px)(A_7 \cos pat + A_8 \sin pat) \quad [12]$$

$$y = (A_9 x + A_{10})(A_{11} t + A_{12}) \quad [13]$$

Note: Of the three solutions, given by [11], [12] and [13], we have to choose that solution which suite the physical nature of the problem and the boundary conditions. Since we may be dealing with problems on vibrations it is clear that y must be a periodic function of x and t. Hence, in the solution, trigonometric terms must be present. So we take up the solution given by [12], i.e., of the form: $y = (c_1 \cos px + c_2 \sin px)(c_3 \cos pat + c_4 \sin pat)$.

(iii) Solution of Laplace equation in two dimension: $\nabla^2 u = 0$, by the method of separation of variables. Two dimensional: $\left(\dfrac{\partial^2 u}{\partial x^2} + \dfrac{\partial^2 u}{\partial y^2}\right) = 0$

Given: $\left(\dfrac{\partial^2 u}{\partial x^2} + \dfrac{\partial^2 u}{\partial y^2}\right) = 0$ [1]

Assume a particular solution of the form

$$u = X(x) \cdot Y(y) = XY \quad [2]$$

where X is a function of x alone and Y is a function of y alone.

then $\dfrac{\partial^2 u}{\partial x^2} = X''Y$, $\dfrac{\partial^2 u}{\partial y^2} = XY''$

Substituting in [1], we have

$$X''y + XY'' = 0$$

$$\Rightarrow \quad \frac{X''}{X} = -\frac{Y''}{Y} \quad [3]$$

the LHS of [3] is a function of x alone and the RHS is a function of y alone. But they are equal for all values of x and y. This is possible only when each is equal to a constant.

Let $\dfrac{X''}{X} = -\dfrac{Y''}{Y} = k.$ [4]

where k is a constant.

If $k = p^2 > 0$, (positive), then from [4] we get two ordinary differential equations,

$$X'' - p^2 X = 0 \text{ and } Y'' + p^2 Y = 0$$

the complete solutions of these equations are

$$X = c_1 e^{px} + c_2 e^{-px} \text{ and } Y = c_3 \cos py + c_4 \sin py$$

where c_1, c_2, c_3, c_4 are arbitrary constants.

Hence, substituting in [2],

$$u = (c_1 e^{px} + c_2 e^{-px})(c_3 \cos py + c_4 \sin py) \qquad [5]$$

If $k = -p^2 < 0$, (negative) then [4] leads to the equations

$$X'' + p^2 X = 0 \text{ and } Y'' - p^2 Y = 0$$

the complete solution of these equations are:

$$X = c_5 \cos px + c_6 \sin px \text{ and } Y = c_7 e^{py} + c_8 e^{-py} \qquad [6]$$

where c_5, c_6, c_7, c_8 are arbitrary constants

Hence from [2]

$$u = (c_5 \cos px + c_6 \sin px)(c_7 e^{py} + c_8 e^{-py})$$

If k=0 [4] reduces to $X'' = 0$ and $Y'' = 0$. On integrations these equations give

$$X = c_9 x + c_{10} \text{ and } Y = c_{11} y + c_{12}$$

So $u = \left((c_9 x + c_{10})(c_{11} y + c_{12})\right)$ [7]

Thus, [5], [6] and [7] are particular solutions of the Laplace equation, obtained by the method of separation of variables. They can also be put in the form

$$\begin{Bmatrix} u = (A_1 \cosh px + A_2 \sinh px)(A_3 \cos py + A_4 \sin py) \\ u = (A_5 \cos px + A_6 \sin px)(A_7 \cosh py + A_8 \sinh py) \\ u = (A_9 x + A_{10})(A_{11} y + A_{12}). \end{Bmatrix}$$

CHAPTER 4

Discrete Fourier Transforms

Introduction

Transform techniques are important tools in the analysis of signals and linear time-invariant systems. It may be noted that the sequences and the linear shift invariant (LSIV) systems are represented in terms of the fourier and Z-transforms, but when the sequences are of finite duration, it forms a special case and we shall see that it is possible to develop an alternative Fourier representation, which we refer to as Discrete Fourier Transform (DFT).

4.1 Discrete Fourier Transforms

It may be noted that DFT itself is a sequence rather than a function of a continuous variable and it corresponds to equally spaced frequency samples of Discrete Time Fourier Transform (DTFT) of a signal. Also, fourier series representation of the periodic sequence corresponds to the (DFT) of the finite length sequence. In short, we can say that DFT is used for transforming discrete-time sequence d_n of finite length into discrete frequency sequence D_k of finite length. Thus, we can draw the conclusion that by using DFT, the discrete time sequence d_n is transformed into corresponding discrete-frequency sequence D_k.

Properties of DFT

Let $\{d_n\}$, $(n = 0,1,2,\cdots,n-1)$ be a finite duration sequence of complex numbers. We define the N-point DFT of the sequence $\{d_n\}$ as:

$$D_k = \sum_{n=0}^{N-1} d_n \cdot e^{-j\left(\frac{2\pi}{N}\right)nk} \qquad [1]$$

$(k = 0,1,2,\cdots,N-1)$

or $$D_k = \sum_{n=0}^{N-1} d_n W^{nk}, \quad (k = 0,1,2,\cdots,N-1) \qquad [2]$$

where $W = e^{-j\left(\frac{2\pi}{N}\right)}$ is called twiddle factor.

The corresponding inverse discrete Fourier transform (IDFT) is expressed as:

$$d_n = \frac{1}{N}\sum_{k=0}^{N-1} D_k e^{j\left(\frac{2\pi}{N}\right)nk} \qquad [1]'$$

or $$d_n = \frac{1}{N}\sum_{k=0}^{N-1} D_k W^{-nk} \qquad [2]'$$

$(n = 0, 1, \cdots, N-1)$

$$\{d_n\} = \{d_0, d_1, \cdots, d_{n-1}\}$$

Properties

P_1: Orthogonality Relationship

Let W be the nth root of unity. The orthogonality relationship is given by

$$\sum_{k=0}^{N-1} W^{ik} \cdot W^{-rk} = \begin{cases} N, & \text{if } j \equiv r (\text{mod}\, r) \\ 0, & \text{otherwise} \end{cases}$$

Proof: Let $W = e^{i\left(\frac{2\pi}{N}\right)}$

$$\therefore \quad W^{jk} \cdot W^{-rk} = e^{i\left(\frac{2\pi}{N}\right)jk} \cdot e^{i\left(\frac{2\pi}{N}\right)(-rk)}$$

$$= e^{i\left(\frac{2\pi}{N}\right)(j-r)k}$$

Hence, $$\sum_{k=0}^{N-1} W^{jk} \cdot W^{-rk} = \sum_{k=0}^{N-1} e^{\frac{2\pi}{N} i(j-r)k}$$

$$= \sum_{k=0}^{N-1} e^{2\pi i \cdot lk}$$

if $j \equiv r (\text{mod}\, N)$

$\Rightarrow \quad j - r \equiv lN$, for some integer l

$$\Rightarrow \quad \sum_{k=0}^{N-1} W^{jk} W^{-rk} = \sum_{k=0}^{N-1} e^{2\pi i lk}$$

$\Rightarrow$ if $\mathrm{j} \equiv r(\bmod N)$

$$= (1+1+1+\cdots+1) = N \qquad [1]$$

Also, $\sum_{k=0}^{N-1} W^{\mathrm{j}k} W^{-rk} = \sum_{k=0}^{N-1} e^{\frac{2\pi}{N}\mathrm{i}(\mathrm{j}-r)k}$

if $\quad \mathrm{j} \neq r(\bmod N) \Rightarrow \mathrm{j}-r=p$, where p is a multiple of N

$$\Rightarrow \quad \sum_{k=0}^{N-1} W^{\mathrm{j}k} W^{-rk} = \sum_{k=0}^{N-1} e^{\mathrm{i}\left(\frac{2\pi}{N}\right)pk}$$

$$= \sum_{k=0}^{N-1} W^{pk}$$

$$W = e^{\mathrm{i}\left(\frac{2\pi}{N}\right)} \qquad = (1+W^p+\left(W^p\right)^2+\cdots+\left(W^p\right)^{N-1}$$

$$= \frac{1-\left(W^p\right)^N}{1-(W^p)} \quad = \frac{1-e^{\frac{2\pi}{N}\mathrm{i}\,p\cdot N}}{1-(W^p)}$$

$$= \frac{1-1}{1-(W^p)} \qquad [2]$$

Equation [1] and [2] prove the result.

P_2: Linearity Property

If (D_k, d_n) and (G_k, g_n) are two DFT pairs, and a, b are any complex constants, then $(aG_k + bD_k)$ and $(ag_n + bd_n)$ is also a transform pair.

P_3: Shifting Property

If d_n is a periodic sequence (period N samples) with DFT D_k, then the periodic sequence d_{n-n_0} has DFT $D_k \cdot e^{\mathrm{j}\left(\frac{2\pi}{N}\right)n_0 k}$ Or

If (G_k, g_n) is a transform pair, Then $\left(W^{n_0 k}, G_k, g_{n-n_0}\right)$ is also a transform pair i.e., DFT of $\left\{W^{n_0 k} G_k\right\}$ is $\left\{g_{n-n_0}\right\}$

P_4: Convolution Property

If (D_k, d_n) and (G_k, g_n) are two pair of DFTs, then

$$\left(\frac{1}{N}\sum_{r=0}^{N-1} G_r D_{j-r},\ g_n d_n\right) \text{ and } G_k D_k, \sum_{r=0}^{N-1} g_n d_{r-n}$$

are also transform pairs.

P_5: (Symmetry Properties)

If $\{d_n\}$ is a real, periodic sequence of period N samples, then D_k satisfies the symmetry conditions.

Note: imaginary part: (i.p.) and real part: (r.p.).

$$(\text{i.p. of})\ D_k = (\text{i.p. of})(D_{n-k})$$

$$(\text{i.p. of})\ D_k = -(\text{i.p. of})(D_{N-k})$$

$$\Rightarrow \quad \begin{Bmatrix} |D_k| = |D_{N-k}| \\ \hat{D}_k = -\hat{D}_{N-k} \end{Bmatrix} \qquad [1]$$

Consider the real periodic sequences d_n and g_n both of period N samples, with N-point DFT's D_k and G_k respectively.

If $\quad Z_n = d_n + \mathrm{j}g_n \qquad [2]$

be a complete sequence, then DFT of Z_n is

$$Z_k = \sum_{n=0}^{N-1} [d_n + \mathrm{j}g_n] e^{-\mathrm{j}\left(\frac{2\pi}{N}\right)nk} \qquad [3]$$

$$Z_k = D_k + \mathrm{j}G_k \qquad [4]$$

Equating real(Re) and imaginary(Im) parts, of Equation [4] gives

$$\mathrm{Re}[Z_k] = \mathrm{Re}[D_k] - \mathrm{Im}[G_k]$$

$$\mathrm{Im}[Z_k] = \mathrm{Im}[D_k] + \mathrm{Re}[G_k]$$

Since the real and imaginary parts of D_k and G_k have even and odd symmetry respectively, it is clear that these components can be recovered by a simple "even-odd" separation.

i.e., $\quad \mathrm{Re}[D_k] = \dfrac{\mathrm{Re}[Z_k] + \mathrm{Re}[Z_{N-k}]}{2}$

$$\left\{\begin{aligned} \text{Im}[G_k] &= \frac{\text{Re}[Z_{N-k}] - \text{Re}[Z_k]}{2} \\ \text{Re}[G_k] &= \frac{\text{Im}[Z_k] + \text{Im}[Z_{N-k}]}{2} \\ \text{Im}[D_k] &= \frac{\text{Im}[Z_k] - \text{Im}[Z_{N-k}]}{2} \end{aligned}\right\} \qquad [5]$$

Thus, a single *N*-point DFT can efficiently transform two *N*-point real sequences at the same time.

Some important Trigonometric form of complex numbers:

1. $(\cos 0 + \text{j}\sin 0 = 1)$

2. $(\cos \pi + \text{j}\sin \pi) = (-1)$

3. $\left(\cos\frac{\pi}{2} + \text{j}\sin\frac{\pi}{2}\right) = \text{i}$

4. $\left(\cos\frac{\pi}{2} + \text{j}\sin\frac{\pi}{2}\right) = -\text{j} = \cos\left(-\frac{\pi}{2}\right) + \text{j}\sin\left(-\frac{\pi}{2}\right)$

5. $(\cos 0 + \text{j}\sin 0)^{\frac{1}{3}} = \left[\cos(2n\pi{+}0) + \text{j}\sin(2n\pi{+}0)\right]^{\frac{1}{3}}$

$$= (\cos 2n\pi + \text{j}\sin 2n\pi)^{\frac{1}{3}}$$

$$= \cos\left(\frac{2n\pi}{3}\right) + \text{j}\sin\left(\frac{2n\pi}{3}\right)$$

putting $n = 0, 1, 2,$ the three cube roots of unity are

$$(\cos 0 + \text{j}\sin 0); \ \left(\cos\frac{2\pi}{3} + \text{j}\sin\frac{2\pi}{3}\right); \ \left(\cos\frac{4\pi}{3} + \text{j}\sin\frac{4\pi}{3}\right)$$

$$\Rightarrow \quad 1; \ \left(\cos\left(\pi - \frac{\pi}{3}\right) + \text{j}\sin\left(\pi - \frac{\pi}{3}\right)\right); \ \left(\cos\left(\pi + \frac{\pi}{3}\right) + \text{j}\sin\left(\pi + \frac{\pi}{3}\right)\right)$$

$$\Rightarrow \quad 1; \ \left(-\cos\frac{\pi}{3} + \text{j}\sin\frac{\pi}{3}\right); \ \left(-\cos\frac{\pi}{3} - \text{j}\sin\frac{\pi}{3}\right)$$

$$\Rightarrow \quad 1; \ \left(-\frac{1}{2} + \text{j}\cdot\frac{\sqrt{3}}{2}\right); \ \left(-\frac{1}{2} - \text{j}\cdot\frac{\sqrt{3}}{2}\right)$$

$$\Rightarrow \quad \left(1; \ \frac{-1 \pm \text{j}\sqrt{3}}{2}\right)$$

6. $\cos\frac{3\pi}{2} + j\sin\frac{3\pi}{2} = 0 + j(-1) = (-j)$

7. $\cos\frac{5\pi}{3} + j\sin\frac{5\pi}{3} = \cos\left(2\pi - \frac{\pi}{3}\right) + j\sin\left(2\pi - \frac{\pi}{3}\right)$

$$= \cos\frac{\pi}{3} - j\sin\frac{\pi}{3} = \left(\frac{1}{2} - j\frac{\sqrt{3}}{2}\right)$$

8. $\left(\cos\frac{3\pi}{4} + j\sin\frac{3\pi}{4}\right) = \cos\left(\pi - \frac{\pi}{4}\right) + j\sin\left(\pi - \frac{\pi}{4}\right)$

$$= -\cos\frac{\pi}{4} - j\sin\frac{\pi}{4}$$

$$= -\frac{1}{\sqrt{2}} + j\cdot\frac{1}{\sqrt{2}} = \left(\frac{-1+j}{\sqrt{2}}\right)$$

9. $\left(\cos\frac{5\pi}{4} + j\sin\frac{5\pi}{4}\right) = \cos\left(\pi + \frac{\pi}{4}\right) + j\sin\left(\pi + \frac{\pi}{4}\right)$

$$= -\cos\frac{\pi}{4} - j\sin\frac{\pi}{4}$$

$$= \left(-\frac{1}{\sqrt{2}} - j\cdot\frac{1}{\sqrt{2}}\right)$$

10. $\left(\cos\frac{7\pi}{4} + j\sin\frac{7\pi}{4}\right) = \cos\left(2\pi - \frac{\pi}{4}\right) + j\sin\left(2\pi + \frac{\pi}{4}\right)$

$$= -\cos\frac{\pi}{4} - j\sin\frac{\pi}{4}$$

$$= \frac{1}{\sqrt{2}} - j\cdot\frac{1}{\sqrt{2}} \quad = \frac{1-j}{\sqrt{2}}$$

(A) EXAMPLES

1. Evaluate the *N*-point DFT of the sequence

$$d_n = \begin{cases} \frac{1}{3}, & 0 \le n \le 2 \\ 0, & \text{otherwise} \end{cases}$$

Solution: $D_k = \sum_{n=0}^{N-1} d_n e^{-j\left(\frac{2\pi}{N}\right)nk}$

Here, given that $N = 3$ therefore we have $[\because \; (0,1,2) = 3 \text{ points}]$

$$= \sum_{n=0}^{2} \left(\frac{1}{3}\right) e^{-j\left(\frac{2\pi}{N}\right)nk}$$

$$= \frac{1}{3}\left[1 + e^{-j\left(\frac{2\pi}{N}\right)k} + e^{-j\left(\frac{2\pi}{N}\right)2k}\right]$$

$$= \frac{1}{3} e^{-j\left(\frac{2\pi}{N}\right)k}\left[1 + e^{j\left(\frac{2\pi}{N}\right)k} + 1 + e^{-j\left(\frac{2\pi}{N}\right)k}\right]$$

$$\therefore \quad D_k = e^{-j\left(\frac{2\pi}{N}\right)k}\left[\frac{1 + 2\cos\left(\frac{2\pi}{N}\right)k}{3}\right]$$

2. Evaluate the N-point DFT of the sequence $d_n = \{1, 0, -1\}$

Solution: We know that DFT is expressed as:

$$D_k = \sum_{n=0}^{N-1} d_n e^{-j\left(\frac{2\pi}{N}\right)nk}$$

Here, given that $N = 3$, therefore, we have $d_0 = 1;\; d_1 = 0;\; d_2 = -1.$

$$\therefore \quad D_k = \sum_{n=0}^{2} d_n e^{-j\left(\frac{2\pi}{3}\right)nk}$$

$$= d_0 + d_1 e^{-j\left(\frac{2\pi}{3}\right)k} + d_2 e^{-j\left(\frac{2\pi}{3}\right)2k}$$

$$= d_0 + d_1 (W)^k + d_2 \left(W^2\right)^k \qquad (\because W = e^{-i})$$

$$= 1 + 0(W)^k + (-1)\left(W^2\right)^k$$

$$= 1 - \left(W^2\right)^k \qquad [1]$$

Now, $D_0 = 1-\left(W^2\right)^0 = 1-1=0$

$$D_1 = 1-\left(W^2\right)^1 = 1-W^2$$

$$D_2 = 1-\left(W^2\right)^2 = 1-W^4 = 1-W\cdot W^3 = 1-W$$

$$\left(\because W^3 = \left[e^{-j\left(\frac{2\pi}{3}\right)}\right]^3 = e^{-j(2\pi)} = \cos 2\pi - i\sin 2\pi = 1-0=1\right)$$

So discrete Fourier transform (DFT) of the given sequence is $(1-W^2, 1-W, 0)$ where W is cube root of unity.

Note: $\cos 0 + j\sin 0; = 1$

$$\cos\frac{2\pi}{3} + j\sin\frac{2\pi}{3} = \cos\left(\pi - \frac{\pi}{3}\right) + j\sin\left(\pi - \frac{\pi}{3}\right)$$

$$\cos\frac{4\pi}{3} + j\sin\frac{4\pi}{3} = \cos\left(\pi + \frac{\pi}{3}\right) + j\sin\left(\pi + \frac{\pi}{3}\right)$$

3. Obtain the DFT of the sequence
$\{d_n\} = \{n, \text{ for } n = 0,1,2,3\}$

***Solution*:** Number of sample point $n=4$.

Hence, we have the 4–point DFT of d_n given by

$$D_k = \sum_{n=0}^{3} d_n e^{j\left(\frac{2\pi}{4}nk\right)}, \qquad k = 0,1,2,3$$

$$= 0 + (1)e\left(-j\frac{\pi}{2}\right)k + (2)e(-j\pi)k + (3)e\left(-j\frac{3\pi}{2}\right)k$$

Now, $D_0 = 1e(0) + 2e(0) + 3e(0)$

$$= 1+2+3 = 6$$

$$D_1 = 1e\left(-j\frac{\pi}{2}\right) + 2e(-j\pi) + 3e\left(-j\frac{3\pi}{2}\right)$$

$$= -2(1-j)$$

$$D_2 = 1e(-j\pi) + 2e(-2j\pi) + 3e(-3j\pi)$$

$$= -1 + 2(1) + 3(-1) \qquad \left(\because \begin{array}{l} \cos 3\pi = (-1)^3 = -1 \\ \cos 2\pi = (-1)^2 = 1 \end{array}\right)$$

$$= -1 + 2 - 3 = -2;$$

$$D_3 = 1e\left(-\frac{3\pi}{2}\mathrm{j}\right) + 2e(-3\pi\,\mathrm{j}) + 3e\left(-\frac{9\pi}{2}\mathrm{j}\right)$$

$$= -2(1+\mathrm{j})$$

So, DFT of the given sequence:

$$6,\ -2(1-\mathrm{j}),\ -2,\ -2(1+\mathrm{j})$$

4. Find the N-point DFT of the sequence

$$d_n = \begin{cases} \frac{1}{3}, & -1 \le n \le 1 \\ 0, & \text{otherwise} \end{cases}$$

Solution: $D_k = \sum_{n=0}^{n=1} d_n e^{-\mathrm{j}\left(\frac{2\pi}{N}nk\right)}$

Here, $d_n = \begin{cases} \frac{1}{3}, & n = 0, 1, N-1 \\ 0, & \text{otherwise} \end{cases}$

Therefore, the N-point DFT of d_n given by

$$D_k = \frac{1}{3}\left(1 + e^{-\mathrm{j}\left(\frac{2\pi}{N}k\right)} + e^{-\mathrm{j}\left(\frac{2\pi(N-1)k}{N}\right)}\right)$$

We add 2π to the exponent of the complex exponential which does not change its value, therefore, we have

$$e^{-\mathrm{j}\left(\frac{2\pi(N-1)}{N}k\right)} = e^{-\mathrm{j}\left[\frac{2(N-1)\pi k}{N} + \left(\frac{2\pi N}{N}\right)k\right]}$$

$$= e^{\mathrm{j}\left[\frac{2\pi k}{N}\right]}$$

$$\therefore \quad D_k = \frac{1}{3}1 + e^{-\mathrm{j}\left(\frac{2\pi k}{N}\right)} + e^{\mathrm{j}\left(\frac{2\pi k}{N}\right)}$$

$$= \left[\frac{1+2\cos\left(\frac{2\pi k}{N}\right)}{3}\right]$$

5. Find the 8-point DFT of the sequence

$$d_n = \begin{cases} 1, & -3 \le n \le 3 \\ 0, & \text{otherwise} \end{cases}$$

Solution: $d_n = \begin{cases} 1, & -3 \le n \le 3 \\ 0, & \text{otherwise} \end{cases}$

$$= \begin{cases} 1, & n = 0, 1, 2, N-1 \\ 0, & \text{otherwise} \end{cases}$$

for N-point DFT

$$\Rightarrow \quad d_n = \begin{cases} 1, & n = 0, 1, 2, 7 \\ 0, & \text{otherwise} \end{cases}$$

(for 8-point DFT)

$$D_k = \sum_{n=0}^{n=1} d_n e^{-j\left(\frac{2\pi}{N}\right)nk}$$

$$= \sum_{n=0,1,2,7} d_n e^{-j\left(\frac{2\pi}{8}\right)nk}$$

$$= 1 + e^{-j\left(\frac{2\pi k}{8}\right)} + e^{-j\left(\frac{4\pi k}{8}\right)} + e^{-j\left(\frac{14\pi k}{8}\right)}$$

$$= 1 + e^{-j\left(\frac{\pi k}{4}\right)} + e^{-j\left(\frac{\pi k}{2}\right)} + e^{-j\left(\frac{7\pi k}{4}\right)}$$

$$= e^{-j\left(\frac{\pi}{8}k\right)}\left[e^{j\left(\frac{\pi}{8}\right)k} + e^{-j\left(\frac{\pi}{8}\right)k}\right] + e^{-j\left(\frac{9\pi}{8}\right)k}\left[e^{j\left(\frac{5\pi}{8}\right)k} + e^{-j\left(\frac{5\pi}{8}\right)k}\right]$$

$$= 2e^{-j\left(\frac{\pi}{8}\right)k}\cos\left(\frac{\pi k}{8}\right) + 2e^{-j\left(\frac{9\pi}{8}k\right)}\cos\left(\frac{5\pi}{8}k\right)$$

6. Find the DFT of d_n if $d_n = \begin{cases} 1, & 2 \le n \le 6 \\ 0, & \text{for } n = 0,1,7,8,9 \text{ otherwise} \end{cases}$

assuming that $N = 10$.

Solution: $D_k = \sum_{n=0}^{n=1} d_n W^{-nk}$ where $W = e^{j\left(\frac{2\pi}{N}\right)}$

$$= \sum_{n=2}^{6} W^{-nk} = \sum_{n=0}^{6} W^{-nk} - \sum_{n=0}^{1} W^{-nk}$$

$$= \left[\frac{1-W^{-7k}}{1-W^{-k}}\right] - \left[\frac{1-W^{-2k}}{1-W^{-k}}\right]$$

$$= \frac{1-W^{-7k}-1+W^{-2k}}{(1-W^{-k})}$$

$$D_k = \frac{(W^{-2k}-W^{-7k}}{(1-W^{-k})} = \frac{e^{-j\left(\frac{\pi}{5}k\right)2} - e^{-j\left(\frac{\pi}{5}k\right)7}}{1-e^{-j\left(\frac{\pi}{5}k\right)(1)}}$$

$$= \frac{e^{-j\left(\frac{9\pi}{10}k\right)}\left[e^{j\left(\frac{\pi}{2}k\right)} - e^{-j\left(\frac{\pi}{2}k\right)}\right]}{e^{-j\left(\frac{\pi k}{10}\right)}\left[e^{j\left(\frac{\pi}{10}k\right)} - e^{-j\left(\frac{\pi}{10}k\right)}\right]}$$

$$D_k = e^{-j\left(\frac{4\pi}{5}k\right)} \frac{\sin\left(\frac{\pi k}{2}\right)}{\sin\left(\frac{\pi k}{10}\right)}$$

7. Find the DFT of the sequence $d_n = \{1,1,0,0\}$ and find the DFT of $D_k = \{1,0,1,0\}$.

Solution: We know that DFT is expressed as:

$$D_k = \sum_{n=0}^{N=1} d_n e^{-j\left(\frac{2\pi}{N}\right)nk}$$

Here, given that $N=4$, therefore, we have

$$D_k = \sum_{n=0}^{3} d_n e^{-j\left(\frac{2\pi}{4}\right)nk}$$

Now, $D_0 = d_0 + d_1 + d_2 + d_3$

or $\quad D_0 = 1+1+0+0 = 2+0 = 2$

$$D_1 = d_0 + d_1 e^{-i\left(\frac{\pi}{2}\right)} = 1 + 1\cdot e^{-i\left(\frac{\pi}{2}\right)}$$

$$= 1 + (0 - i) = 1 - j$$

$$D_2 = d_0 + d_1 e^{-i\pi} = 1 + (-1) = 1 - 1 = 0$$

$$D_3 = d_0 + d_1 e^{-i\left(\frac{3\pi}{2}\right)} = 1 + j$$

Hence $D_k = [2, 1-j, 0, 1+j]$

8. Obtain the DFT of the sequence $d_n = \{1,1,1,1,0,0,0,0\}$

Solution: Method (1), We know that

$$D_k = \sum_{n=0}^{N-1} d_n e^{-j\left(\frac{2\pi}{N}\right)nk}, \qquad k = 0,1,2,\cdots N-1$$

Here, given that $N=8$, therefore, we have

$$D_k = \sum_{n=0}^{7} d_n e^{-j\left(\frac{2\pi}{N}\right)nk} \qquad k = 0,1,2,3,4,5,6,7$$

$$= \sum_{n=0}^{7} d_n W^{nk} \qquad = (1 + W^k + W^{2k} + W^{3k})$$

where $W = e^{\left(\frac{-j2\pi}{N}\right)} = e^{\left(\frac{-j2\pi}{8}\right)} = e^{\left(\frac{-j\pi}{4}\right)}$

$$= \cos\left(\frac{\pi}{2}\right) + j\sin\left(\frac{\pi}{4}\right) = \frac{1+j}{\sqrt{2}}$$

with $k = 0,1,2,\cdots,7$ in the D_k equation, we get

$$D_0 = (1+1+1+1) = 4$$

$$D_1 = \left\{1 + e^{\left(\frac{-j\pi}{4}\right)} + e^{\left(\frac{-j\pi}{2}\right)} + e^{\left(\frac{-j3\pi}{4}\right)}\right\} = \left[1 - j\left(\sqrt{2}+1\right)\right]$$

Similarly, $D_2 = 0$; $D_3 = \left(1 - \frac{j}{\sqrt{2}}\right)$; $D_4 = (0)$; $D_5 = \left(1 + \frac{j}{\sqrt{2}}\right)$

$D_6 = 0$; $D_7 = \left[1 + j\left(\sqrt{2}+1\right)\right]$.

9. Obtain the DFT of the sequence: $\{d_n\} = \{1,1,1,1,0,0,0,0\}$.

Solution: The twiddle factor W_8^n, $n = 0(n)7$ lie on the circumference of the circle of unit radius as shown in Figure 4.1.

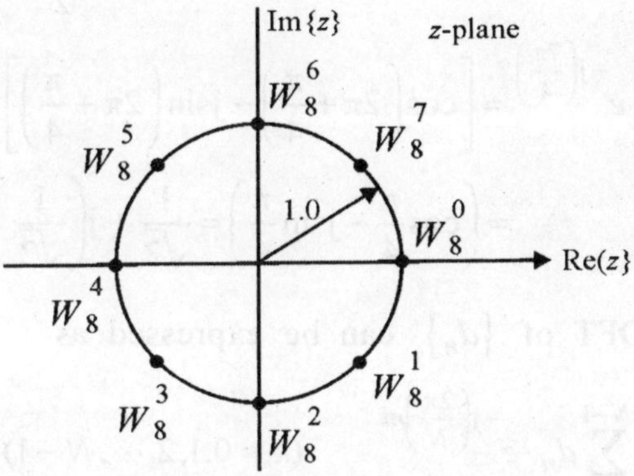

Figure 4.1: sequence W_8^n for $0 \le n \le 8$.

$$N = 8; \quad \therefore \; W_8^n = e^{-j\frac{2\pi n}{8}} = \left(\cos\frac{2n\pi}{8} - j\sin\frac{2n\pi}{8}\right) = \left(\cos\frac{n\pi}{4} - j\sin\frac{n\pi}{4}\right)$$

$(n = 0,1,2,3,4,5,6,7)$

$$W_8^0 = e^{-0} = (\cos 0 - j\sin 0) = \cos 0 = 1$$

$$W_8^1 = e^{-j\left(\frac{\pi}{4}\right)} = \left(\cos\frac{\pi}{4} - j\sin\frac{\pi}{4}\right) = \frac{1}{\sqrt{2}} - j\frac{1}{\sqrt{2}} = \left(\frac{1-j}{\sqrt{2}}\right)$$

$$W_8^2 = e^{-j\left(\frac{\pi}{2}\right)} = \left(\cos\frac{\pi}{2} - j\sin\frac{\pi}{2}\right) = 0 - j(1) = -j$$

$$W_8^3 = e^{-j\left(\frac{3\pi}{4}\right)} = \left(\cos\frac{3\pi}{4} - j\sin\frac{3\pi}{4}\right) = -\frac{1}{\sqrt{2}} - j\frac{1}{\sqrt{2}}$$

$$W_8^4 = e^{-j\pi} = (\cos\pi - j\sin\pi) = -1 - j(0) = -1$$

$$W_8^5 = e^{-j\left(\frac{5\pi}{4}\right)} = \left(\cos\frac{5\pi}{4} - j\sin\frac{5\pi}{4}\right) = \left(\cos\left(\pi + \frac{\pi}{4}\right) - j\sin\left(\pi + \frac{\pi}{4}\right)\right)$$

$$= \left(-\cos\frac{\pi}{4} + j\sin\frac{\pi}{4}\right) = -\frac{1}{\sqrt{2}} + j\left(\frac{1}{\sqrt{2}}\right)$$

$$W_8^6 = e^{-j\left(\frac{3\pi}{2}\right)} = \left(\cos\frac{3\pi}{2} - j\sin\frac{3\pi}{2}\right) = \left[\cos\left(\pi + \frac{\pi}{2}\right) - j\sin\left(\pi + \frac{\pi}{2}\right)\right]$$

$$= -\cos\frac{\pi}{2} + j\sin\frac{\pi}{2} = -0 + j(1) = j$$

$$W_8^7 = e^{-j\left(\frac{7\pi}{4}\right)} = \left[\cos\left(2\pi + \frac{\pi}{4}\right) - j\sin\left(2\pi + \frac{\pi}{4}\right)\right]$$

$$= \left(\cos\frac{\pi}{4} - j\sin\frac{\pi}{4}\right) = \frac{1}{\sqrt{2}} + j\left(\frac{1}{\sqrt{2}}\right)$$

Now, the DFT of $\{d_n\}$ can be expressed as

$$D_k = \sum_{n=0}^{N-1} d_n \cdot e^{-j\left(\frac{2\pi}{N}\right)nk} \qquad (k = 0, 1, 2, \cdots, N-1)$$

$$= \text{DFT}\{d_n\}$$

or, $$D_k = \sum_{n=0}^{7} d_n W_8^{nk} \qquad \left(\because\ W_8 = e^{-j\left(\frac{2\pi}{N}\right)}\right)$$

$$= 1(1)\cdot W_8^{k} + (1)\cdot W_8^{2k} + (1)\cdot W_8^{3k}$$

$$= 1 + W_8^{k} + W_8^{2k} + W_8^{3k}; \qquad k = 0, 1, 2, 3, 4, 5, 6, 7.$$

$$D_0 = 1 + 1 + 1 + 1 = 4$$

$$D_1 = 1 + W_8^1 + W_8^2 + W_8^3 = 1 + \left(\frac{1}{\sqrt{2}} - \frac{j}{\sqrt{2}}\right) + (-j) + \left(-\frac{1}{\sqrt{2}} - \frac{j}{\sqrt{2}}\right)$$

$$= 1 - j2 \cdot 414$$

$$D_2 = 1 + W_8^2 + W_8^4 + W_8^6 = (1 + (-\text{j}) + (-1) + \text{j})$$

$$D_3 = 1 + W_8^3 + W_8^6 + W_8^1 = \left(1 - \frac{1}{\sqrt{2}} - \text{j}\frac{1}{\sqrt{2}} + \text{j} + \frac{1}{\sqrt{2}} - \frac{\text{j}}{\sqrt{2}}\right) = 1 - \text{j}0.414$$

$$D_4 = 1 + W_8^4 + W_8^0 + W_8^4 = (1 - 1 + 1 - 1) = 0$$

$$D_5 = 1 + W_8^5 + W_8^2 + W_8^7 = \left(1 - \frac{1}{\sqrt{2}} + \frac{\text{j}}{\sqrt{2}} - \text{j} + \frac{1}{\sqrt{2}} + \frac{\text{j}}{\sqrt{2}}\right) = 1 + \text{j}0.414$$

$$D_6 = 1 + W_8^6 + W_8^4 + W_8^2 = (1 + \text{j} - 1 - \text{j}) = 0$$

$$D_7 = 1 + W_8^7 + W_8^6 + W_8^5 = \left(1 + \frac{1}{\sqrt{2}} + \frac{\text{j}}{\sqrt{2}} + \text{j} - \frac{1}{\sqrt{2}} + \frac{\text{j}}{\sqrt{2}}\right) = 1 + \text{j}2.414$$

Hence $D_k \equiv \begin{bmatrix} 4, & 1-\text{j}2.414, & 0, & 1-\text{j}0.414 \\ 0, & 1+\text{j}0.414, & 0, & 1+\text{j}2.414 \end{bmatrix}$

10. Find the DFT of the sequence defined by $d_n = (-1)^n$ for

(a) $N = 3$; (b) $N = 4$; (c) N odd and (d) N even.

Solution: (a) Given that $N = 3$;

We know that DFT of d_n can be expressed as

$$D_k = \sum_{n=0}^{N-1} (-1)^n \cdot e^{-\text{j}\left(\frac{2\pi}{N}\right)nk}$$

$$= \sum_{n=0}^{N-1} (-1)^n \cdot W_N^{nk} \qquad \left(\because W_N = e^{-\text{j}\left(\frac{2\pi}{N}\right)}\right)$$

$$= \sum_{n=0}^{N-1} \left[W_N^{nk}\right]^n$$

$$= \frac{1-(-1)^N}{1+W_N^k}, \text{ for } W_N^k \neq -1$$

$$D_k = \frac{2}{1+W_3^k} = \frac{2}{1+\cos\left(\frac{2\pi}{3}\right)k - \text{j}\sin\left(\frac{2\pi}{3}\right)k}, \quad (0 \le k \le 2)$$

(b) Here, given that $N = 4$, therefore, we have

$$D_k = \sum_{n=0}^{N-1} (-1)^n \cdot W_4^{2n} = 0, \quad \text{for } W_4^k \neq -1 \text{ or } k \neq 2.$$

with $k = 2$, we get $D_2 = \sum_{n=0}^{3} (-1)^n W_4^{2n}$

$$= 1 - W_4^2 + W_4^4 - W_4^6$$

$$= 1 - (-1) + (-1)^2 - (-1)^3 = 4.$$

Hence, $D_k = 4\delta(k-2)$.

(c) We know that $W_N^k = e^{-j\left(\frac{2\pi}{N}\right)k}$

if $N = 2k$, we get $W_N^k = -1$.

Here, given that N = odd, no k exists.

$\Rightarrow \quad W_N^k \neq -1 \quad$ for all k from 0 to $N - 1$

$$\therefore \quad D_k = \frac{2}{1 + W_N^k}, \; 0 \le k \le N-1$$

$$= \frac{2}{1 + \cos\frac{2\pi}{N}k - j\sin\frac{2\pi}{N}k}$$

(d) Here, given that N even: $W_N^k = -1$, if $k = \dfrac{N}{2}$

$$D_k = 0 \text{ for } k \neq \frac{N}{2}$$

and $$D_{\frac{N}{2}} = \sum_{n=0}^{N-1} \left[-W_N^{N/2} \right]^n$$

$$= \sum_{n=0}^{N-1} (1) = N.$$

Hence, $D_k = N\delta\left(k - \dfrac{N}{2}\right)$

P_6: Parseval's theorem

Let $\{d_r\}$ be a set of complex number i.e., $\{d_1, d_2, \cdots, d_{n-1}\}$ is a sequence for $(r = 1, 2, \cdots, n-1)$ and $\{D_j\}$ be DFT of $\{d_r\}$,

then $\frac{1}{n}\sum_{j=0}^{n-1}\left|D_j\right|^2 = \sum_{r=0}^{n-1}\left|d_r\right|^2$

Proof: By definition of DFT, we have

$$D_j = \sum_{r=0}^{n-1} d_r W^{jr},\ j = 0, 1, 2, \cdots, n-1 \qquad [1]$$

$$G_j = \sum_{r=0}^{n-1} g_r W^{jr},\ j = 0, 1, 2, \cdots, n-1 \qquad [2]$$

$$\Rightarrow \bar{G}_j = \sum_{r=0}^{n-1} \bar{g}_r W^{-jr},\ j = 0, 1, \cdots, n-1$$

on taking conjugate on both sides

Now, $\sum_{j=1}^{n-1} D_j \cdot \bar{G}_j = \sum_{j=1}^{n-1} D_j \sum_{r=0}^{n-1} \bar{g}_r W^{-jr}$

$$= \sum_{r=0}^{n-1} \bar{g}_r \sum_{r=0}^{n-1} D_j W^{-jr} = \sum_{r=0}^{n-1} \bar{g}_r\, ndr$$

Using $dr = \frac{1}{N}\sum_{j=0}^{n-1} D_j W^{-jr}$

$r = 0, 1, 2, \cdots, n-1$

$$\Rightarrow \frac{1}{n}\sum_{j=1}^{n-1} D_j \cdot \bar{G}_j = \sum_{r=0}^{n-1} \bar{g}_r\, dr \qquad [3]$$

Putting $\bar{G}_j = \bar{D}_j$

and $\bar{g}_r = \bar{d}_r$

i.e., by taking the pair $(\bar{D}_j, \bar{d}_r)$ in place of $G_j; g_r$

we have, $\frac{1}{n}\sum_{j=0}^{n-1} D_j \cdot \bar{D}_j = \sum_{r=0}^{n-1} \bar{g}_r\, dr$

$$=\frac{1}{n}\sum_{j=1}^{n-1}\left|D_j\right|^2=\sum_{r=0}^{n-1}\left|d_r\right|^2$$

Hence the theorem.

11. Verify the equation $d_r=\frac{1}{N}\sum_{j=0}^{n-r} D_j W^{-jr}$, $(r=0,1,2,\cdots,n-1)$

Using the equation: $D_j=\sum_{r=0}^{n-1} d_r W^{jr}$

where $W=e^{\frac{2\pi i}{n}}$ is the nth root of unity.

Solution: We can write the given equation as:

$$D_j=\sum_{l=0} d_l W^{jl} \qquad [1]$$

$(j=0,1,2,\cdots,n-1)$

Substituting the value of D_j in the RHS of the equation:

$$d_r=\frac{1}{n}\sum_{j=0}^{n-1} D_j W^{-jr}$$

$$=\frac{1}{n}\sum_{j=0}^{n-1}\left(\sum_{l=0}^{n-1} d_l W^{jl}\right)W^{-jr}$$

$$=\frac{1}{n}\sum_{l=0}^{n-1}\sum_{j=0}^{n-1} d_l W^{jl}\,W^{-jr}$$

$$=\frac{1}{n}\left(\sum_{l=0}^{n-1} d_l \sum_{j=0}^{n-1} W^{jl}\cdot W^{-jr}\right)$$

$$=\frac{1}{n}\cdot\left[d_r\cdot n\right]=d_l$$

using orthogonality relationship

Note: D_j and d_r form a transform pair, i.e., $\{D_j\}$ is the DFT of $\{d_r\}$. Also D_j and d_r are periodic with period *n*. So we may consider D_j and d_r defined for all *j* and *r* with

$$D_{j+r}N=D_j \qquad \text{and} \qquad d_{l+r}N=d_l$$

12. Find the DFT of the sequence $\{d_r\} = \{1, 0, -1\}$

Solution: Given, $n = 3$; so by the definition,

$$D_1 = \sum_{r=0}^{2} d_r W^{ir}$$

where $W = e^{\frac{2\pi i}{3}}$

$$= 1 + 0 + (-1)W^2$$

$$= 1 - W^2$$

$$D_2 = \sum_{r=0}^{2} d_r W^{2r}$$

$$= 1 + 0 + (-1)W^4$$

$$= 1 - W^3 \cdot W = 1 - W$$

$$D_3 = \sum_{r=0}^{2} d_r W^{3r}$$

$$= 1 + 0 - W^6 \quad = 1 - \left(W^3\right)^2$$

$$= 1 - 1 = 0$$

So DFT is $= (1 - W^2, 1 - W, 0)$, where W is cube root of unity.

Discrete Fourier cosine transform

Let $\{d_r\}$ be a sequence of real numbers $(k = 0, 1, 2, \cdots, n-1)$, then we have

$$D_j = \sum_{r=0}^{n-1} d_r \cos\frac{2\pi jr}{n} \qquad [1]$$

$(j = 0, 1, 2, \cdots n-1)$

is called DFT of $\{d_r\}$. and cosine inverse discrete transform (CIDFT) of $\{D_j\}$

$(j = 0, 1, 2, \cdots n-1)$ by

$$d_r = \frac{1}{n} \sum_{j=0}^{n-1} D_j \cos\frac{2\pi jr}{n} \qquad [2]$$

we can compute CDFT of a given sequence $\{d_r\}$ of real number by [1], and discrete inverse cosine transform of (DICT) $\{D_j\}$ by Equation [2].

13. Compute the N-point DFT for $\{d_n\} = a^n$ for $0 < a < 1$.

Solution: We know that N-point DFT is expressed

as $$D_k = \sum_{n=0}^{N-1} d_n e^{-j\left(\frac{2\pi}{N}\right)nk}, \quad k = 0, 1, \cdots, N-1$$

or $$D_k = \sum_{n=0}^{N-1} a^n e^{-j\left(\frac{2\pi}{N}\right)nk}$$

or $$D_k = \sum_{n=0}^{N-1} \left(a e^{-j\frac{2\pi k}{N}} \right)^n$$

or $$D_k = \frac{1 - \left(a e^{-j\frac{2\pi k}{N}} \right)^n}{1 - a e^{-j\frac{2\pi k}{N}}}$$

or $$D_k = \frac{1 - a^N}{1 - a e^{-j\frac{2\pi k}{N}}}, \qquad k = 0, 1, \cdots, N-1.$$

Note: (i) Sum of the geometric progression $a + ar + ar^2 + \cdots$ to n terms $= \dfrac{a(1-r^n)}{1-r}$ or $\dfrac{a(r^n - 1)}{r-1}$ when $r \neq 1$; if $r = 1$, the formula fails, However when $r = 1$, Sum $= a + a + a + \cdots$ to n terms $= na = ax \times$ numbers n terms. Sum of the Nth power of the n, n^{th} roots of unity is $S_n = (1)^N + (\alpha)^N + (\alpha^2)^N + \cdots$ to n terms which is a geometric progression with common ratio $r = \alpha^N$, If N is a multiple of n, let $N = kn$.

Then the common ratio $\alpha^N = \left(\text{cis} \dfrac{2\pi}{n} \right)^{kn} = cis\, 2k\pi = 1$

$\therefore \quad S_n = 1 + 1 + 1 + \cdots$ to n terms $= n$.

It N is not a multiple of n, common ratio $\alpha^N \neq 1$

$$\therefore \quad S_n = \frac{1\cdot\left[1-\left(\alpha^N\right)^n\right]}{1-\alpha^N} = \frac{1}{1-\alpha^N}\left[1-\left((\alpha)^n\right)^N\right]$$

$$= \left[1-\frac{1}{1-\alpha^N}\left\{\left(cis\frac{2\pi}{n}\right)^n\right\}^N\right] = \frac{1}{1-\alpha^N}\left[1-(cis\,2\pi)^N\right]$$

$$= \frac{1}{1-\alpha^N}[1-1] = 0$$

(ii) $$1+\alpha+\alpha^2+\cdots+\alpha^{n-1} = \frac{1\cdot(1-\alpha^n)}{1-\alpha} = \frac{1-\left(cis\frac{2\pi}{n}\right)}{1-\alpha}$$

$$= \frac{1}{1-\alpha}[1-cis\,2\pi] = \frac{1}{1-\alpha}[1-1] = 0.$$

14. Find the N-point DFT of the sequence (infinite): $d_n = a^n u(n),$ for all n.

Solution: DFT of $\{d_n\}$ can be expressed as:

$$D_k = \sum_{n=0}^{\infty} a^n \cdot e^{-j\left(\frac{2\pi}{N}\right)nk}, \quad k = 0,1,2,\cdots\infty$$

$$= \sum_{n=0}^{\infty}\left(a e^{-j\left(\frac{2\pi}{N}\right)k}\right)^n$$

$$= 1+\left(a e^{-j\left(\frac{2\pi}{N}\right)k}\right)+\left(a e^{-j\left(\frac{2\pi}{N}\right)k}\right)^2+\cdots$$

$$= \frac{1}{1-a e^{-j\left(\frac{2\pi}{N}\right)k}}, |a|<1$$

$$D_k = \left(1-e^{-j\left(\frac{2\pi k}{n}\right)}\right)^{-1}, \quad 0\le k\le N-1$$

is the DFT sequence

15. Find the DFT of the sequence $d_n = an, 0 \le n \le N-1$

Solution: The DFT of the given sequence can be expressed as:

$$D_k = \sum_{n=0}^{N-1} d_n e^{-j\left(\frac{2\pi}{N}\right)nk}$$

$$= a\sum_{n=0}^{N-1} nW_N^{nk}, \quad 0 \le k \le N-1 \qquad [1]$$

Let $\quad S = \sum_{n=0}^{N-1} b^n = \dfrac{b^N - 1}{b-1}, b \ne 1 \qquad [2]$

Differentiating w.r.t. b, we get

$$\sum_{n=0}^{N-1} nb^{n-1} = \frac{(b-1)Nb^{N-1} - (b^N - 1)\times 1}{(b-1)^2}$$

$$\Rightarrow \quad \sum_{n=0}^{N-1} nb^n = \frac{b(Nb^N - Nb^{N-1} - b^N + 1)}{(b-1)^2}$$

$$= \frac{b(b^N - Nb^{N-1} - b^N + 1)}{(b-1)^2}$$

Let $b = W_N^k$ in the above expression, we get

$$\sum_{n=0}^{N-1} nW_N^{kn} = \frac{W_N^k\left[W_N^{kN}(N-1) - NW_N^{k(N-1)} + 1\right]}{\left[W_N^k - 1\right]^2}W$$

$$= \frac{W_N^k\left[N - 1 - NW_N^{-k} + 1\right]}{\left[W_N^k - 1\right]^2}$$

$$\left(\because \; W_N^{kN} = \left(e^{-j\left(\frac{2\pi}{N}\right)kN}\right) = e^{-j2\pi k} = 1\right)$$

$$= \frac{N\left[W_N^k - 1\right]}{\left[W_N^k - 1\right]^2} = \frac{N}{\left[W_N^k - 1\right]}, \quad k \ne 0.$$

$$\therefore \quad D_0 = a\sum_{n=0}^{N-1} n = \frac{aN(N-1)}{2}$$

Hence, $$D_k = \begin{cases} \dfrac{aN(N-1)}{2}, & k=0 \\ \dfrac{aN}{W_N^k - 1}, & k \neq 0 \end{cases}$$

16. For $N=5$, complete the DFT of $\{d_n\} = (1,1,1,0,0)$ and compare the result with the DFT of $\{d_n\} = (1,1,1)$ for $N=3$

Solution: Given, $d_n = (1,1,1,0,0)$ [1]

$N = 5$

We know that the DFT of $\{d_n\}$ can be expressed as:

$$D_k = \sum_{n=0}^{N-1} d_n \cdot W_5^{kn}, \; 0 \le k \le 4$$

where $W = e^{-j\left(\frac{2\pi}{5}\right)}$

$$W_5^0 = e^{-j\left(\frac{2\pi}{5}\right)^0} = e^{-0} = 1$$

$$W_5^1 = e^{-j\left(\frac{2\pi}{5}\right)^1} = \cos\left(\frac{2\pi}{5}\right) - j\sin\left(\frac{2\pi}{5}\right) = 0.309 - j0.951$$

$$W_5^2 = e^{-j\left(\frac{2\pi}{5}\right)^2} = \cos\left(\frac{4\pi}{5}\right) - j\sin\left(\frac{4\pi}{5}\right) = -0.809 - j0.587$$

$$W_5^3 = e^{-j\left(\frac{2\pi}{5}\right)^3} = \cos\left(\frac{6\pi}{5}\right) + j\sin\left(\frac{6\pi}{5}\right) = -0.809 - j0.587$$

$$W_5^4 = e^{-j\left(\frac{2\pi}{5}\right)^4} = \cos\left(\frac{8\pi}{5}\right) + j\sin\left(\frac{8\pi}{5}\right) = 0.309 + j0.951$$

Hence, $$D_k = \text{DFT}\{d_n\} = \sum_{n=0}^{4} d_n W_5^{kn}, \; 0 \le k \le 4$$

$$= 1 + W_5^k + W_5^{2k}$$

$$\left[\begin{array}{l} \Rightarrow \quad D_0 = 1+1+1=3 \\ D_1 = 1+W_5^1+W_5^2 = 0.5-\mathrm{j}\,1.538 \\ D_2 = 1+W_5^2+W_5^4 = 0.5-\mathrm{j}\,0.364 \\ D_3 = 1+W_5^3+W_5^1 = 0.5-\mathrm{j}\,0.364 \\ D_4 = 1+W_5^4+W_5^3 = 0.5-\mathrm{j}\,1.538 \end{array}\right]$$

Let us denote the sequence (1,1,1)

by $\{d'_n\} = (1,1,1)$ [2]

Here, given that $N = 3$

Now, $W_N = e^{-\mathrm{j}\left(\frac{2\pi}{N}\right)}$

$$W_3 = e^{-\mathrm{j}\left(\frac{2\pi}{3}\right)}$$

$$\Rightarrow \quad W_3^0 = 1$$

$$W_3^1 = -0.5-\mathrm{j}0.866$$

$$W_3^2 = -0.5-\mathrm{j}0.866$$

Hence, the corresponding DFT of $\{d'_n\}$ is denoted by D'_k.

$$\therefore \quad D'_k = DFT\{d'_n\}$$

$$= \sum_{n=0}^{2} d'_n W_N^{kn}, \quad 0 \le k \le 2$$

$$= W_3^{0k} + W_3^{k} + W_3^{2k}$$

$$= 1 + W_3^{k} + W_3^{2k}$$

$$\Rightarrow \quad D'_0 = 1+1+1 = 3$$

$$D'_1 = 1+W_3^1+W_3^2 = 1-0.5-\mathrm{j}0.866+(-0.5+\mathrm{j}0.866)$$

$$= 1-1 = 0$$

$$D'_2 = 1+W_3^2+W_3^4 = 0$$

$$\Rightarrow \quad D_k \ne D'_k$$

17. Find the DFT of the following periodic signals:

$$d_{nT} = \begin{cases} 1, & n = 3,7 \\ 0, & n = 1,2,4,5,6,8,9 \end{cases}$$

***Solution*:** $D_k = \sum_{n=0}^{N-1} d_{nT}\, e^{-j\left(\frac{2\pi}{N}\right)nk}$

$$= \sum_{n=0}^{N-1} 1 \cdot W^{-nk}; \qquad \left(W = e^{+j\left(\frac{2\pi}{N}\right)} \right)$$

$$= \sum_{n=3,7} W^{-nk}$$

$$= W^{-3k} + W^{-7k}$$

$$= e^{-j\left(\frac{3\pi k}{5}\right)} + e^{-j\left(\frac{7\pi k}{5}\right)}$$

$$= e^{-j\pi k}\left[e^{j\left(\frac{2\pi k}{5}\right)} - e^{-j\left(\frac{2\pi k}{5}\right)} \right]$$

$$D_k = 2e^{-j\pi k}\cos\left(\frac{2\pi k}{5}\right)$$

18. Compute the 4-point DFT of the sequence $d_n = \cos\dfrac{n\pi}{4}$

***Solution*:** Here, given that $N = 4$.
Therefore, we have

$$d_n = \left[\cos(0),\ \cos\left(\frac{\pi}{4}\right),\ \cos\left(\frac{\pi}{2}\right),\ \cos\left(\frac{3\pi}{4}\right) \right]$$

or $\quad d_n = [1,\ 0.707,\ 0,\ -0.707]$

We know that *N*-point DFT of the sequence is expressed as:

$$D_k = \sum_{n=0}^{N-1} d_n\, e^{-j\left(\frac{2\pi}{N}\right)nk}, \quad k = 0,1,\cdots,N-1$$

the DFT is $D_k = \sum_{n=0}^{3} d_n \cdot e^{-j\left(\frac{2\pi}{4}\right)nk}$, $k = 0,1,2,3$

$$= \sum_{n=0}^{3} d_n \cdot e^{-j\left(\frac{\pi}{2}\right)nk}$$

$$D_k = \sum_{n=0}^{3} d_n \cdot e^{-j\frac{\pi nk}{2}}, \quad k = 0,1,2,3$$

$$\therefore \quad D_0 = \sum_{n=0}^{3} d_n = d_0 + d_1 + d_2 + d_3 = 1 + 0.707 + 0 - 0.707 = 1$$

$$D_1 = \sum_{n=0}^{3} d_n e^{-j\left(\frac{n\pi}{2}\right)}$$

$$= 1 + (0.707)e^{-j\frac{\pi}{2}} + (0)(-0.707)e^{-3j\frac{\pi}{2}}$$

$$= 1 + (0.707)(-j) + 0 - (0.707)(j)$$

$$\therefore \quad D_1 = 1 - j(1.414)$$

$$D_2 = \sum_{n=0}^{N-1} d_n e^{-j\pi n}$$

$$D_2 = \sum_{n=0}^{3} d_n e^{-j\pi n}$$

$$= 1 + (0.707)e^{-j\pi} + 0 + (-0.707)e^{-j3\pi}$$

$$D_2 = 1 + (0.707)(-1) + 0 + (-0.707)(-1) = 1$$

$$D_3 = \sum_{n=0}^{3} d_n e^{-j\pi(3)\frac{n}{2}}$$

$$= 1 + (0.707)e^{-\frac{3\pi}{2}} + 0 + (-0.707)e^{-j\frac{9\pi}{2}}$$

$$= 1 + (0.707)(j) + (-0.707)(-j)$$

$$= 1 + 1.414j$$

Hence $D_k = [1,\ 1 - j1.414,\ 1,\ 1 + j1.414]$.

19. Compute the DFT of the sample data sequence: $d_n = \{1,1,2,2,3,3\}$

Solution: We know that the N-point DFT of a finite duration sequence d_n is defined as

$$D_k = \sum_{n=0}^{N-1} d_n e^{-j\left(\frac{2\pi}{N}\right)nk} \qquad \text{Here } N = 6 \text{ and } k = 0, 1, \cdots N-1$$

$$\therefore \quad D_0 = \sum_{n=0}^{5} d_n e^{-j\left(\frac{2\pi}{6}\right)(0)n}$$

$$= \sum_{n=0}^{5} d_n = 1+1+2+2+3+3 = 12$$

$$D_1 = \sum_{n=0}^{5} d_n e^{-j\left(\frac{2\pi}{6}\right)n(1)}$$

$$= \sum_{n=0}^{5} d_n e^{-j\left(\frac{\pi}{3}\right)n}$$

$$= 1 + e^{-j\frac{\pi}{3}} + 2e^{-j\frac{2\pi}{3}} + 2e^{-j\pi} + 3e^{-j\frac{4\pi}{3}} + 3e^{-j\frac{5\pi}{3}}$$

$$= 1 + 0.5 - j0.866 + 2(-0.5 - j0.866) + 2(-1) + 3(-0.5 + j0.866) + 3(0.5 + j0.866)$$

$$= -1.5 + j2.598$$

$$D_2 = \sum_{n=0}^{5} d_n e^{\frac{-j2\pi(2)n}{6}}$$

$$= \sum_{n=0}^{5} d_n e^{\frac{-2j\pi n}{3}}$$

$$= 1 + e^{\frac{-j2\pi}{3}} + 2e^{\frac{-j4\pi}{3}} + 2e^{-j2\pi} + 3e^{\frac{-j8\pi}{3}} + 3e^{\frac{-j10\pi}{3}}$$

$$= 1 + (-0.5) - j0.866 + 2(-0.5 + j0.866) + 2(1) + 3(-0.5 - j0.866) + 3(-0.5 + j0.866)$$

$$= -1.5 + j0.866$$

$$D_3 = \sum_{n=0}^{5} d_n e^{\frac{-j2\pi(3)n}{6}}$$

$$= \sum_{n=0}^{5} d_n e^{-j\pi n}$$

$$= 1 + e^{-j\pi} + 2e^{-j2\pi} + 2e^{-j3\pi} + 3e^{-j4\pi} + 3e^{-j5\pi}$$

$$= 1 - 1 + 2(1) + 2(-1) + 3(1) + 3(-1) = 0$$

$$D_4 = \sum_{n=0}^{5} d_n e^{\frac{-j2\pi(4)n}{6}}$$

$$= \sum_{n=0}^{5} d_n e^{\frac{-j4\pi n}{3}}$$

$$= 1 + e^{\frac{-j4\pi}{3}} + 2e^{\frac{-j8\pi}{3}} + 2e^{-j4\pi} + 3e^{\frac{-j16\pi}{3}} + 3e^{\frac{-j20\pi}{3}}$$

$$= 1 + (-0.5 + j0.866) + 2(-0.5 - j0.866) + 2(1) + 3(-0.5 + j0.866) + 3(-0.5 - j0.866)$$

$$= -1.5 - j0.866$$

$$D_5 = \sum_{n=0}^{5} d_n e^{\frac{-j2\pi(5)n}{6}}$$

$$= \sum_{n=0}^{5} d_n e^{\frac{-j5\pi n}{3}}$$

$$= 1 + e^{\frac{-j5\pi}{3}} + 2e^{\frac{-j10\pi}{3}} + 2e^{-j5\pi} + 3e^{\frac{-j20\pi}{3}} + 3e^{\frac{-j25\pi}{3}}$$

$$= 1 + (-0.5 + j0.866) + 2(-0.5 + j0.866) + (-1) + 3(-0.5 - j0.866) + 3(0.5 - j0.866)$$

$$= -1.5 - j2.598$$

Hence $D_k = [12, -1.5 + j2.598, -1.5 + j0.866, 0, -1.5 - j0.866, -15 - j2.598]$

(1) To find amplitude spectrum of D_k:

$$|D_k| = \left[\sqrt{12 \times 12}, \sqrt{(-1.5)^2 + (-2.598)^2}, \sqrt{(-1.5)^2 + (0.866)^2}, 0, \sqrt{(-1.5)^2 + (0.866)^2}, \sqrt{(-1.5)^2 + (-2.598)^2}\right]$$

$$|D_k| = [12, 2.999, 1.732, 0, 1.732, 2.999]$$

(2) To find phase spectrum of D_k:

$$\hat{D}_k = \left[\tan^{-1}(0), \tan^{-1}\left(\frac{2.598}{-1.5}\right), \tan^{-1}\left(\frac{0.866}{-1.5}\right), \tan^{-1}(0), \tan^{-1}\left(\frac{-0.866}{-1.5}\right), \tan^{-1}\left(\frac{-2.598}{-1.5}\right)\right]$$

or $$\hat{D}_k = \left[0, -\frac{\pi}{3}, -\frac{\pi}{6}, 0, \frac{\pi}{6}, \frac{\pi}{3}\right]$$

20. Complete the N-point of the sequence $d_n = \cos(nW_0)$, $W_0 = \frac{2\pi}{N}k_0$, $0 \le n \le N-1$

Solution: We know that

$$\boxed{\cos\theta = \frac{1}{2}\left(e^{j\theta} + e^{-j\theta}\right)} \qquad [1]$$

Here, given that $d_n = \cos(nW_0)$

$$\Rightarrow \quad d_n = \frac{1}{2}e^{+jnW_0} + \frac{1}{2}e^{-jnW_0} \qquad \text{(using [1])}$$

$$= \frac{1}{2}e^{jn\left(\frac{2\pi}{N}\right)k_0} + \frac{1}{2}e^{-jn\left(\frac{2\pi}{N}\right)k_0}$$

$$= \frac{1}{2}e^{-j\left(\frac{2\pi}{N}\right)(-k_0 n)} + \frac{1}{2}e^{-j\left(\frac{2\pi}{N}\right)(k_0 n)}$$

$$= \frac{1}{2}W_N^{-k_0 n} + \frac{1}{2}W_N^{k_0 n}$$

Hence, $D_k = DFT\{d_n\}$, $0 \le k \le N-1$

$$= \frac{1}{2}\sum_{n=0}^{N-1} W_N^{k-k_0 n} + \frac{1}{2}\sum_{n=0}^{N-1} W_N^{(k+k_0)n}$$

$$= \frac{1}{2}S_1 + \frac{1}{2}S_2$$

To find S_1 and S_2:

$$\boxed{S_1 \sum_{n=0}^{N-1} W_N^{(k-k_0)n}} = \frac{W_N^{(k-k_0)N} - 1}{W_N^{(k-k_0)} - 1} = \frac{W_N^{kn}\, W_N^{-k_0 N} - 1}{W_N^{(k-k_0)} - 1}$$

$$= \frac{1\times 1-1}{W_N^{(k-k_0)}-1} = 0, \qquad k \neq k_0$$

When $k = k_0$, we get

$$S_1 = \sum_{n=0}^{N-1} (1)^n = N$$

Hence, $S_1 = \sum_{n=0}^{N-1} W_N^{(k-k_0)n}$

$$= \begin{cases} 0, & k \neq k_0 \\ N, & k = k_0 \end{cases}$$

or, $\quad S_1 = N\delta(k-k_0)$

Similarly, $S_2 = N\delta(k-k_0)$

$$= N\delta[k-(N-k_0)]$$

thus, $\quad D_k = \frac{1}{2}S_1 + \frac{1}{2}S_2$

21. Find the N-point DFT of the following sequences:

(i) $d_n = \delta(n)$ (ii) $d_{n-n_0} = \delta(n-n_0)$

Solution: (i) $D_k = \text{DFT}\{d_n\}$

$$= \sum_{n=0}^{N-1} d_n W_N^{kn}$$

$$= \sum_{n=0}^{N-1} \delta(n) W_N^{kn}, \quad 0 \le k \le N-1$$

Using shifting property, we get

$$D_k = W_N^{kn}\Big]_{n=0} = 1$$

(ii) $\quad D_k = \text{DFT}\{d_{n-n_0}\}$

$$= \sum_{n=0}^{N-1} \delta(n-n_0)\, N_N^{kn}$$

Using shifting property, we get

$$D_k = W_N^{kn}\Big|_{n=n_0}$$

$$= W_N^{kn_0}, \quad 0 \le k \le N-1$$

$$= e^{-j\left(\frac{2\pi}{N}\right)kn_0}$$

22. Find the N-point DFT of the sequences: $d_n = e^{jwmn}, \quad 0 \le n \le N-1$

Solution: $D_k = \text{DFT}\{d_n\}$

$$= \sum_{n=0}^{N-1} d_n \cdot e^{-j\left(\frac{2\pi}{N}\right)nk}$$

$$= \sum_{n=0}^{N-1} e^{jwnn - j\left(\frac{2\pi}{N}\right)nk}$$

$$= \sum_{n=0}^{N-1} W_N^{-mn} W_N^{kn}$$

$$= \sum_{n=0}^{N-1} W_N^{(k-m)n}$$

$$= \frac{W_N^{(k-m)N} - 1}{W_N^{k-m} - 1}, \quad k \ne m \qquad \left(\because \sum_{n=0}^{N-1} r^n = \frac{r^N - 1}{r - 1}, \ r \ne 1\right)$$

Hence, $D_k = \dfrac{W_N^{kN} \cdot W_N^{-mN} - 1}{W_N^{k-m} - 1}$

$$= \frac{1 \times 1 - 1}{W_N^{k-m} - 1} = 0, \qquad k \ne m.$$

When $k = m$,

$$D_m = \sum_{n=0}^{N-1} 1 = N$$

Hence, $D_k = \begin{cases} 0, & k \ne m \\ N, & k = m \end{cases}$ or $D_k = N\delta(k-m), \qquad (0 \le m \le N-1)$

4.2 Fast Fourier Transform and its properties

Fast Fourier Transform algorithm

The fast Fourier Transform (FFT) refers to algorithms that complete the (DFT) in a computationally efficient manner. This approach depends on the decomposition of an *N*-point DFT into successively smaller size DFTs. When *N* is factorised as $N = r_1 r_2 r_3 \cdots r_L$ where $r_1 = r_2 = r_3 = \cdots r_L = r$, then $N = r^L$. Therefore, the DFT will be of size '*r*' is called the radius of the FFT algorithm.

Here, we develop the FFT algorithm for *N* when it is product of two integer i.e., when $N = r_1 r_2$

By definition,

$$D_j = \sum_{k=0}^{N-1} d_k W^{jk}, \quad j = 0, 1, \cdots, N-1 \qquad [1]$$

Let $j = j_1 + r_1 j_2; \quad k = k_2 + r_2 k_1$

then, for $j_1, k_1 = 0$ to $r_1 - 1$ and for j_2, k_2 to $r_2 - 1$

both *j* and *k* run their required range 0 to N – 1

Now, $W^{jk} = W^{(j_1 + r_1 j_2)(k_2 + r_2 k_1)}$

$$= W^{j_1 k_2 + j_1 k_1 r_2 + r_1 j_2 k_2 + r_1 r_2 j_2 k_1}$$

$$= W^{j_1 k_2 + j_1 k_1 r_2 + r_1 j_2 k_2} \qquad \left(\text{Since} \quad W^{r_1 r_2 j_2 k_1} = W^{j_2 k_1 N} = 1\right)$$

Substituting in [1], we get

$$D_j = \sum_{k_2=0}^{r_2-1} \sum_{k_1=0}^{r_1-1} d_k W^{j_1 k_1 r_2} \cdot W^{(r_1 j_2 + j_1) k_2}$$

This can be split into the following form:

$$F_1(j_1 k_2) = \sum_{k_1=0}^{r_1-1} d_k W^{j_1 k_1 r_2} \qquad [A]$$

$$\therefore \quad D_j = F_2(j_1, j_2) = \sum_{k_2=0}^{r_2-1} F(j_1 k_2) W^{(r_1 j_2 + j_1) k_2} \qquad [B]$$

[A] and [B] are the two step algorithm for the FFT.

Decimation in Frequency radius-2 FFT (or DIT-FFT algorithm)

In this case, the given sequence $\{d_n\}$ is decimated (broken) into two sequence each of length $\left(\frac{N}{2}\right)$ samples, i.e., the first sequence $\{x_1(n)\}$ consists of the first $\left(\frac{N}{2}\right)$ point of $\{d_n\}$, whereas the second sequence $\{x_2(n)\}$ consists of the last $\left(\frac{N}{2}\right)$ point of $\{d_n\}$.

Thus

$$\left.\begin{aligned} x_1(n) &= d_n, \qquad n = 0, 1, \cdots, \left(\frac{N}{2} - 1\right) \\ x_2(n) &= d_{n+\frac{N}{2}}, \quad n = 0, 1, \cdots, \left(\frac{N}{2} - 1\right) \end{aligned}\right] \qquad [1]$$

The N-point DFT of $\{d_n\}$ can be expressed as:

$$D_k = \sum_{n=0}^{\left(\frac{N}{2}-1\right)} d_n W_N^{nk} + \sum_{\left(n=\frac{N}{2}\right)}^{N-1} d_n W_N^{nk} \qquad [2]$$

$$= \sum_{n=0}^{\left(\frac{N}{2}-1\right)} x_1(n) W_N^{nk} + \sum_{n=0}^{\left(\frac{N}{2}-1\right)} x_2(n) W_N^{\left(n+\frac{N}{2}\right)k} \qquad [3]$$

$$D_k = \sum_{n=0}^{\left(\frac{N}{2}-1\right)} \left[x_1(n) + e^{-j\pi k} \cdot x_2(n)\right] W_N^{nk} \qquad [4]$$

$$\left(\text{where } W^{\frac{Nk}{2}} = e^{-j\pi k}\right) \qquad [5]$$

If we now consider the even and odd samples of the DFT separately, we have

$$D_{2k} = \sum_{n=0}^{\left(\frac{N}{2}-1\right)} \left[x_1(n) + x_2(n)\right] \left(W_N^2\right)^{nk}$$

$$\boxed{D_{2k} = \sum_{n=0}^{\left(\frac{N}{2}-1\right)} \left[x_1(n) + x_2(n)\right] \left(W_{\frac{N}{2}}^{nk}\right)} \qquad [A]$$

$$D_{2k+1} = \sum_{n=0}^{\left(\frac{N}{2}-1\right)} [x_1(n) - x_2(n)] W_N^{n(2k+1)}$$

$$\boxed{D_{2k+1} = \sum_{n=0}^{\left(\frac{N}{2}-1\right)} \left\{[x_1(n) - x_2(n)] W_N^{n}\right\} W_{\frac{N}{2}}^{nk}} \qquad [B]$$

'Even' and 'odd' valued samples of the DFT are given by the equation [A] and [B] respectively.

Now, let us denote the sequences $f(n)$ and $g(n)$ as:

$$f(n) = x_1(n) + x_2(n) \qquad n = 0,1,\cdots,\left(\frac{N}{2}-1\right)$$

$$g(n) = [x_1(n) - x_2(n)] W_N^{n}, \qquad n = 0,1,\cdots,\left(\frac{N}{2}-1\right)$$

Hence, the problem of obtaining the N-point DFT has been reduced to one of obtaining two $\left(\frac{N}{2}\right)$ point DFT's.

FFT for n when n is product of there integers: $n = r_1 \times r_2 \times r_3$

Let $\quad n = r_1 \times r_2 \times r_3$

Using definition of F.D.T of $\{d_k\}$, $\quad k = 0,1,2,\cdots n-1,$

we have $\{D_j\}$, $\quad j = 0,1,2,\cdots n-1,$ such that

$$D_j = \sum_{k=0}^{N-1} d_k\, w^{jk}, \; j = 0,1,2,\cdots n-1 \qquad [1]$$

Let $\qquad \mathrm{j} = \mathrm{j}_1 + r_1\mathrm{j}_2 + r_1r_2\mathrm{j}_3$

and $\qquad k = k_3 + r_3k_2 + r_3r_2k_1$

Then for $\mathrm{j}_1, k_1 = 0$ to $r_1 - 1$, for $\mathrm{j}_2, k_2 = 0$ to $r_2 - 1$

and $\qquad \mathrm{j}_3,\; k_3 = 0$ to $r_1 - 1$,

both j and k run their required ranges 0 to $n - 1$

Now putting in [1] we get

$$D_j = \sum_{k=0}^{N-1} d_k\, w^{(j_1 + r_1 j_2 + r_1 r_2 j_3)(k_3 + r_3 k_2 + r_2 r_3 k_1)}$$

$$D_j = \sum_{k=0}^{N-1} d_k\, w^{(j_1 + r_1 j_2 + r_1 r_2 j_3) k_3 + (j_1 r_3 + j_2 r_1 r_3) k_2} \times w^{j_1 r_2 r_3 k_1}$$

The term of the form $w^{r_1 r_2 r_3 k_1}$ becomes 1.

$$= \sum_{k_3=0}^{r_3-1} \left[\sum_{k_2=0}^{r_2-1} \left(\sum_{k_1=0}^{r_1-1} d_k\, w^{j_1 r_2 r_3 k_1} \right) w^{(j_1 r_3 + j_2 r_1 r_3) k_2} \times w^{(j_1 + r_1 j_2 + r_1 r_2 j_3) k_3} \right]$$

with k_1 appearing only in the inner sum and k_2 appearing in middle. Thus, the three steps of the FFT algorithm for $n = r_1 \times r_2 \times r_3$ is given by the equation [2], [3] and [4] as shown below.

$$F_1(j_1, k_2, k_3) = \sum_{k_1=0}^{r_1-1} d_k\, w^{j_1 r_2 r_3 k_1} \qquad [2]$$

$$F_2(j_1, j_2, k_3) = \sum_{k_2=0}^{r_2-1} F_1(j_1, k_2, k_3)\, w^{(j_1 r_3 + j_2 r_1 r_3) k_2} \qquad [3]$$

$$d_k = F_2(j_1, j_2, j_3) = \sum_{k_3=0}^{r_3-1} F_1(j_1, j_2, k_3)\, w^{(j_1 + r_1 j_2 + r_1 r_2 j_3) k_3} \qquad [4]$$

(B) EXAMPLES

1. Find the DIFT of the DFT given by $\{D_j\} = \{0, 2\sqrt{3}i, 0, 0, 0, -2\sqrt{3}i\}$ $(j = 0, 1, 2, 3, 4, 5)$

Solution: Given, $n = 6$; $\{D_j\} = \{0, 2\sqrt{3}i, 0, 0, 0, -2\sqrt{3}i\}$

$\therefore$ Using $d_k = \dfrac{1}{6} \sum_{j=0}^{5} D_j\, w^{-jk}$, $k = 0, 1, 2, \cdots 5.$ [1]

$$d_0 = \frac{1}{6} \sum_{j=0}^{5} D_j\, w^0 = 0$$

$$d_1 = \frac{1}{6} \sum_{j=0}^{5} D_j\, w^{-j}$$

$$= \frac{1}{6} \cdot \left[2\sqrt{3}i \frac{1}{w} - \frac{2\sqrt{3}i}{w^5} \right]$$

$$= \frac{2\sqrt{3}i}{6}\left[w^{-1} - w^{-5}\right] = \frac{2\sqrt{3}i \times -2\sqrt{3}i}{6} = 1$$

$$d_2 = \frac{1}{6}\sum_{k=0}^{5} D_j\, w^{-2j} = \frac{2\sqrt{3}i}{6}\left[w^{-2} - w^{-10}\right] = 1$$

$$d_3 = \frac{1}{6}\sum_{j=0}^{5} D_j\, w^{-3j}$$

$$= \frac{2\sqrt{3}i}{6}\left[w^{-3} - w^{-15}\right]$$

$$= \frac{2\sqrt{3}i}{6}\left[w^{-3} - \left(w^{-16}\right)^2 w^{-3}\right]$$

$$= \frac{2\sqrt{3}i}{6}\left[w^{-3} - w^{-3}\right] = 0$$

$$d_4 = \frac{1}{6}\sum_{j=0}^{5} D_j\, w^{-4j} = \frac{2\sqrt{3}i}{6}\left[w^{-4} - w^{-20}\right]$$

$$= \frac{2\sqrt{3} \times \sqrt{3}i^2}{6} = -1$$

$$d_5 = \frac{1}{6}\sum_{j=0}^{5} D_j\, w^{-5j}$$

$$= \frac{2\sqrt{3}i}{6}\left[w^{-5} - w^{-25}\right]$$

$$= \frac{2\sqrt{3}}{6} j\left[w^{-5} - w^{-1}\right] = \frac{2\sqrt{3}i}{6}\sqrt{3}\, j = -1$$

$$d_6 = \frac{1}{6}\sum_{j=0}^{5} D_j\, w^{-6j} = \frac{2\sqrt{3}i}{6} \cdot \left[w^{-6} - w^{-30}\right] = 0$$

Hence, the inverse transforms are given by

$$\{d_k\} = \{0, 1, 1, 0, -1, -1, 0\}$$

Illustration:

$$\text{Given: } \begin{cases} f(n) = x_1(n) + x_2(n), & n = 0,1,\cdots,\left(\dfrac{N}{2}-1\right) \\ g(n) = \left[x_1(n) - x_2(n)\right] W_N^n, & n = 0,1,\cdots\left(\dfrac{N}{2}-1\right) \end{cases} \quad [1]$$

The procedure is illustrated in Figure 4.2 and 4.3 for $N = 8$

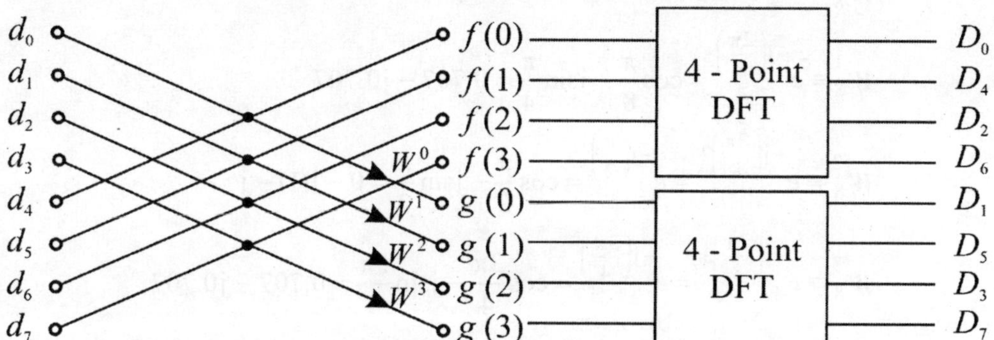

Figure 4.2 An 8-point DFT reduced to two 4-point DFTs by DIF.

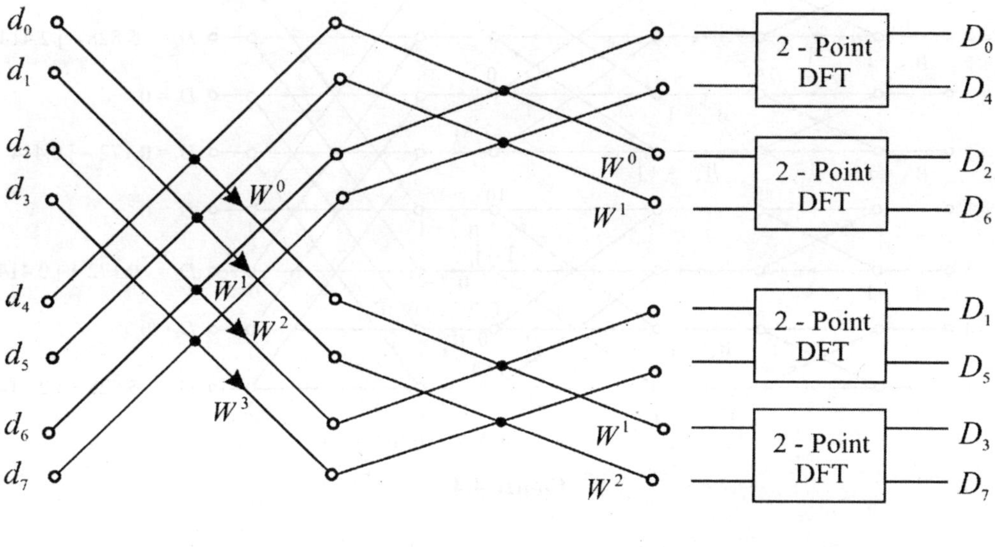

Figure 4.3 An 8-point DFT reduced to four 4-point DFTs by DIFT.

2. Find the 8-point DFT of the sequence, $d_n = (0,1,2,3,4,5,6,7)$ Using DIT-FFT algorithm.

***Solution*:** Here it is given that N=8 and the twiddle factors $W_N^r = e^{-6j\left(\frac{2\pi}{N}\right)r}$ are as follows:

$$W_8^0 \equiv 1 \equiv e^{-j\left(\frac{2\pi}{8}\right)0}$$

$$W_8^1 \equiv e^{-j\left(\frac{2\pi}{8}\right)} = \cos\frac{\pi}{8} - \text{j}\sin\frac{\pi}{4} = 0.707 - \text{j}0.707$$

$$W_8^2 \equiv e^{-j\left(\frac{2\pi}{8}\right)2} = e^{-j\left(\frac{\pi}{2}\right)} = \cos\frac{\pi}{2} - \text{j}\sin\frac{\pi}{2} = 0 - \text{j}(1) - \text{j}$$

$$W_8^3 \equiv e^{-j\left(\frac{2\pi}{8}\right)3} = e^{-j\left(\frac{3\pi}{4}\right)} = \cos\frac{3\pi}{4} - \text{j}\sin\frac{3\pi}{4} = 0.707 - \text{j}0.707$$

Using DIT-FFT algorithm, we can find D_k from the given sequence d_n as shown in Figure 4.4

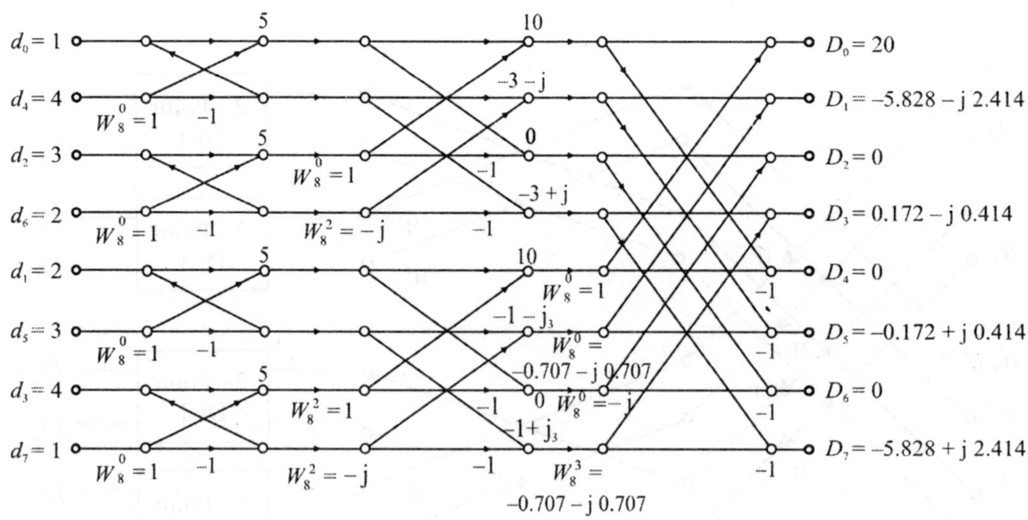

Figure 4.4

3. Find the DFT of the following sequence d_n using the DIT-FFT algorithm $d_n = (1,-1,-1,-1,1,1,1,-1)$

Solution: Here, given that $N = 8$ and the twiddle factors: $W_8^r = e^{-j\left(\frac{2\pi}{8}\right)r}$.

$$\Rightarrow \quad W_8^0 \equiv e^{-j\left(\frac{2\pi}{8}\right)0} = 1$$

$$W_8^1 \equiv e^{-j\left(\frac{2\pi}{8}\right)} = \cos\frac{\pi}{4} - j\sin\frac{\pi}{4} = \frac{1}{\sqrt{2}} - j\left(\frac{1}{\sqrt{2}}\right)$$

$$W_8^2 \equiv e^{-j\left(\frac{\pi}{2}\right)} = \cos\frac{\pi}{2} - j\sin\frac{\pi}{2} = -j$$

$$W_8^3 \equiv e^{-j\left(\frac{2\pi}{8}\right)3} = e^{-j\left(\frac{3\pi}{4}\right)} = \left[\cos\left(\frac{3\pi}{4}\right) - j\sin\left(\frac{3\pi}{4}\right)\right] = -\frac{1}{\sqrt{2}} - j\left(\frac{1}{\sqrt{2}}\right).$$

Using DIT-FFT algorithm, we can find D_k from the given sequence d_n as shown in Figure 4.5

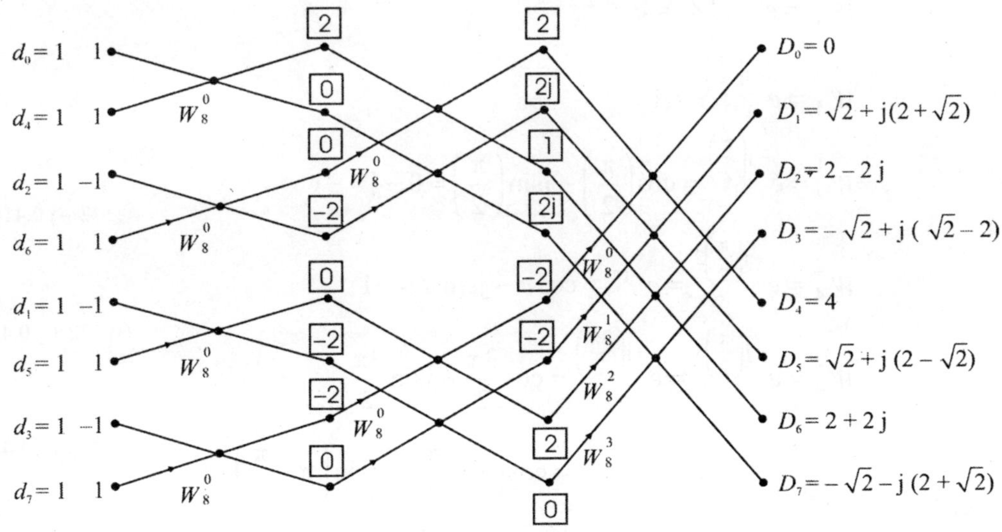

Figure 4.5

(Bit reversed)	D_k	$\lvert D_k \rvert$	Arg D_k (radians)
$d_0 = 1$	0	0	–
$d_4 = 1$	$-\sqrt{2} + j(2+\sqrt{2})$	3.6955	1.9635
$d_2 = -1$	$2-2j$	2.8284	– 0.7854
$d_6 = 1$	$\sqrt{2} + j(\sqrt{2}-2)$	1.5307	– 0.3927
$d_1 = -1$	4	4	0
$d_5 = 1$	$\sqrt{2} + j(2-\sqrt{2})$	1.5307	0.03927
$d_3 = -1$	$2 + j2$	2.8284	0.7854
$d_7 = -1$	$-\sqrt{2} - j(2+\sqrt{2})$	3.6955	– 1.9635

4. Find the DFT of the following sequence $\{d_n\}$ using the DIT- FFT algorithm $d_n = \{1, 0, 1, 0\}$

Solution: Here, given that $N = 4$ and the twiddle factors

$$W_4^r = e^{-j\left(\frac{2\pi}{4}\right)r} = e^{-j\left(\frac{\pi}{2}\right)r}$$

$$W_4^0 = e^{-j\left(\frac{\pi}{2}\right)0} = 1$$

$$W_4^1 = e^{-j\left(\frac{\pi}{2}\right)} = \cos\left(\frac{\pi}{2}\right) - j\sin\left(\frac{\pi}{2}\right) = 0 - j = -j$$

$$W_4^2 = e^{-j\left(\frac{\pi}{2}\right)2} = e^{-j\pi} = \cos\pi - j\sin\pi = -1$$

$$W_4^3 = e^{-j\left(\frac{\pi}{2}\right)3} = e^{-j\left(\frac{3\pi}{2}\right)} = \cos\frac{3\pi}{2} - j\sin\frac{3\pi}{2}$$

$$= \cos\left(\pi + \frac{\pi}{2}\right) - j\sin\left(\pi + \frac{\pi}{2}\right)$$

$$= -\cos\left(\frac{\pi}{2}\right) - j\left(-\sin\frac{\pi}{2}\right)$$

$$= -0 + j(1) = j$$

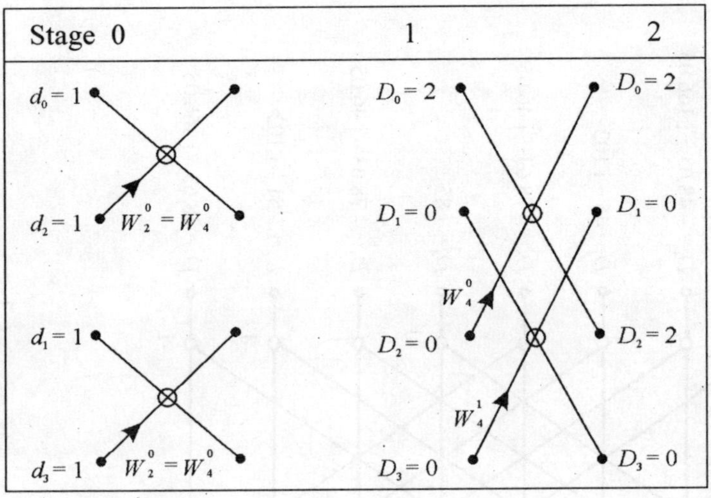

Figure 4.6 A 4-point DFT. Computation flow graph

Hence $D_k = [2, 0, 2, 0]$.

5. Given $d_n = 2^n$ and $N = 8$, find $X(k)$ using DIT-FFT algorithm.

Solution: Here, given that $N = 8$ and the twiddle factors:

$$W_4^r = e^{-j\left(\frac{2\pi}{8}\right)r} = e^{-j\left(\frac{\pi}{4}\right)r} \qquad [1]$$

Now,

$$\boxed{\begin{aligned} \{d_n\} &= 2^n \\ d_0 &= 1 \\ d_1 &= 2 \\ d_2 &= 4 \\ d_3 &= 8 \\ d_4 &= 16 \\ d_5 &= 32 \\ d_6 &= 64 \\ d_7 &= 128 \end{aligned}}$$

$$\therefore \{d_n\} = [1, 2, 4, 8, 16, 32, 64, 128]$$

Using (1), we have,

$$W_8^0 = 1; \quad W_8^1 = 0.707 - j0.707$$

$$W_8^2 = -j; \quad W_8^3 = -0.707 - j0.707$$

The procedure is illustrated in Figure 4.7 below

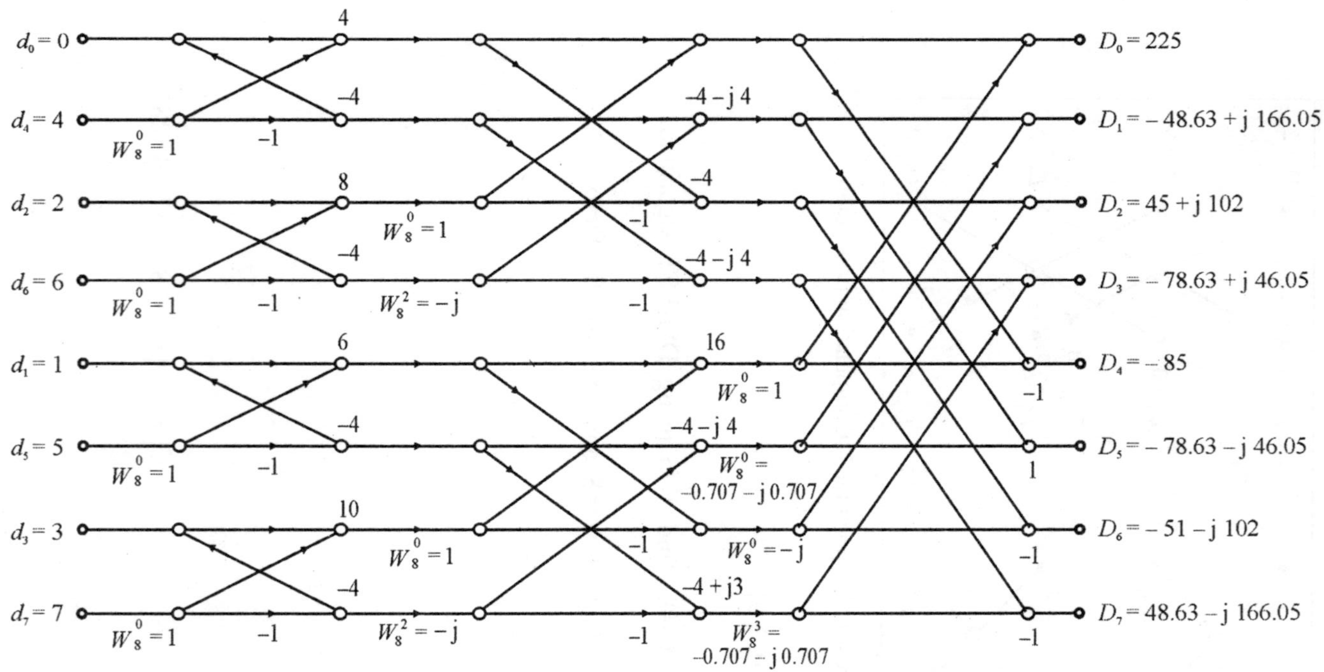

Figure 4.7

$$\text{Hence,}\quad D_k = \begin{bmatrix} 255, 48.63 + j166.05, -45 + j102, -78.63 + j46.05, -85, \\ -78.63 - j46.05, -51 - j102, 48.63 - j166.05 \end{bmatrix}$$

6. Find the 6-point DFT of the sequence, $d_n = (0,1,1,0,-1,-1,0)$ using FFT algorithm.

Solution: Here, given that $N = 6 = 2 \times 3 = r_1 \times r_2$

So $(r_1 = 2;\ r_2 = 3)$; $(\mathrm{j} + \mathrm{j}_1 + 2\mathrm{j}_2,\ k = k_2 + 3k_1)$

and also, $(d_0 = 0,\ d_1 = 1,\ d_2 = 1,\ d_3 = 0,\ d_4 = -1,\ d_5 = -1,\ d_6 = 0)$

We find the values of $F_1(\mathrm{j}_1, k_2)$ using equation (A):

$$F_1(\mathrm{j}_1, k_2) = \sum_{k_1=0}^{1} d_{k_2+3k_1} W^{3\mathrm{j}_1 k_1} \qquad [1]$$

where W is the 6th root of unity, $W = e^{\mathrm{i}\left(\frac{2\pi}{6}\right)} = e^{\mathrm{i}\left(\frac{\pi}{3}\right)}$

$$F_1(0,0) = F_1(\mathrm{j}_1 = 0, k_2 = 0) = \sum_{k_1=0}^{1} d_{3k_1} = d_0 + d_3 = 0 + 0 = 0$$

$$F_1(0,1) = F_1(\mathrm{j}_1 = 0, k_2 = 1) = \sum_{k_1=0}^{1} d_{1+3k_1} = d_1 + d_4 = 1 - 1 = 0.$$

$$F_1(0,2) = F_1(\mathrm{j}_1 = 0, k_2 = 2) = \sum_{k_1=0}^{1} d_{2+3k_1} = d_2 + d_5 = 1 - 1 = 0.$$

$$F_1(1,0) = F_1(\mathrm{j}_1 = 1, k_2 = 0) = \sum_{k_1=0}^{1} d_{3k_1} W^{3k_1} = d_0 + d_3 W^3 = d_0 - d_3 = 0 - 0 = 0$$

$$\left(\because\ W^3 = e^{\mathrm{i}(\pi)} = -1\right)$$

$$F_1(1,1) = F_1(\mathrm{j}_1 = 1, k_2 = 1) = \sum_{k_1=0}^{1} d_{1+3k_1} W^{3k_1} = d_1 + d_4 W^3 = 1 - 1(-1) = 2$$

$$F_1(1,2) = F_1(\mathrm{j}_1 = 1, k_2 = 2) = \sum_{k_1=0}^{1} d_{2+3k_1} W^{3k_1} = d_2 + d_5 W^3 = d_2 - d_5 = 1 - (-1) = 2$$

Next, we find the values $F_2(\mathrm{j}_1, \mathrm{j}_2)$ using equation (B):

$$\boxed{D_j = F_2(\mathrm{j}_1, \mathrm{j}_2) = \sum_{k_2=0}^{2} F_1(\mathrm{j}_1, k_2) W^{\mathrm{j}_1 k_2 + 2\mathrm{j}_2 k_2}} \qquad [2]$$

$$D_0 = F_2(0,0) = F_1(0,0) + F_1(0,1) + F_1(0,2) = 0 + 0 + 0 = 0$$

$$D_1 = F_2(1,0) = \sum_{k_2=0}^{2} F_1(1, k_2) W^{k_2}$$

$$= F_1(1,0) + F_1(1,1)W + F_1(1,2)W^2$$

$$= 0 + 2W + 2W^2 = 2\left(e^{\frac{\pi i}{3}} + e^{\frac{2\pi i}{3}} \right)$$

$$= 2\left[\cos\frac{\pi}{3} + i\sin\frac{\pi}{3} + \cos\frac{2\pi}{3} + i\sin\frac{2\pi}{3} \right]$$

$$= 2\left[\frac{1}{2} + i\frac{\sqrt{3}}{2} + \left(-\frac{1}{2}\right) + i\left(\frac{\sqrt{3}}{2}\right) \right]$$

$$= 2\left[2i\frac{\sqrt{3}}{2} \right] = 2i\sqrt{3}$$

$$D_2 = F_2(0,1) = \sum_{k_2=0}^{2} F_1(0,k_2)W^{2k_2}$$

$$= F_1(0,0) + F_1(0,1)W^2 + F_1(0,2)W^4$$
$$= 0 + 0 + 0 + = 0.$$

$$D_3 = F_2(1,1) = \sum_{k_2=0}^{2} F_1(1,k_2)W^{3k_2}$$

$$= F_1(1,0) + F_1(1,1)W^3 + F_1(1,2)W^6$$

$$= 0 + 2W^3 + 2W^6 = 2W^3 + 2\left(W^3\right)^2$$

$$= -2 + 2 = 0; \qquad (W^3 = -1)$$

$$D_4 = F_2(0,2) = \sum_{k_2=0}^{2} F_1(0,k_2)W^{4k_2}$$

$$= F_1(0,0) + F_1(0,1)W^4 + F_1(0,2)W^8$$
$$= 0 + 0 + 0 + = 0$$

$$D_5 = F_2(1,2) = \sum_{k_2=0}^{2} F_1(1,k_2)W^{5k_2}$$

$$= F_1(1,0) + F_1(1,1)W^5 + F_1(1,2)W^{10}$$

$$= 0 + 2W^5 + 2W^{10}$$

$$= 2\left[e^{\frac{10\pi i}{6}} + e^{\frac{20\pi i}{6}}\right] = 2\left[-\frac{\sqrt{3}}{2}i - \frac{\sqrt{3}}{2}i\right] = 2\sqrt{3}i$$

$$\therefore \quad D_j = FDT\{d_k\} \quad = \left[0,\ 2\sqrt{3}i,\ 0,\ 0,\ 0,\ -2\sqrt{3}i\right]$$

7. Find the 8-point DFT of a real sequence $d_n = (1,2,2,2,1,0,0,0)$ using decimation-in-frequency FFT algorithm.

Solution: Here, given that $N = 8$ and twiddle factors.

$$W_8^r = e^{-j\left(\frac{2\pi}{8}\right)r} = e^{-j\left(\frac{\pi}{4}\right)r}$$

$$\Rightarrow \quad W_8^0 = e^{-0} = 1$$

$$W_8^1 = e^{-j\left(\frac{\pi}{4}\right)} = \cos\left(\frac{\pi}{4}\right) - j\sin\left(\frac{\pi}{4}\right) = \frac{1}{\sqrt{2}} - j\frac{1}{\sqrt{2}}$$

$$W_8^2 = e^{-j\left(\frac{\pi}{2}\right)} = \cos\left(\frac{\pi}{2}\right) - j\sin\left(\frac{\pi}{2}\right) = 0 - j(1) = -j$$

$$W_8^3 = e^{-j\left(\frac{3\pi}{4}\right)r} = \cos\left(\frac{3\pi}{4}\right) - j\sin\left(\frac{3\pi}{4}\right)$$

$$= \cos\left(\pi - \frac{\pi}{4}\right) - j\sin\left(\pi - \frac{\pi}{4}\right)$$

$$= -\cos\frac{\pi}{4} - j\left(\sin\frac{\pi}{4}\right)$$

$$= -\frac{1}{\sqrt{2}} - j\frac{1}{\sqrt{2}} \qquad (\because\ \sin(\pi - \theta) = \sin\theta;\ \cos(\pi - \theta) = -\cos\theta)$$

Using DIT-FFT algorithm, we can find D_k from the given sequence d_n as shown in Figure 4.7

Stage 1

$$D_0 = 1 + 1 = 2$$

$$D_4 = 2 + 0 = 2$$

$$D_2 = 2 + 0 = 2$$

$$D_6 = 2 + 0 = 2$$

$$D_1 = (1-1)W_8^0 = 0$$

$$D_5 = (2-0)W_8^1 = 2\left[\frac{1}{\sqrt{2}} - j\frac{1}{\sqrt{2}}\right] = \sqrt{2} - j\sqrt{2}$$

$$D_3 = (2-0)W_8^2 = 2(-j) = j2$$

$$D_7 = (2-0)W_8^3 = 2\left[-\frac{1}{\sqrt{2}} - j\frac{1}{\sqrt{2}}\right] = -\sqrt{2} - j\sqrt{2}$$

8. Perform the linear convolution of the two sequences below in time domain and compare the result obtained by using the DFT relations:

$$h_n = \begin{cases} n, & 0 \le n \le 3 \\ n, & \text{otherwise} \end{cases}$$

Solution: $f_n - g_n * h_n$

$f_{-1} = 0;$

$f_0 = 1$

$f_1 = 2$

$f_2 = 4$

$f_3 = 2$

$f_4 = 3$

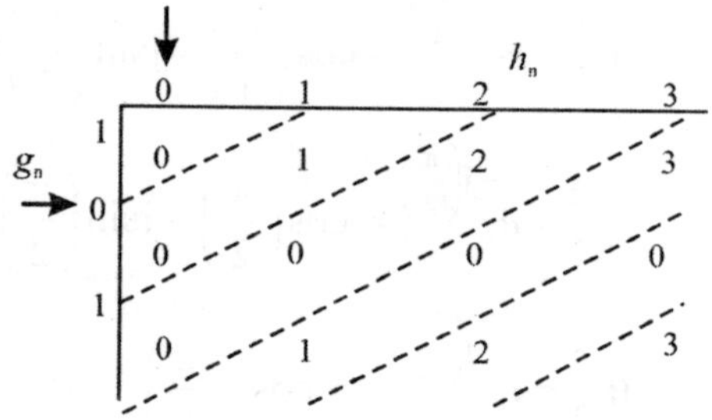

Figure 4.8

$$f_n = \{0, 1, 2, 4, 2, 3\} \qquad [1]$$
$$\uparrow$$

[1] is the result of convolution (linear) by "overlap and add method". We shall now find the linear convolution using the DFT technique.

$$\therefore \quad H_k = \sum_{n=0}^{3} h_n e^{-j\left(\frac{2\pi}{N}\right)nk}$$

$$= \sum_{n=0}^{3} h_n e^{-j\left(\frac{2\pi}{6}\right)nk} \qquad (\because\ N = 6)$$

$$= e^{-j\left(\frac{\pi}{3}\right)k} + 2e^{-j\left(\frac{2\pi}{3}\right)k} + 3e^{-j(\pi k)}$$

Here, $N_G = 4;\ \ N_H = 3$

$$\therefore \quad N_G + N_H - 1 = N = 6.$$

$$H_0 = 6$$

$$H_1 = e^{-j\left(\frac{\pi}{3}\right)} + 2e^{-j\left(\frac{2\pi}{3}\right)} + 3e^{-j(\pi)}$$

$$H_2 = e^{-j\left(\frac{2\pi}{3}\right)} + 2e^{-j\left(\frac{4\pi}{3}\right)} + 3e^{-j(2\pi)}$$

$$H_3 = e^{-j\pi} + 2e^{-j(2\pi)} + 3e^{-j(3\pi)}$$

$$H_4 = e^{-j\left(\frac{4\pi}{3}\right)} + 2e^{-j\left(\frac{8\pi}{3}\right)} + 3e^{-j(4\pi)}$$

$$H_5 = e^{-j\left(\frac{5\pi}{3}\right)} + 2e^{-j\left(\frac{10\pi}{3}\right)} + 3e^{-j(5\pi)}$$

$$G_k = \sum_{n=-1}^{1} g_n e^{-j\left(\frac{2\pi}{N}\right)nk}$$

$$= e^{j\left(\frac{2\pi}{6}\right)k} + 0 + e^{-j\left(\frac{2\pi}{6}\right)k}$$

$$G_k = e^{j\left(\frac{2\pi}{6}\right)k} + e^{-j\left(\frac{2\pi}{6}\right)k}$$

$$G_0 = 1 + 1 = 2$$

$$G_1 = e^{j\left(\frac{2\pi}{6}\right)} + e^{-j\left(\frac{2\pi}{6}\right)}$$

$$G_2 = e^{j\left(\frac{2\pi}{3}\right)} + e^{-j\left(\frac{2\pi}{3}\right)}$$

$$G_3 = e^{j\pi} + e^{-j(\pi)}$$

$$G_4 = e^{j\left(\frac{4\pi}{3}\right)} + e^{-j\left(\frac{4\pi}{3}\right)}$$

$$G_5 = e^{j\left(\frac{5\pi}{3}\right)} + e^{-j\left(\frac{5\pi}{3}\right)}$$

Now, $F_k = H_k G_k$

$$F_0 = H_0 G_0 = 12$$

$$F_1 = H_1 G_1 = \left[e^{-j\left(\frac{\pi}{3}\right)} + 2e^{-j\left(\frac{2\pi}{3}\right)} + 3e^{-j(\pi)}\right]\left[e^{j\left(\frac{2\pi}{6}\right)} + e^{-j\left(\frac{2\pi}{6}\right)}\right]$$

$$= 1 + 6e^{-j\left(\frac{2\pi}{3}\right)} + 2e^{-j(\pi)} + 3e^{-j\left(\frac{4\pi}{6}\right)}$$

$$F_2 = H_2G_2 = \left[e^{-j\left(\frac{2\pi}{3}\right)} + 2e^{-j\left(\frac{4\pi}{3}\right)} + 3e^{-j(2\pi)}\right]\left[e^{j\left(\frac{2\pi}{3}\right)} + e^{-j\left(\frac{2\pi}{3}\right)}\right]$$

$$= 1 + 4e^{-j\left(\frac{4\pi}{3}\right)} + 2e^{-j\left(\frac{2\pi}{3}\right)} + 2e^{-j(2\pi)} + 3e^{-j\left(\frac{8\pi}{3}\right)}$$

$$F_3 = H_3G_3 = \left[e^{-j(\pi)} + 2e^{-j(2\pi)} + 3e^{-j(3\pi)}\right]\left[e^{j(\pi)} + 2e^{-j(2\pi)} + e^{-j(\pi)}\right]$$

$$= 1 + 2e^{-j\pi} + 4e^{-j2\pi} + 2e^{-j3\pi} + 3e^{-j4\pi}$$

$$F_4 = H_4G_4 = \left[e^{-j\left(\frac{4\pi}{3}\right)} + 2e^{-j\left(\frac{8\pi}{3}\right)} + 3e^{-j(4\pi)}\right]\left[e^{j\left(\frac{4\pi}{3}\right)} + e^{-j\left(\frac{4\pi}{3}\right)}\right]$$

$$= 1 + 4e^{-j\left(\frac{8\pi}{3}\right)} + 2e^{-j\left(\frac{4\pi}{3}\right)} + e^{-j\left(\frac{16\pi}{3}\right)} + 2e^{-j\left(\frac{12\pi}{3}\right)}$$

$$F_5 = H_5G_5 = \left[e^{-j\left(\frac{5\pi}{3}\right)} + 2e^{-j\left(\frac{10\pi}{3}\right)} + 3e^{-j(5\pi)}\right]\left[e^{j\left(\frac{5\pi}{3}\right)} + e^{-j\left(\frac{5\pi}{3}\right)}\right]$$

$$= 1 + 4e^{-j\left(\frac{10\pi}{3}\right)} + 2e^{-j\left(\frac{5\pi}{3}\right)} + e^{-j\left(\frac{15\pi}{3}\right)} + 3e^{-j\left(\frac{20\pi}{3}\right)}.$$

The Inverse Discrete Fourier Transform (IDFT) is given by

$$f_n = \frac{1}{N}\sum_{k=0}^{N-1} F_k\, e^{-j\left(\frac{2\pi}{N}\right)nk}, \qquad 0 \le n \le N-1$$

Here, $N = 6$, $f_n = \frac{1}{6}\sum_{k=0}^{5} F_k\, e^{j\left(\frac{2\pi nk}{6}\right)}$

Thus, f_n can be evaluated using the above expression and output may be sketched as shown in Figure 4.9

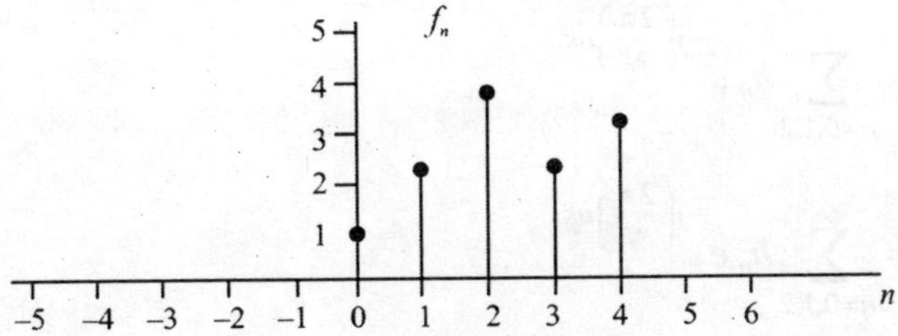

Figure 4.9

Given $g_n = \begin{cases} 1, & n=0 \\ 0.5, & n=1 \\ 0, & \text{otherwise} \end{cases}$ and $h_n = \begin{cases} 0.5, & n=0 \\ 1, & n=1 \\ 0, & \text{otherwise} \end{cases}$

9. Evaluate $f_n = g_n * h_n$ using DFT techniques.

Solution: By data, $N_G = 2$ (Sample point of g_n)

$N_H = 2$ (Sample point of h_n)

$$N_F = N_G + N_H - 1 = 2 + 2 - 1 = 3.$$

Now, we shall convert the 2-sample input sequences into three sample sequences by "padding with zero".

The 3-point DFT sequences are given by

$$G_k = \sum_{n=0,1,2} g_n W^{-nk}$$

$$= \sum_{n=0,1,2} g_n e^{-j\left(\frac{2\pi}{N}\right)nk}$$

Since $N = 3$

$$\therefore \quad G_k = g_0 e^{-j\left(\frac{2\pi}{3}\right)(0)k} + g_1 e^{-j\left(\frac{2\pi}{3}\right)k} + g_2 e^{-j\left(\frac{2\pi}{3}\right)2k}$$

$$G_k = 1 + 0.5 e^{-j\left(\frac{2\pi}{3}\right)k}$$

$D_0 = 1 + 0.5 = 1.5$	$D_1 = 1 + 0.5 e^{-j\left(\frac{2\pi}{3}\right)}$	$D_2 = 1 + 0.5 e^{-j\left(\frac{4\pi}{3}\right)}$

$$H_k = \sum_{n=0,1,2} h_n e^{-j\left(\frac{2\pi}{N}\right)nk}$$

$$= \sum_{n=0,1,2} h_n e^{-j\left(\frac{2\pi}{N}\right)nk}$$

$$H_k = 0.5 + e^{-j\left(\frac{2\pi}{3}\right)k}$$

$$H_0 = 1.5$$

$$H_1 = 0.5 + e^{-j\left(\frac{2\pi}{3}\right)}$$

$$F_k = H_k G_k, \text{ for } 0 \le k \le 2.$$

Since a time domain a convolution is equivalent to a Frequency domain Multiplication

$$F_0 = (1.5)(1.5) = 2.25$$

$$F_1 = \left(0.5 + e^{-j\left(\frac{2\pi}{3}\right)}\right)\left(1 + 0.5e^{-j\left(\frac{2\pi}{3}\right)}\right)$$

$$= 0.5 + 0.25e^{-j\left(\frac{2\pi}{3}\right)} + e^{-j\left(\frac{2\pi}{3}\right)} + 0.5e^{-j\left(\frac{4\pi}{3}\right)}$$

$$F_1 = 0.5 + 1.25e^{-j\left(\frac{2\pi}{3}\right)} + 0.5e^{-j\left(\frac{4\pi}{3}\right)} + 0.5e^{-j\left(\frac{4\pi}{3}\right)}$$

$$F_2 = 0.5 + 0.25e^{-j\left(\frac{4\pi}{3}\right)} + e^{-j\left(\frac{4\pi}{3}\right)} + 0.5e^{-j\left(\frac{8\pi}{3}\right)}$$

$$= 0.5 + 1.25e^{-j\left(\frac{4\pi}{3}\right)} + 0.5e^{-j\left(\frac{8\pi}{3}\right)}$$

The IDFT is given by

$$f_n = \frac{1}{N}\sum_{k-0}^{N-1} F_n e^{j\left(\frac{2\pi nk}{3}\right)}, \qquad 0 \le n \le N-1$$

$$f_n = \frac{1}{3}\sum_{k-0}^{2} F_n e^{j\left(\frac{2\pi nk}{3}\right)} \qquad (\because N = 3, \text{ meaning 3-point DFT})$$

$$f_n = \frac{1}{3}\left[F_0 e^0 + F_1 e^{j\left(\frac{2\pi n}{3}\right)} + F_1 e^{\left(\frac{4\pi n}{3}\right)}\right]$$

$$= \frac{1}{3}\left[2.25 + \left[0.5 + 1.25e^{-j\left(\frac{2\pi}{3}\right)} + 0.5e^{-j\left(\frac{4\pi}{3}\right)}\right] e^{j\left(\frac{2\pi n}{3}\right)}\right]$$

$$+ \left[0.5 + 1.25e^{-j\left(\frac{4\pi}{3}\right)} + 0.5e^{-j\left(\frac{8\pi}{3}\right)}\right] e^{\left(\frac{4\pi n}{3}\right)}$$

$$= \frac{1}{3}\left[2.25 + 0.5e^{j\left(\frac{2\pi n}{3}\right)} + 0.25e^{-j\left(\frac{2\pi}{3}\right)(1-n)} + 0.5e^{j\left(\frac{2\pi}{3}\right)(2-n)}\right.$$

$$\left. + 0.5e^{j\left(\frac{4\pi n}{3}\right)} + 1.25e^{-j\frac{4\pi}{3}(1-n)} + 0.5e^{-j\frac{4\pi}{3}(2-n)}\right]$$

$$= \frac{1}{3}\left[2.25 + 0.5 + 1.25e^{-j\left(\frac{2\pi}{3}\right)} + 0.5e^{-j\left(\frac{4\pi}{3}\right)} + 0.5 + 1.25e^{-j\left(\frac{8\pi}{3}\right)}\right]$$

$$= \frac{1}{3}\left[3.25 + 1.25e^{-j\left(\frac{2\pi}{3}\right)} + 1.75e^{-j\left(\frac{4\pi}{3}\right)} + 4.5e^{-j\left(\frac{8\pi}{3}\right)}\right]$$

$$= \frac{1}{3}\left[3.25 + 1.75\left(-0.5 - j\frac{\sqrt{3}}{2}\right) + 1.75\left(-0.5 + j\frac{\sqrt{3}}{2}\right)\right]$$

$$\Rightarrow \quad f_n = \frac{1}{3}[3.25 - 1.75]$$

$$f_0 = 0.5$$

Similarly, we can find: $f_1 = 1.25;\quad f_2 = 0.5$

We shall now check the output sample values f_0, f_1 and f_2 by "overlap and add method".

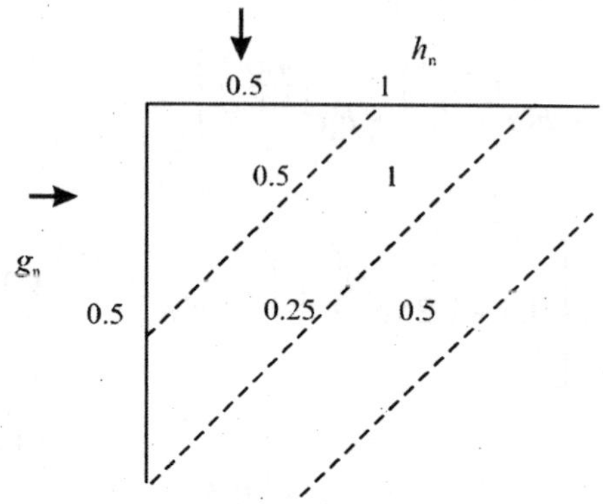

Figure 4.10

From the above table,

$$\left.\begin{aligned} f_0 &= 0.5 \\ f_1 &= 1.25 \\ f_2 &= 0.5 \end{aligned}\right\} \qquad \therefore f_n = \begin{cases} 0.5, & n=0 \\ 1.25, & n=1 \\ 0.5, & n=2 \\ 0, & \text{otherwise} \end{cases}$$

EXERCISES

1. Find the DFT of $\{d_r\} = \{1, -1, 0\}$, $r = 0, 1, 2$

$\left[\textit{Ans.}\ \{1 - w^2, 0, 1 - w\} \quad \text{where } w^3 = 1\right]$

2. State and prove convolution formula for the DFT

3. If (G_j, g_r) and (H_j, h_r) one two pair of DFT, prove that

$$\frac{1}{N}\sum_{r=0}^{N-1} G_r H_{r-j} = \frac{1}{N}\sum_{r=0}^{N-1} G_{j+r} + H_r$$
$$= g_r h_{-r}$$

4. Apply FFT, algorithm for finding DFT for the following sequence $\{d_r\} = \{0, 0, 1, 1, 1, 0\}$ $r = 0, 1, 2, 3, 4, 5$

$\left[\textit{Ans.}\ D_j = \{3, -2, 0, -1, 0, -2\}\right]$

CHAPTER 5

Orthogonal Set of Functions

Introduction

In this chapter, we have introduced the corresponding concepts and notations of orthogonal set of functions. The concept of an orthogonal set of function is a natural generalization of the concept of an orthogonal set of vectors. The orthogonal complement of a vector V (or of a subset S) of a vector space in the set of all vectors of the space which are orthogonal to V (or to each vector of S).

5.1 Orthogonal Set of Functions

The orthogonal complement of a vector in three dimensions is the set of all vectors perpendicular to the given vector i.e., the set of the all linear combinations of any two linearly independent vectors perpendicular to the given vector. In fact, a function can be considered as a generalized vector so that the fundamental properties of the set of functions are suggested by the analogous properties of the set of vectors.

Some Basic definitions:

Definition (1): (Inner Product) The inner product of two functions $f(x)$ and $g(x)$ is the number defined by the equation

$$(f,g)=\int_a^b f(x)g(x)dx, \quad \text{[on the interval } a\le x\le b]. \qquad [1]$$

i.e., the integral in [1] is the inner product (f_n, g_m) of f_n and g_m.

Definition (2): (Orthogonal function) The condition that the two function be orthogonal is written as

$$(f,g)=0;\ \text{i.e.,}\ \int_a^b f(x)g(x)dx=0 \qquad [2]$$

The norm of the function $f(x)$ is defined as non-negative square root of (f, f) and is generally denoted by $\|f\|$.

Thus, norm of $f(x)=\|f\|=\sqrt{(f,f)}$

$$= \sqrt{\left(\int_a^b f^2(x)dx\right)} \qquad [3]$$

$f(x)$ is said to be normalized when

$$\|f\| = 1 \text{ i.e., } \int_a^b f^2(x)\,dx = 1 \qquad [4]$$

Definition (3): (Orthogonal set of functions) A set of real functions $\phi_1, \phi_2, \phi_3, \cdots$ is orthogonal on an interval $a \le x \le b$ if

$$\int_a^b \phi_m(x)\phi_n(x)dx = 0, \qquad \text{whenever } m \ne n$$

and is called an orthogonal set of functions on that interval.

Definition (4): (Orthogonal set of functions with respect to a weight function) Consider a set $\{\phi_n(x)\}$ $(n = 1,2,3,\cdots)$ of real functions defined on an interval $a \le x \le b$. For some function $p(x) \ge 0$, the set $\{\phi_n(x)\}$ is said to be an orthogonal set of functions on the interval $a \le x \le b$ if $\int_a^b p(x)\phi_m(x)\phi_n(x)dx = 0$ whenever $m \ne n$.

Definition (5): (Orthogonal set of functions)

Let $\{g_m(x)\}$ $(m = 1,2,3,\cdots)$ be a orthogonal set of functions. If none of the function g_m can be normalized by dividing it by the positive constant $\|g_m\|$. The new set $\{\phi_m(x)\}$ so formed where $\phi_m(x) = \|g_m\|^{-1} g_m(x)$ $(m = 1,2,3,\cdots)$ is called an orthogonal set of functions on the interval $a \le x \le b$.

Clearly, we have

$$\int_a^b \phi_m(x)\phi_m(x)dx = \begin{cases} 0 & \text{when } m \ne n \ (m = 1,2,\cdots) \\ 0 & \text{when } m = n \end{cases}$$

Evidently, from an orthogonal set we may obtain an orthonormal set by dividing each function by its norm on the interval under consideration. The method of orthonormalization is called Gram-Schmidt orthogonalization method.

(A) EXAMPLES

Show that the following functions are orthogonal on the intervals indicated against them.

1. $\{\sin(mx)\}, (m = 1,2,3,\cdots)$ on $(-\pi \le x \le \pi)$

2. The functions $1, \cos x, \sin x, \cos 2x, \sin 2x, \cdots$ on $(-\pi \le x \le \pi)$

3. $\{\cos nx\}$, $(n = 0,1,2,\cdots)$ on $(0, \pi)$

4. The functions $\sin x, \sin 2x, \sin 3x, \cdots 1, \cos x, \cos 2x, \cos 3x, \cdots$ on $(-\pi, \pi)$

5. The functions $f_1(x) = 1$ and $f_2(x) = x$ on $(-1,1)$ and determine the constants A and B so that the function $f_3(x) = 1 + Ax + Bx^2$ is orthogonal to both f_1 and f_2 on that interval.

Solutions:

1. Let $\phi_m(x) = \sin mx,\ m = 1,2,3,\cdots$

$$\therefore \quad (\phi_m, \phi_n) = \int_{-\pi}^{\pi} \sin(mx)\sin(nx)dx$$

$$= \frac{1}{2}\int_{-\pi}^{\pi} \cos(m-n)xdx - \frac{1}{2}\int_{-\pi}^{\pi} \cos(m+n)xdx = 0 \ (m \neq n)$$

The norm $\|\phi_m\|$ equals $\sqrt{\pi}$

$$\because \quad \|\phi_m\|^2 = \int_{-\pi}^{\pi} \sin^2 mx\, dx = \pi \qquad (m = 1,2,\cdots)$$

Hence, the corresponding orthonormal set consists of the functions:

$$\frac{\sin x}{\sqrt{\pi}}, \frac{\sin 2x}{\sqrt{\pi}}, \frac{\sin 3x}{\sqrt{\pi}}, \cdots$$

2. Given: $1, \cos x, \sin x, \cos 2x, \sin 2x, \cdots$ on $(-\pi, \pi)$

Hence, $\displaystyle\int_{-\pi}^{\pi} \sin(mx)\sin(nx)dx = \frac{1}{2}\int_{-\pi}^{\pi} \sin(m+n)x\,dx - \frac{1}{2}\int_{-\pi}^{\pi} \sin(m-n)x\,dx = 0$

for all $m, n = 0, 1, \cdots$

the corresponding orthonormal set is

$$\frac{1}{\sqrt{2\pi}}, \frac{\cos x}{\sqrt{\pi}}, \frac{\sin x}{\sqrt{\pi}}, \frac{\cos 2x}{\sqrt{\pi}}, \frac{\sin 2x}{\sqrt{\pi}}, \cdots$$

3. Given: $\{\cos nx\}\ (n = 0,1,2,\cdots)$

for $m \neq n$, we have

$$\int_0^{\pi} \cos mx \cos nx\, dx = \frac{1}{2}\int_0^{\pi}[\cos(m+n)x + \cos(m-n)]dx.$$

$$= \frac{1}{2}\left[\frac{\sin(m+n)x}{m+n} + \frac{\sin(m-n)x}{m-n}\right]_0^{\pi} = 0.$$

This shows that the given set of functions is orthogonal. We now obtain norm of each function of the set. By definition, we have:

For $n = 0$, $\|\cos nx\| = \|1\| = \sqrt{\left(\int_0^\pi 1^2\,dx\right)} = \sqrt{\pi}$

for $n \neq 0$, $\|\cos nx\| = \sqrt{\left(\int_0^\pi \cos^2 nx\,dx\right)}$

$$= \left[\int_0^\pi \frac{1+\cos 2nx}{2}\,dx\right]^{\frac{1}{2}}$$

$$= \left[\int_0^\pi \left(x + \frac{\sin 2nx}{2n}\right)_0^\pi\right]^{\frac{1}{2}} = \sqrt{\frac{\pi}{2}}.$$

Hence the corresponding orthonormal set is

$$\frac{1}{\sqrt{\pi}}, \sqrt{\left(\frac{2}{\pi}\right)}\cos x, \sqrt{\left(\frac{2}{\pi}\right)}\cos 2x, \sqrt{\left(\frac{2}{\pi}\right)}\cos 3x, \cdots$$

$$\frac{1}{\sqrt{\pi}}, \left\{\sqrt{\left(\frac{2}{\pi}\right)}\cos nx\right\} \qquad (n = 1, 2, 3, \cdots)$$

4. For $m \neq n$, we have the following results:

(i) $\int_{-\pi}^{\pi} \sin mx \sin nx\,dx$

$= 2\int_0^\pi \sin mx \sin nx\,dx$ ($\because$ $\sin mx \sin nx$ is an even function of x)

$$= \int_0^\pi [\cos(m-n)x - \cos(m+n)x]\,dx$$

$$= \left[\frac{\sin(m-n)x}{m-n} - \frac{\sin(m+n)x}{m+n}\right]_0^\pi = 0$$

(ii) $\int_{-\pi}^{\pi} \sin mx \sin nx\,dx$

$= 0$ ($\because$ $\sin mx \cos nx$ is an odd function of x)

(iii) $\int_{-\pi}^{\pi} \cos mx \cos nx\,dx$

$= 2\int_0^\pi \cos mx \sin nx\,dx$ ($\because$ $\cos mx \cos nx$ is an even function of x)

$$= \int_0^\pi [\cos(m+n)x + \cos(m-n)x]\,dx$$

$$= \left[\frac{\sin(m+n)x}{m+n} + \frac{\sin(m+n)x}{m+n} \right]_0^{\pi} = 0$$

(iv) $\int_{-\pi}^{\pi} 1 \cdot \sin mx \, dx = 0$ ($\because \sin mx$ is an odd function)

(v) $\int_{-\pi}^{\pi} 1 \cdot \cos nx \, dx = 2\int_0^{\pi} \cos nx \, dx = \left[\frac{\sin nx}{n} \right]_0^{\pi} = 0.$

(i), (ii), (iii), (iv), and (v) together show that the given set of functions are orthogonal.

Now, $\|\sin nx\| = \sqrt{\left(\int_{-\pi}^{\pi} \sin^2 nx \, dx \right)} = \left[2\int_0^{\pi} \sin^2 nx \, dx \right]^{\frac{1}{2}}$

($\because \sin^2 nx$ is an even function of x)

$$= \left[\int_0^{\pi} 1 - \cos 2nx \, dx \right]^{\frac{1}{2}} = \left[\left(x - \frac{\sin 2nx}{2n} \right)_0^{\pi} \right]^{\frac{1}{2}} = \sqrt{\pi},$$

$\|\cos nx\| = \sqrt{\left(\int_{-\pi}^{\pi} \cos^2 nx \, dx \right)} = \left[2\int_0^{\pi} \cos^2 nx \, dx \right]^{\frac{1}{2}}$

($\because \cos^2 nx$ is an even function of x)

$$= \left[\int_0^{\pi} 1 + \cos 2nx \, dx \right]^{\frac{1}{2}} = \left[\left(x + \frac{\sin 2nx}{2n} \right)_0^{\pi} \right]^{\frac{1}{2}} = \sqrt{\pi}$$

and $\|1\| = \sqrt{\left(\int_{-\pi}^{\pi} 1^2 \, dx \right)} = \sqrt{(2\pi)}.$

Hence the required orthonormal set is

$$\frac{\sin x}{\sqrt{\pi}}, \frac{\sin 2x}{\sqrt{\pi}}, \frac{\sin 3x}{\sqrt{\pi}}, \dots \frac{1}{\sqrt{(2\pi)}}, \frac{\cos x}{\sqrt{\pi}}, \frac{\cos 2x}{\sqrt{\pi}}, \frac{\cos 3x}{\sqrt{\pi}}, \dots$$

5. Given: $f_1(x) = 1, f_2(x) = x, f_3(x) = 1 + Ax + Bx^2$ [1]

$\therefore \int_{-1}^{1} f_1(x) f_2(x) \, dx = \int_{-1}^{1} x dx$, by [1]

$$= \left[\frac{x^2}{2} \right]_{-1}^{1} = 0.$$

This shows that the functions $f_1(x)$ and $f_2(x)$ are orthogonal on the interval $(-1, 1)$.

Let $f_3(x)$ be orthogonal to both $f_1(x)$ and $f_2(x)$ on $(-1, 1)$. Then by definition, we must have

$$\int_{-1}^{1} f_1(x) f_3(x)\, dx = 3 \text{ and } \int_{-1}^{1} f_1(x) f_3(x)\, dx = 0$$

i.e., $\int_{-1}^{1} (1 + Ax + Bx^2)\, dx = 0$ and $\int_{-1}^{1} (x + Ax^2 + Bx^3)\, dx = 0$, by [1]

i.e., $\left[x + \frac{A}{2}x^2 + \frac{B}{3}x^3\right]_{-1}^{1} = 0$ and $\left[\frac{1}{2}x^2 + \frac{A}{3}x^3 + \frac{B}{4}x^4\right]_{-1}^{1} = 0$

i.e., $2 - \frac{2}{3}B = 3$ and $\frac{2}{3}A = 0$, giving $A = 0$ and $B = -3$.

Hence $A = 0$ and $B = -3$ give the required values of A and B.

Hermite Orthogonality of a set of complex-valued functions

A set $\{f_n(x)\}$, $(n = 1, 2, 3, \cdots)$ of complex-valued functions of a real variable x is orthogonal in the Hermite sense on an interval (a,b) if

$$\int_a^b f_n(x)\overline{f_m(x)}\, dx = 0,\ m \neq n,$$

where $\overline{f_m(x)}$ denotes the complex conjugate of function $f_m(x)$.

Also, norm of $\quad f_n(x) = \sqrt{\left(\int_a^b f_n(x)\overline{f_n(x)}\, dx\right)}$

6. Establish the Hermite orthogonality of the exponential functions $\{e^{inx}\}$, $(n = 0, \pm 1, \pm 2, \pm 3, \cdots)$ on the interval $(-\pi, \pi)$. Also show that norm of each function is $\sqrt{(2\pi)}$.

Solution: Let $f_n(x) = e^{inx}$, $f_m(x) = e^{imx}$ and $\overline{f_m(x)} = e^{-imn}$

Here, $\quad \int_{-\pi}^{\pi} f_n(x)\overline{f_m(x)}\, dx = \int_{-\pi}^{\pi} e^{inx}\, e^{-imx}\, dx$

$$= \int_{-\pi}^{\pi} e^{ix(n-m)} dx = \frac{1}{i(n-m)}\left[e^{ix(n-m)}\right]_{-\pi}^{\pi}$$

$$=\frac{1}{\mathrm{i}(n-m)}\left[e^{\mathrm{i}\pi(n-m)}-e^{-\mathrm{i}\pi(n-m)}\right]$$

$$=\frac{1}{(n-m)}\times 2\sin(n-m)\pi=0,\ n\neq m$$

$\therefore$ The given set of functions are orthogonal in the Hermite sense.

Norm of $e^{\mathrm{i}nx}$, $(n=0,1,2,\cdots)$

$$=\sqrt{\left(\int_{-\pi}^{\pi}e^{\mathrm{i}nx}\overline{\left(e^{-\mathrm{i}nx}\right)}dx\right)}=\sqrt{\left(\int_{-\pi}^{\pi}e^{\mathrm{i}nx}e^{-\mathrm{i}nx}dx\right)}$$

$$=\sqrt{\left(\int_{-\pi}^{\pi}dx\right)}=\sqrt{(2\pi)}.$$

7. Show that the set of functions $\left\{\sin\frac{n\pi x}{c}\right\}(n=1,2,3,\cdots)$ is orthogonal on the interval $(0,c)$. Normalize the set.

Solution: For $m\neq n$, we have

$$\int_0^c \sin\frac{m\pi x}{c}\sin\frac{n\pi x}{c}dx=\frac{1}{2}\int_0^c\left[\cos\left\{\frac{\pi x}{c}(m-m)\right\}-\cos\left\{\frac{\pi x}{c}(m+n)\right\}\right]dx$$

$$=\frac{1}{2}\left[\frac{c}{\pi(m-n)}\sin\left\{\frac{\pi x}{c}(m-n)\right\}-\frac{c}{\pi(m+n)}\sin\left\{\frac{\pi x}{c}(m+n)\right\}\right]_0^c=0$$

This shows that the given set of functions is orthogonal.

Now, $$\left\|\sin\frac{n\pi x}{c}\right\|=\left[\int_0^c\sin^2\frac{n\pi x}{c}dx\right]^{\frac{1}{2}}=\left[\frac{1}{2}\int_0^c\left(1-\cos\frac{2n\pi x}{c}\right)\right]^{\frac{1}{2}}$$

$$=\left[\frac{1}{2}\left(x-\frac{c}{2n\pi}\sin\frac{2n\pi}{c}\right)_0^c\right]^{\frac{1}{2}}=\sqrt{\left(\frac{c}{2}\right)}.$$

Hence the required normalized set is

$$\left\{\sqrt{\left(\frac{2}{c}\right)}\sin\frac{n\pi x}{c}\right\}\ (n=1,2,3,\cdots)$$

Working Rule: Gram Schmidt orthonormalization method

Suppose $\{f_n(x)\}$ $(n=1,2,3,\cdots)$ be a given set of orthogonal functions. If none of the function $f_n(x)$ have zero norms, each function $f_n(x)$ can be normalized by dividing it by the positive constant $\|f_n(x)\|$. The new set $\{\phi_n(x)\}$ so formed, where

$$\phi_n(x)=\frac{f_n(x)}{\|f_n(x)\|} \qquad (n=1,2,3,\cdots)$$

is normalized and orthogonal. Hence $\{\phi_n(x)\}$ so constructed will be the required orthogonal set of functions.

Gram-Schmidt process of Orthonormalization

Suppose that $\{f_n(x)\}$, $(n=1,2,3,\cdots)$ be a set of a linearly independent real functions for each of which norm $\|f_n\|$ exists and is non-zero. Then we wish to obtain an orthonormal set $\{\phi_n(x)\}$ $(n=1,2,3,\cdots)$ so that

$$\int_a^b \phi_n(x)\phi_m(x)\,dx=\begin{cases}0 & \text{if } m\neq n,\\ 1 & \text{if } m=n.\end{cases}$$

Procedure: We select f_1 and obtain $\phi_1=\dfrac{f_1}{\|f_1\|}$.

We next choose f_2 and write $F_2=f_2+c\phi_1$, [1]

where c is chosen in such a manner so that F_2 may be orthogonal to ϕ_1, i.e.,

$$\int_a^b F_2\phi_1\,dx=0 \qquad \Rightarrow\int_a^b f_2+c\phi_1\,\phi_1\,dx=0,$$

$$\Rightarrow \quad \int_a^b f_2\phi_1\,dx+c\int_a^b \phi_1^2\,dx=0 \qquad \Rightarrow\int_a^b f_2\phi_1\,dx+c=0, \text{ by [1]—[2]}$$

Hence $c=-\int_a^b f_2\phi_1\,dx$. With this value of c, [2] reduces to

$$F_2=f_2-\phi_1\int_a^b f_2\phi_1\,dx \tag{2}$$

we now take $\phi_2=\dfrac{F_2}{\|F_2\|}$ [3]

In the third step, we write $F_3=f_3+c_1\phi_1+c_2\phi_2$ [4]

where c_1 and c_2 are chosen in such a manner so that F_2 may be orthogonal to ϕ_1 and ϕ_2, i.e.,

$$\int_a^b (f_3 + c_1\phi_1 + c_2\phi_2)\phi_1 dx = 0 \qquad [5]$$

and $$\int_a^b (f_3 + c_1\phi_1 + c_2\phi_2)\phi_2 dx = 0. \qquad [6]$$

Using [1], [5] and [6] reduce to

$$\int_a^b (f_3\phi_1\, dx + c_1 = 0 \text{ and } \int_a^b (f_3\phi_2\, dx + c_2 = 0$$

$$\therefore \quad c_1 = -\int_a^b (f_3\phi_1\, dx \text{ and } c_2 = -\int_a^b (f_3\phi_2\, dx. \qquad [7]$$

Using [7], [4] reduce to

$$F_3 = f_3 - \phi_1 \int_a^b (f_3\phi_1\, dx - \phi_2 \int_a^b (f_3\phi_2\, dx. \qquad [8]$$

So we take $\phi_3 = \dfrac{F_3}{\|F_3\|}$. [9]

By repeated application of the above process, the nth required normalized function ϕ_n is given by

$$\phi_n = \frac{F_n}{\|F_n\|}, \qquad [10]$$

where $F_n = f_n - \phi_1 \int_a^b (f_n\phi_1\, dx - \phi_2 \int_a^b (f_n\phi_2\, dx \cdots - \phi_{n-1} \int_a^b (f_n\phi_{n-1}\, dx, \cdots$ [11]

The above process would fail if and only if at a stage

$F_r = 0$ for $1 \le r \le n$. [12]

But $F_r = 0$

$\Rightarrow \quad f_r - \phi_1 \int_a^b f_r\phi_1\, dx - \phi_2 \int_a^b (f_r\phi_2\, dx - \phi_{r-1} \int_a^b (f_r\phi_{r-1}\, dx = 0$

$\Rightarrow \quad f_r$ is a linear combination of $\phi_1, \phi_2, \cdots, \phi_{r-1}$

$\Rightarrow \quad f_r$ is a linear combination of $f_1, f_2, \cdots, f_{r-1}$

(by our construction of $\phi_1, \phi_2, \cdots, \phi_{r-1}$)

$\Rightarrow \quad f_1, f_2, \cdots, f_r$ is a linearly dependent set for $1 \le r \le n$.

But this is a contradiction because $\{f_n\}$ $(n = 1, 2, \cdots, n)$ is linearly independent and so its every subset would also be linearly independent. The existence of this contradiction shows that $F_r \ne 0$ for $1 \le r \le n$ and hence we would always get a set $\{\phi_n(x)\}$ $(n = 1, 2, \cdots, n)$ which would be orthonormal.

8. With the help of 1, x, x^2, $\cos x$ at three function g_0, g_1 and g_2 which are orthogonal over $-1 \le x \le 1$.

Solution: Choosing $g_0 = 1$. [1]

We next choose x and write

$$g_1 = x + c,\ 1 = x + c. \qquad [2]$$

Let g_1 be orthogonal to g_0 so that

$$\int_{-1}^{1} g_0 g_1\, dx = 0 \text{ i.e., } \int_{-1}^{1} x + c\, dx = 0.$$

$$\therefore \quad \left[\frac{1}{2}x^2 + cx\right]_{-1}^{1} = 0 \text{ or } 2c = 0 \text{ or } c = 0.$$

$\therefore$ From [2], $g_1 = x$. [3]

In the third step, we take

$$g_2 = x^2 + c_1 g_0 + c_2 g_1 \text{ i.e., } g_2 = x^2 + c_1 + c_2 x. \qquad [4]$$

Let g_2 be orthogonal to both g_0 and g_1, so that

$$\int_{-1}^{1} g_2 g_0\, dx = 0 \text{ and } \int_{-1}^{1} g_2 g_1\, dx = 0$$

$$\int_{-1}^{1} (x^2 + c_1 + c_2 x)\, dx = 0 \text{ and } \int_{-1}^{1} (x^2 + c_1 + c_2 x)\, x\, dx = 0$$

i.e., $$\left[\frac{1}{3}x^3 + c_1 x + \frac{1}{2}c_2 x^2\right]_{-1}^{1} = 0 \text{ and } \left[\frac{1}{4}x^4 + \frac{1}{2}c_1 x^2 + \frac{1}{3}c_2 x^3\right]_{-1}^{1} = 0$$

i.e., $$\frac{2}{3} + 2c_1 = 0 \text{ and } c_2 \frac{2}{3} = 0, \text{ So } c_1 = -\frac{1}{3} \text{ and } c_2 = 0.$$

with these values of c_1 and c_2, [4] reduces to

$$g_2(x) = x^2 - \left(\frac{1}{3}\right).$$

Hence the required functions are

$$g_0 = 1,\ g_1 = x,\ g_2 = x^2 - \left(\frac{1}{3}\right).$$

9. With the help of $1, x, x^2$ construct three functions g_0, g_1 and g_2 which are orthogonal over $0 \le x \le \infty$ w.r.t. e^{-x}

Solution: Choosing $g_0 = 1$. [1]

We next choose x and write

$$g_1 = x + c,\ g_0 = x + c. \qquad [2]$$

Let g_1 be orthogonal to go with respect to weight function e^{-x}

$$\therefore \quad \int_0^\infty (x+c)e^{-x}dx = 0$$

or $\quad \left[(x+c)(-e)^{-x} - (1)(e^{-x})\right]_0^\infty = 0, \quad$ (by parts)

or $\quad \lim\limits_{x\to\infty} \dfrac{x+c}{e^x} + c + 1 = 0$

or $\quad c + 1 = 0$ i.e., $c = -1$. $\qquad \left(\because \lim\limits_{x\to\infty} \dfrac{x+c}{e^x} = \lim\limits_{x\to\infty} \dfrac{1}{e^x} = 0 \right.$ Using L'Hospital's rule $\left.\right)$

with this value of c, [2] reduces to

$$g_1 = x - 1. \qquad [3]$$

Finally, we take

$$g_2 = x^2 + c_1 g_0 + c_2 g_1 = x^2 + c_2 x + c_1 - c_2. \qquad [4]$$

Let g_2 be orthogonal to g_0 and g_1 with respect to weight function e^{-x}.

$$\therefore \quad \int_0^\infty e^{-x} g_2 g_0\, dx = 0 \text{ i.e., } \int_0^\infty x^2 + c_2 x + c_1 - c_2\, e^{-x} dx = 0 \qquad [5]$$

and $\quad \int_0^\infty e^{-x} g_2 g_1\, dx = 0,$

i.e., $\quad \int_0^\infty x^2 + c_2 x + c_1 - c_2\,(x-1)e^{-x} dx = 0.$

or $\quad \int_0^\infty [x^3 + x^2(c_2 - 1) + x(c_1 - 2c_2) + (c_2 - c_1)]e^{-x} dx = 0 \qquad [6]$

Applying chain rule of integration by parts, [5] gives

$$\left[(x^2 + c_2 x + c_1 - c_2) - (2x + c_2)(e^{-x}) + (2)(-e^{-x})\right]_0^\infty = 0$$

i.e., $\quad (c_1 - c_2) + c_2 + 2 = 0$ so that $c_1 = -2$.

Similarly, [6] reduces to

$$\Big[\{x^3 + x^2(c_2 - 1) + x(c_1 - 2c_2) + (c_2 - c_1)(e^{-x})$$
$$-\{3x^2 + 2x(c_2 - 1) + (c_1 - 2c_2)\}(-e^{-x})$$
$$+\{6x + 2(c_2 - 1)\}(-e^{-x}) - (6)(e^{-x})\Big]_0^\infty = 0$$

or $\quad c_2 - c_1 + c_1 - 2c_2 + 2(c_2 - 1) + 6 = 0$ so that $c_2 = -4.$

Putting values of c_1 and c_2 in (4), we get

$$g_2 = x^2 - 4x + 2. \qquad [7]$$

Hence the required functions are

$$g_0 = 1,\ g_1 = x - 1 \text{ and } g_2 = x^2 - 4x + 2.$$

10. Given the set of function $1, x, x^2, x^3, \cdots$, obtain from these a set of functions which are mutually orthonormal in $(-1,1)$.

Solution: Suppose $\{\phi_n(x)\}$ be the required orthonormal set of functions so that

$$\int_{-1}^{1} \phi_n\, \phi_m\, dx = \begin{cases} 0 & \text{if } m \neq n, \\ 1 & \text{if } m = n. \end{cases} \qquad [1]$$

Now, $\|1\| = \left[\int_{-1}^{1} 1^2\, dx\right]^{\frac{1}{2}} = \sqrt{2}$

Choose $\quad \phi_1 = \dfrac{1}{\|1\|} = \dfrac{1}{\sqrt{2}}. \qquad [2]$

We next choose x and write

$$F_2 = x + c\phi_2. \qquad [3]$$

Let F_2 be orthogonal is ϕ_1 so that

$$\int_{-1}^{1} (x + c\phi_1)\phi_1\, dx = 0 \ \Rightarrow \ = \int_{-1}^{1} x\phi_1\, dx + c\int_{-1}^{1} \phi_1^2\, dx = 0$$

$$\Rightarrow \quad \int_{-1}^{1} \frac{x}{\sqrt{2}} dx + c \cdot 1 = 0 \text{ by [1]} \ \Rightarrow \ \frac{1}{\sqrt{2}}\left[\frac{x^2}{2}\right]_{-1}^{1} + c = 0 \text{ so that } c = 0.$$

So from [3], $F_2 = x.$

Now, $\|F_2\| = \left[\int_{-1}^{1} x^2\, dx\right]^{\frac{1}{2}} = x\sqrt{\dfrac{2}{3}}.$

We take $\phi_2 = \frac{F_3}{\|F_3\|} = x\sqrt{\left(\frac{3}{2}\right)}$. [5]

Next we choose x^2 and write

$$F_3 = x^2 + c_1\phi_1 + c_2\phi_2. \qquad [6]$$

Let F_3 be orthogonal to both ϕ_1 and ϕ_2 so that

$$\int_{-1}^{1}(x^2 + c_1\phi_1 + c_2\phi_2)\phi_1\, dx = 0 \qquad [7]$$

and $\int_{-1}^{1}(x^2 + c_1\phi_1 + c_2\phi_2)\phi_2\, dx = 0$ [8]

Using [1], [7] and [8] reduces to

$$\int_{-1}^{1} x^2\phi_1\, dx + c_1 = 0 \Rightarrow \int_{-1}^{1} x^2\left(\frac{1}{\sqrt{2}}\right)dx + c_1 = 0, \text{ by [2]}$$

and $\int_{-1}^{1} x^2\phi_2\, dx + c_2 = 0 \Rightarrow \int_{-1}^{1} x^2\left(\frac{\sqrt{3}}{\sqrt{2}}\right)x\, dx + c_2 = 0$, by [5]

these give $c_1 = -\frac{1}{\sqrt{2}}\int_{-1}^{1} x^2\, dx = -\frac{1}{\sqrt{2}}\left[\frac{x^3}{3}\right]_{-1}^{1} = -\frac{2}{3\sqrt{2}}$

and $c_2 = \left(\frac{3}{2}\right)^{\frac{1}{2}}\int_{-1}^{1} x^3\, dx = -\left(\frac{3}{2}\right)^{\frac{1}{2}}\left[\frac{x^4}{4}\right]_{-1}^{1} = 0.$

Substituting these values of c_1 and c_2 and using [2] and [5], [6] reduces to

$$F_3 = x^2 + \left(-\frac{2}{3\sqrt{2}}\right)\left(\frac{1}{\sqrt{2}}\right) = \frac{3x^2 - 1}{2}$$

$$\therefore \quad \|F_3\| = \left[\int_{-1}^{1}\left(\frac{3x^2 - 1}{2}\right)^2 dx\right]^{\frac{1}{2}} = \left[\frac{1}{4}\int_{-1}^{1}(9x^4 - 6x^2 + 1)\, dx\right]^{\frac{1}{2}}$$

$$= \left[\frac{1}{4}\left\{9\cdot\frac{x^5}{5} - 6\cdot\frac{x^3}{3} + x\right\}_{-1}^{1}\right]^{\frac{1}{2}} = \sqrt{\frac{2}{5}}.$$

So we take $\phi_3 = \frac{F_3}{\|F_3\|} = \sqrt{\frac{5}{2}}\left(\frac{3x^2-1}{2}\right)$.

Continuing likewise, we obtain

$$\phi_4(x) = \sqrt{\frac{7}{2}}\left(\frac{5x^3-3x}{2}\right),\ \phi_5(x) = \sqrt{\frac{9}{2}}\left(\frac{5x^4-30x^2+3}{8}\right) \text{ and so on.}$$

Generalised Fourier Series

Let $f(x)$ be any given function and if $\{\phi_n\}$ $(n = 1,2,3,\cdots)$ be an orthonormal set on $[a,b]$, then, we have

$$f(x) = c_1\phi_1(x) + c_2\phi_2(x) + \cdots + c_n\phi_n(x) + \cdots \qquad [1]$$

Such series are called orthonormal series and they are generalization of Fourier series. Since $\{\phi_n(x)\}$ is an orthonormal set, we have

$$\int_a^b \phi_n(x)\phi_m(x)\,dx = \begin{cases} 0 & \text{if } n \neq m, \\ 1 & \text{if } n = m. \end{cases} \qquad [2]$$

Assume that the series on the RHS of [1] is convergent. Assume also that the series obtained by multiplying all terms of RHS of [1] by $\phi_n(x)$ is integrable. So multiplying both sides of [1] by $\phi_n(x)$ and then integrating w.r.t. 'x' from a to b, we get

$$\int_a^b f(x)\phi_n(x)\,dx = c_1\int_a^b \phi_1(x)\phi_n(x)\,dx + c_2\int_a^b \phi_2(x)\phi_n(x)\,dx + \cdots$$

$$+ c_n\int_a^b \phi_n^2(x)\,dx + \cdots$$

or $\quad \int_a^b f(x)\phi_n(x)\,dx = c_1 \times 0 + c_2 \times 0 + \cdots + c_n \times 1 + \cdots$, by [2]

Thus $c_n = \int_a^b f(x)\phi_n(x)\,dx.$ $\qquad$ [3]

The constants c_n given by [3] are called Fourier constants and the series in equation [1] with c_n given by [3] is called the generalized fourier series corresponding to $f(x)\cdots$. The fact is expressed as

$$f(x) \sim \sum_{n=1}^{\infty} c_n\,\phi_n(x). \qquad [4]$$

Theorem: The Fourier constants of a function $f(x)$ with respect to the functions $\phi_1, \phi_2, \cdots, \phi_m$ of an orthonormal set are those coefficients for which a linear combination of those m functions is the best approximation in the mean to $f(x)$, on the fundamental interval.

Proof: Let $\{\phi_n\}$ $(n = 1, 2, 3, \cdots)$ be an orthogonal set. then

$$\int_a^b \phi_n(x)\phi_m(x)\,dx = \begin{cases} 0 & \text{if } m \neq n, \\ 1 & \text{if } m = n. \end{cases} \qquad [1]$$

We know that the Fourier constants c_n of a given function $f(x)$ are given by

$$c_n = \int_a^b f(x)\phi_n(x)\,dx. \qquad [2]$$

Consider function in the space of sectionally continuous functions on an interval $a \le x \le b$. Let $S_m(x)$ be a linear combination of m functions $\phi_1, \cdots, \phi_m$, i.e., $S_m(x) = k_1\phi_1(x) + k_2\phi_1(x) + \cdots + k_m\phi_m(x)$ [3]

Then the mean square error E of approximation [3] is given by

$$E = \frac{1}{b-a}\int_a^b \left[f(x) - S_m(x)\right]^2 dx. \qquad [4]$$

We are interested in finding k_n (constants) $(1 \le n \le m)$ so that S_m is the best approximation in the mean to $f(x)$.

Now, $f(x) - S_m(x) = f(x) - (k_1\phi_1 + k_2\phi_1 + \cdots + k_m\phi_m)$.

On squaring both sides, we get

$$[f(x) - S_m(x)]^2 = [f(x)]^2 + k_1^2\phi_1^2 + k_2^2\phi_2^2 + \cdots + k_m^2\phi_m^2$$
$$- 2\sum_{n=1}^{m} k_n f(x)\phi_n(x) - 2\sum_{i=1}^{m}\sum_{j=1}^{m} k_i k_j \phi_i \phi_j \qquad [5]$$

Integrating both sides of [5] w.r.t. 'x' from a to b and using [1] and [2], we have

$$\int_a^b \left[f(x) - S_m(x)\right]^2 dx = \int_a^b [f(x)]^2\,dx + k_1^2 + k_2^2 + \cdots + k_m^2$$
$$-2(k_1c_1 + k_2c_2 + \cdots + k_nc_n)$$
$$= \int_a^b [f(x)]^2\,dx + (k_1 - c_1)^2 + \cdots + (k_m - c_m)^2$$
$$-\left(c_1^2 + c_2^2 + \cdots + c_m^2\right) \qquad [6]$$

From [4] and [6] we have

$$E = \frac{1}{b-a}\left[\int_a^b [f(x)]^2\,dx + (k_1 - c_1)^2 + \cdots + (k_m - c_m)^2 - \left(c_1^2 + c_2^2 + \cdots + c_m^2\right)\right] \quad [7]$$

From [7] we see that the mean square error will be the least when $(k_n - c_n)^2 = 0$ for each n i.e., when $k_n = c_n$ for each n.

Corollary (1): To prove that

$$\sum_{n=1}^{m} c_n^2 \le \int_a^b [f(x)]^2\,dx$$

where c_n are Fourier constants of a function f.

Proof: Since $k_n = c_n$ therefore [7] becomes

$$E = \frac{1}{b-a}\left[\int_a^b [f(x)]^2\,dx - \sum_{n=1}^{m} c_n^2\right]$$

Since $E \ge 0$, [9] reduces to

$$\int_a^b [f(x)]^2\,dx - \sum_{n=1}^{m} c_n^2 \ge 0$$

so that $\displaystyle\sum_{n=1}^{m} c_n^2 \le \int_a^b [f(x)]^2\,dx.$

Corollary (2): Fourier constants c_n always approach to zero as $n \to \infty$.

Proof: From corollary 1, we have

$$\sum_{n=1}^{\infty} c_n^2 \le \int_a^b [f(x)]^2\,dx.$$

We note that the RHS does not depend on m. Hence, taking the limit as $m \to \infty$, the above inequality reduces to

$$\sum_{n=1}^{\infty} c_n^2 \le \int_a^b [f(x)]^2\,dx. \quad [10]$$

From [2] it follows that if the RHS of [10] exists, then the series on LHS of [10] must be convergent. From the theory of convergence of infinite series, if $\sum_{n=1}^{\infty} c_n^2$ is convergent, then the nth term c_n^2, and with it c_n, must tend to zero as $n \to \infty$, i.e.,

$$\lim_{n\to\infty} c_n = 0. \qquad [11]$$

Definition: (Closed set) An orthonormal set of functions $\{\phi_n(x)\}$ $(n=1,2,3,\cdots)$ is said to be closed in the sense of mean convergence if for each function $f(x)$ of the function space is being considered as

$$\sum_{n=1}^{\infty} c_n^2 = \int_a^b [f(x)]^2\, dx,$$

where $\quad c_n = \int_a^b f(x)\phi_n(x)\, dx.$

Theorem: If an orthonormal set $\{\phi_n(x)\}$ is closed, it is complete.

Proof: Let $\{\phi_n(x)\}$ $(n=1,2,3,\cdots)$ be an orthonormal set and let it be closed in the sense of mean convergence. Then for each $f(x)$ of the function space being considered, we have (by definition),

$$\sum_{n=1}^{\infty} c_n^2 = \int_a^b [f(x)]^2\, dx, \qquad [1]$$

where $\quad c_n = \int_a^b f(x)\phi_n(x)\, dx. \qquad [2]$

If possible, let $\{\phi_n(x)\}$ be not complete. Then by definition there exists a function $g(x)$ in the function space, such that

$$\text{norm of } g(x) = \| g(x) \| > 0 \qquad [3]$$

and that $g(x)$ is orthogonal to each function $\phi_n(x)$ of the given set $\{\phi_n(x)\}$,

i.e., $\quad \int_a^b g(x)\phi_n(x)\, dx = 0. \qquad [4]$

Since [1] and [2] are satisfied by each function $f(x)$ of the function space, these will also be satisfied by $g(x)$ so that

$$\sum_{n=1}^{\infty} c_n^2 = \int_a^b [g(x)]^2\, dx, \qquad [5]$$

where $\quad c_n = \int_a^b g(x)\phi_n(x)\,dx.$

Using [4], [6] reduces to $c_n = 0$, Hence [5] reduces to

$$\int_a^b [g(x)]^2\,dx = 0 \text{ or } \qquad \left[\int_a^b [g(x)]^2\,dx\right]^{\frac{1}{2}} = 0$$

or $\quad \| g(x) \| = 0$, by definition of norm which contradicts [3].

Since this contradiction arises by assuming that the set $\{\phi_n(x)\}$ is not complete it follows that $\{\phi_n(x)\}$ is complete.

Example: Show why the set $\{\sin n\pi x\};\, n = 1,2,3\cdots$ is orthogonal on the interval $-1 \le x \le 1$, but not complete even in the space of all continuous functions on the interval.

***Solution*:** For $m \ne n$, we have

$$\int_{-1}^{1} \sin n\pi x \sin m\pi x\,dx = \int_{-1}^{1} \frac{\cos(n-m)\pi x - \cos(n+m)\pi x}{2}\,dx$$

$$= \frac{1}{2}\left[\frac{\sin(n-m)\pi x}{(n-m)\pi} - \frac{\sin(n+m)\pi x}{(n+m)\pi}\right]_{-1}^{1} = 0.$$

This shows that the given set of functions is orthogonal.

The norm of $\sin n\pi x$

$$= \left[\int_{-1}^{1} \sin^2 n\pi x\,dx\right]^{\frac{1}{2}} = \left[\int_{-1}^{1} \frac{1-\cos 2n\pi x}{2}\,dx\right]^{\frac{1}{2}}$$

$$= \left[\frac{1}{2}\left\{x - \frac{\sin 2n\pi x\,dx}{2n\pi}\right\}_{-1}^{1}\right]^{\frac{1}{2}} = 1.$$

So norm of each function $\sin n\pi x$ is unity.

From these facts we see that $\{\sin n\pi x\}$ is an orthonormal set.

Let the function space under consideration consists of all continuous functions on $-1 \le x \le 1$. Let k be a non-zero constant function defined on $-1 \le x \le 1$. Then we have

$$\int_{-1}^{1} k \sin n\pi x\,dx = k\left[-\frac{\cos n\pi x}{n\pi}\right]_{-1}^{1} = 0 \text{ for } n = 1,2,\cdots \qquad [1]$$

Furthermore, the norm of the function k

$$=\| k \|=\left[\int_{-1}^{1} k^2\, dx\right]^{\frac{1}{2}} = \sqrt{2}\, k.$$

Since $k \neq 0$, so $\| k \| > 0$. [2]

From [1] and [2] we see that there exists function k in the function space being considered, with positive norm, which is orthogonal to each of the functions $\sin n\pi x$ of the set $\{\sin n\pi x\}$. Hence by definition, the set $\{\sin n\pi x\}$ is not complete.

Sturm-Liouville-Problem

The second ODE of the form:

$$\left[R(x)y'\right]' + \left[Q(x) + \lambda P(x)\right]y = 0 \qquad [1]$$

on some internal [a,b] satisfying conditions of the form:

$$\text{(i)} \qquad \left.\begin{aligned} a_1 y + a_2 y' = 0 \ \text{ at } x = a \\ b_1 y + b_2 y' = 0 \ \text{ at } x = b \end{aligned}\right\} \qquad [2]$$

here λ is a real parameter and a_1, a_2, b_1, b_2 are given real constants at least one in each condition, [2] being different from zero.

The equation [1] is known as Sturm Liouville equation. The condition [1] and [2] are called boundary conditions.

The boundary problem given by [1] and [2] is called a Sturm Liouville problem.

The solution $y \neq 0$ are called the characteristic function or eigen function of the problem, and the values of λ for which such solution exists are called characteristic values or eigen values of the problem.

Theorem (1):

Let $\left.\begin{aligned} y_m = y_m(x); \\ y_n = y_n(x) \end{aligned}\right]$ be given function of Sturm Liouville problem corresponding to distinct eigen values λ_m and $\lambda_n(x)$ respectively and let the derivatives $y_m(x), y_n(x)$ be continuous on the interval. Then y_m and y_n are orthogonal on that interval w.r.t. the weight function P. (P,Q,R are all assumed to be real and continuous on [a,b].

Proof: Since y_m and y_n are the solutions of Sturm Liouville problem

$$[R(x)y']' + [Q(x) + \lambda P(x)]y = 0 \qquad [1]$$

$$\text{on } [a,b] \ni \left.\begin{array}{l}\text{(i) } a_1 y + a_2 y' = 0 \text{ at } x = a \\ \text{(ii) } b_1 y + b_2 y' = 0 \text{ at } x = b\end{array}\right\} \qquad [2]$$

$$\therefore \text{ we have } \left.\begin{array}{l}(Ry'_m)' + (Q + \lambda_m P)y_m = 0 \\ \text{and } (Ry'_n)' + (Q + \lambda_n P)y_n = 0\end{array}\right\} \qquad [3]$$

Multiply [2] by y_n and [3] by y_m, and on subtracting, we have

$$(\lambda_m - \lambda_n)Py_m y_n = y_m (Ry'_n)' - y_n (Ry'_m)'$$

or $(\lambda_m - \lambda_n)Py_m y_n = (Ry'_n y_m - Ry'_m y_n)'$

Integrating from a to b, (both sides) we get

$$(\lambda_m - \lambda_n)\int_a^b Py_m y_n \, dx = (Ry'_n y_m - Ry'_m y_n)_a^b$$

$$= R(b)[y'_n(b)y_m(b) - y'_m(b)y_n(b)]$$

$$-R(a)[y'_n(a)y_m(a) - y'_m(a)y_n(a)] \qquad [4]$$

Five different cases arise:

Case (i): If $R(a) = R(b) = 0$ (each)

$\therefore$ from [4] we have

$$(\lambda_m - \lambda_n)\int_a^b Py_m y_n \, dx = 0$$

$$\Rightarrow \int_a^b Py_m y_n \, dx = 0 \rightarrow (\lambda_m \neq \lambda_n)$$

$\Rightarrow$ y_m and y_n are orthogonal w.r.t. the weight functions P.

Case (ii): Let $R(a) = 0$; and $R(b) \neq 0$

$\therefore$ Equation [4] becomes

$$(\lambda_m - \lambda_n)\int_a^b Py_m y_n \, dx = R(b)[y'_n(b)y_m(b) - y'_m(b)y_n(b)] \qquad [5]$$

Also from [2] we have

$$b_1 y_m(b) + b_2 y'_m(b) = 0 \qquad [6]$$

and $\quad b_1 y_n(b) + b_2 y_n'(b) = 0$ [7]

If $b_2 \neq 0$; then multiplying [6] by $y_n(b)$ and [7] by $y_m(b)$ and on subtracting, we have

$$b_2\left[y_m'(b)y_n(b) - y_n'(b)y_m(b)\right] = 0$$

and $\quad y_m'(b)y_n(b) - y_n'(b)y_m(b)$

$\therefore$ From [5], we have

$$\int_a^b P y_m y_n \, dx = 0$$

$\Rightarrow$ y_m and y_n are orthogonal w.r.t. weight function P.

If $b_2 = 0$, then assuming $b_1 \neq 0$ and multiplying [6], by $y_n'(b)$ and [7] by $y_m'(b)$. Also on subtracting,

we have

$$b_1\left[y_m(b)y_n'(b) - y_m'(b)y_n(b)\right] = 0$$

$$y_m(b)y_n'(b) - y_m'(b)y_n(b) = 0$$

$\therefore$ From [5], we have $\int_a^b P y_m y_n \, dx = 0$

$\Rightarrow$ y_m and y_n are orthogonal w.r.t. the weight function P.

Case (iii): Let $R(b) = 0$; and $R(a) \neq 0$

$\therefore$ Equation [4] becomes

$$(\lambda_m - \lambda_n)\int_a^b P y_m y_n \, dx = -R(a)\left[y_n'(a)\cdot y_m(a) - y_m'(a)y_n(a)\right] \quad [8]$$

Also, from (i), we have

$$a_1 y_m(a) + a_2 y_m'(a) = 0 \quad [9]$$

and $\quad a_1 y_n(a) + a_2 y_n'(a) = 0$ [10]

If $a_2 \neq 0$, then multiplying [9] by $y_n(a)$ and [10] by $y_m(a)$ and then on subtracting, we have

$$a_2\left[y_m'(a)y_n(a) - y_n'(a)y_m(a)\right] = 0$$

or, $\quad y_m'(a)y_n(a) - y_n'(a)y_m(a) = 0$

$\therefore$ From equation [8], we have

$$\int_a^b P y_m y_n \, dx = 0$$

$\therefore$ y_m and y_n are orthogonal w.r.t. the weight function P.

If $a_2 = 0$, then we assume $a_1 \neq 0$.

Then multiplying [9] by $y_n'(a)$ and [10] by $y_m'(a)$ and subtracting, we have

$$a_1\left[y_n'(a)y_m(a) - y_m'(a)y_n(a)\right] = 0$$

or $\quad y_n'(a)y_m(a) - y_m'(a)y_n(a) = 0$

$\therefore$ From [8], we have

$$\int_a^b P y_m y_n \, dx = 0.$$

$\therefore$ y_m and y_n are orthogonal w.r.t. the weight function P.

Case (iv): Let $R(a) \neq 0$; $R(b) \neq 0$

Thus, using (i) and (ii) and proceeding as in case (ii) and (iii), we have, from [4],

$$\int_a^b P y_m y_n \, dx = 0$$

Case (v): Let $R(a) = R(b)$

$\therefore$ From [iv], we have

$$(\lambda_m - \lambda_n)\int_a^b P y_m y_n \, dx = R(b)\left[y_n'(b)\cdot y_m(b) - y_m'(b)y_n(b) - y_n'(a)y_m(a) + y_m'(a)y_n(a)\right] \qquad [11]$$

and $\quad a_1 y_m(a) + a_2 y_m'(a) = 0 \qquad [12]$

$$a_1 y_n(a) + a_2 y_n'(a) = 0 \qquad [13]$$

$$b_1 y_m(b) + b_2 y_m'(b) = 0 \qquad [14]$$

$$b_1 y_n(b) + b_2 y_n'(b) = 0 \qquad [15]$$

If $a_1 \neq 0$, $b_1 \neq 0$, multiplying [12] by $y_n(a)$ and [13] by $y_m'(a)$ and on subtracting, we have

$$y_m'(a)y_n(a) - y_m(a)y_n'(a) = 0$$

Also multiplying [14] by $y_n(b)$ and [15] by $y_m(b)$ and on subtracting, we have

$$y_m'(b)y_n(b) - y_n'(b)y_m(b) = 0$$

$\therefore$ From [11], we have

$$\int_a^b P y_m y_n \, dx = 0.$$

$\Rightarrow$ y_m and y_n are orthogonal w.r.t. the weight function P. (proved)

Theorem (2): Eigen values of the Sturm Liouville problem are all real.

Example: Find the characteristic functions of $X'' + \lambda X = 0, X(0)$ and $X'(L) = 0$.

Solution: Given $\quad X'' + \lambda X = 0 \quad$ [1]
with boundary conditions $X(0) = 0,\ X'(L) = 0$. [2]

Case (i): Let $\lambda = 0$. Then solution of [1] is

$$X(x) = Ax + B. \quad [3]$$

From [3], $X'(x) = A$. [4]

Replacing x by 0 in [3] and using [2], we get $B = 0$. Again, replacing x by L in [4] and using [2], we get $A = 0$. With $A = 0$ and $B = 0$, [3] reduces to $X(x) = 0$. So there is no eigen function corresponding to $\lambda = 0$.

Case (ii): Let $\lambda = -\mu^2$, where $\mu \neq 0$ (i.e., λ is –ive). Then [1] reduces to

$$X'' - \mu^2 X = 0$$

whose solution is $X(x) = Ae^{\mu x} + Be^{-\mu x}$ [5]

From [5], $X'(x) = A\mu e^{\mu x} - B\mu e^{-\mu e}$ [6]

Using [2], [5] and [6] reduce to

$$0 = A + B \text{ and } A\mu e^{\mu L} - B\mu e^{-\mu L} = 0. \quad [7]$$

Solving [7], we find $A = B = 0$. Hence [5] gives $X(x) = 0$. Since $X(x) \neq 0$, so there is no eigen function corresponding to $\lambda = \mu^2$

Case (iii): Let $\lambda = \mu^2$ where $\mu \neq 0$ (i.e., λ is +tive). Then (1) reduces to

$$X'' - \mu^2 X = 0$$

whose solution is $X(x) = A\cos\mu x + B\sin\mu x$ [8]

From [8], $X'(x) = A\mu\sin\mu x + B\mu\cos\mu x$. [9]

Using [2], [8] and [9] reduce to

$0 = A$ and $0 = B\mu\cos\mu L$.

Thus $A = 0$ and $B\cos\mu L = 0$. $(\because \mu \neq 0)$ [10]

Now consider $B\cos\mu L = 0$. [11]

If $B = 0$, then with $A = 0$, [8] reduces to $X(x) = 0$, which is not an eigen function. So we take $B \neq 0$, for the existence of eigen functions. Since $B \neq 0$, [11] reduces to

$$\cos\mu L = 0 \text{ so that } \mu L = (2n-1)\frac{\pi}{2}, n = 1, 2, 3, \cdots$$

$$\therefore \quad \mu = \frac{\pi}{2L}(2n-1), n = 1, 2, 3, \cdots$$

Using $A = 0$ and [12], [8] reduces to

$$X(x) = B\sin\left\{\frac{\pi}{2L}(2n-1)\right\}, n = 1, 2, 3, \cdots$$

and then $\lambda = \mu^2 = \dfrac{\pi^2}{4L^2}(2n-1)^2,\ n = 1, 2, 3, \cdots$

Hence the required eigen function $X_n(x)$ with corresponding eigen values λ_n are given by

$$X_n(x) = B_n \sin\left\{\frac{\pi}{2L}(2n-1)\right\}, n = 1, 2, 3, \cdots$$

and $$\lambda_n = \frac{\pi^2}{4L^2}(2n-1)^2, \ n = 1, 2, 3, \cdots$$

Example 1: Find all the eigen values and functions of the Sturm Liouville problem $y'' + \lambda y = 0$ with $y(0) + y'(0) = 0$ and $y(1) + y'(1) = 0$.

Solution: Given $y'' + \lambda y = 0$ [1]

with boundary conditions $y(0) + y'(0) = 0$ [2]

and $y(1) + y'(1) = 0.$ [3]

Case (i): Let $\lambda = 0$. Then solution of [1] is

$$y(x) = Ax + B \quad [4]$$

From [4], $y'(x) = A$. [5]

From [4] and [5], $y(0) = B$, $y'(0) = A$. With these values, [2] reduces to

$$B + A = 0. \qquad [6]$$

From [4] and [5], $y(1) = A + B$ and $y'(1) = A$. With these values, [3] reduces to

$$2A + B = 0. \qquad [7]$$

Solving [6] and [7], $A = B = 0$. Hence [4] reduces to $X(x) = 0$, which is not an eigen function and so $\lambda = 0$ is not an eigen value.

Case (ii): Let $\lambda = -\mu^2$, where $\mu \neq 0$. Then [1] reduces to $y'' - \mu^2 y = 0$ whose solution is

$$y(x) = Ae^{\mu x} + Be^{-\mu x} \qquad [8]$$

From [8],
$$y'(x) = A\mu e^{\mu x} + B\mu e^{-\mu x} \qquad [9]$$

From [8] and [9], $y(0) = A + B$ and $y'(0) = \mu(A - B)$. [10]

Using [10], [12] reduces to

$$A + B + \mu(A - B) = 0 \text{ i.e., } A(1+\mu) + B(1-\mu) = 0. \qquad [11]$$

Again, from [8] and [9], we have

$$y(1) = Ae^{\mu} + Be^{-\mu},\ y'(1) = \mu(Ae^{\mu} - Be^{-\mu}) \qquad [12]$$

Using [12], [3] reduces to

$$Ae^{\mu} + Be^{-\mu} + \mu(Ae^{\mu} - Be^{-\mu}) = 0$$

i.e.,
$$Ae^{\mu}(1+\mu) + Be^{-\mu}(1-\mu) = 0. \qquad [13]$$

We now use the theory of determinants for solving [11] and [13]. For non-trivial solution of these equation, we must have

$$\begin{vmatrix} 1+\mu & 1-\mu \\ e^{\mu}(1+\mu) & e^{-\mu}(1-\mu) \end{vmatrix} = 0$$

i.e.,
$$(1+\mu)(1-\mu)(e^{-\mu} - e^{\mu}) = 0,$$

giving $\mu = -1$, and $\mu = 1$.

When $\mu = -1$, [11] and [13] give $B = 0$, while A will be arbitrary. So [8] reduces to $y(x) = Ae^{-x}$.

Example 2: Find the characteristic functions of $X'' + \lambda X = 0, X'(0) = 0$ and $X'(L) = 0$.

***Solution*:** Given $X'' + \lambda X = 0$ [1]

with boundary conditions $X'(0) = 0$ and $X'(L) = 0$. [2]

Case (i): Let $\lambda = 0$. Then solution of [1] is

$$X(x) = Ax + B. \quad [3]$$

From [3], $X'(x) = A$. [4]

Using [2], [4] gives $0 = A$ and $0 \neq A$

These give $A = 0$, while B is arbitrary. Hence [3] reduces to $X(x) = B$ which is non-zero. Hence taking $B = 1$. $X(x) = 1$ is an eigen function with $\lambda = 0$ as the corresponding eigen value.

Case (ii): Let $\lambda = -\mu^2$, where $\mu \neq 0$. Then [11] reduces to $X'' - \mu^2 X = 0$ whose solution is

$$X(x) = Ae^{\mu x} + Be^{-\mu x} \quad [5]$$

From [5], $X(x) = A\mu e^{\mu x} - B\mu e^{-\mu x}$

Using [2], [6] reduces to

$$0 = A\mu + B\mu \text{ and } A\mu e^{\mu L} - B\mu e^{-\mu L} = 0$$

i.e., $A - B = 0$ and $Ae^{\mu L} - Be^{\mu L} = 0$, as $\mu \neq 0$. [6]

Solving equations [6], we get $A = B = 0$. Hence [5] gives $X(x) = 0$, which is not an eigen function.

Case (iii): Let $\lambda = \mu^2$, where $\mu \neq 0$. Then [1] reduces to $X'' + \mu^2 X = 0$ whose solution is

$$X(x) = A\cos\mu x + B\sin\mu x. \quad [7]$$

From [7], $X'(x) = A\mu\sin\mu x + B\mu\cos\mu x$. [8]

Using [2], [8] reduces to

$$0 = B\mu \text{ and } 0 = -A\mu\sin\mu L + B\mu\cos\mu L$$

i.e., $B = 0$ and $A\sin\mu L = 0$, as $\mu \neq 0$. [9]

We now consider $A\sin\mu L = 0$, [10]

If $A=0$ then with $B=0$, [8] reduces to $X(x)=0$ which is not an eigen function. So we take $A\neq 0$, for the existence as eigen function. Since [10] reduces to

$$\sin\mu L=0 \text{ so that } \mu L=n\pi,\ n=1,2,3,\cdots$$

$$\therefore \quad \mu=\frac{n\pi}{L},\ n=1,2,3,\cdots \tag{11}$$

($n=0$ is omitted as $n=0\Rightarrow\mu=0\Rightarrow\lambda=0$, which has already been considered in case (i)).

Using $B=0$ and [11], [7] reduces to

$$X(x)=A\cos\frac{n\pi x}{L},\ n=1,2,3,\cdots \tag{12}$$

and then $\lambda=u^2=\dfrac{n^2\pi^2}{L},\ n=1,2,3,\cdots$ [13]

Hence the required eigen functions $X_n(x)$ with the corresponding eigen values λ_n are given by (taking $A=1$)

$$X_n(x)=\cos\frac{n\pi x}{L},\ n=1,2,3,\cdots \tag{14}$$

and
$$\lambda_n=\frac{n^2\pi^2}{L^2},\ n=1,2,3,\cdots \tag{15}$$

Example 3: Find the eigen values and the corresponding eigen function of the boundary value problem:

$$y''+2y'+(1+\lambda)y=0,\ y(0)=0, y'(a)=0$$

***Solution*:** Given $(D^2+2D+1+\lambda)y=0,\ D\equiv\dfrac{d}{dx}.$ [1]

with $y(0)=0,\ y'(a)=0.$ [2]

Case (i): Let $\lambda=0$. Then solution of [1] is

$$y(x)=(A+Bx)e^{-x} \tag{3}$$

From [3], $\quad y'(x)=Be^{-x}-(A+Bx)e^{-x}.$ [4]

Using [2], [3] and [4] reduce to

$$0=A \text{ and } Be^{-a}-(A+Bx)e^{-a}=0,$$

i.e., $A = 0,\ Be^{-a}(1-a) = 0.$

These give $A = B = 0$. So [3] reduces to $y(x) = 0$, which is not an eigen function.

Case (ii): Let $\lambda = -\mu^2$, where $\mu \neq 0$. Then [1] reduces to

$$[D^2 + 2D + 1 - \mu^2]y = 0 \qquad [5]$$

whose auxiliary equation is $D^2 + 2D + (1-\mu^2) = 0$.

This gives $D = \dfrac{-2 \pm \sqrt{[4 - 4(1-\mu^2)]}}{2} = -1 \pm \mu.$

Hence solution of [5] is

$$y(x) = Ae^{(-1+\mu)x} + Be^{-(1+\mu)x} \qquad [6]$$

so that $y'(x) = A(-1+\mu)e^{(-1+\mu)x} - B(1+\mu)e^{-(1+\mu)x}$ [7]

Using [2], [6] and [7] reduce to

$$0 = A + B \text{ and } 0 = A(-1+\mu)e^{(-1+\mu)a} - B(1+\mu)e^{-(1+\mu)a}$$

i.e., $A + B = 0.$ and $A(\mu - 1)e^{\mu a} - B(\mu+1)e^{-\mu a} = 0.$

These give $A + B = 0$. So [6] reduces to $y(x) = 0$, which is not an eigen function.

Case (iii): Let $\lambda = \mu^2$, where $\mu \neq 0$. Then [1] reduces to

$$(D^2 + 2D + 1 + \mu^2)y = 0. \qquad [8]$$

whose auxiliary equation is $D^2 + 2D + 1 + \mu^2 = 0$.

This gives $D = \dfrac{-2 \pm \sqrt{[4 - 4(1+\mu^2)]}}{2} = -1 \pm i\mu.$

Hence solution of [8] is

$$y(x) = e^{-x}(A\cos\mu x + B\sin\mu x). \qquad [9]$$

From [9], $y'(x) = -e^{-x}(A\cos\mu x + B\sin\mu x)$

$$+ e^{-x}(-A\mu\sin\mu x + B\mu\cos\mu x). \qquad [10]$$

Using [2], [9] and [10] reduce to

$$0 = A,\ 0 = e^{-a}(A\cos\mu a + B\sin\mu a) + \mu e^{-a}(-A\sin\mu a + B\cos\mu a)$$

$\therefore \qquad A = 0$ and $B(\mu \cos \mu a - \sin \mu a) = 0.$ [11]

Consider $\qquad B(\mu \cos \mu a - \sin \mu a) = 0.$

If $B = 0$ then with $A = 0$, [9] reduces to $y(x) = 0$, which is not an eigen function. So we take $B \neq 0$ for the existence of eigen functions. Since $B \neq 0$, [12] reduces to

$$\mu \cos \mu a - \sin \mu a = 0 \quad \text{or} \quad \sin \mu a = \mu \cos \mu a,$$

i.e., $\qquad \tan \mu a = \mu.,$

which is a trigonometrical equation in μ. Let $\mu_n\ (n = 1, 2, 3, \cdots)$ be the positive roots of [13]. With $A = 0$, [9] reduces to

$$y(x) = Be^{-x} \sin \mu x.$$

Thus the required eigen function $y(x)$ with the corresponding eigen values λ_n are given by (taking $B = 1$)

$$y_n(x) = e^{-x} \sin \mu_n x,\ \lambda_n = \mu_n^2 \qquad (n = 1, 2, 3, \cdots)$$

where μ_n are the positive roots of [13].

EXERCISES

1. Define (i) Inner product (ii) Norm (iii) Orthogonal function.

2. Show that the functions $1 - x$, $1 - 2x + \frac{1}{2}x^2$ and $1 - 3x + \frac{3}{2}x^2 - \frac{1}{6}x^3$ are orthogonal with respect to e^{-x} on $0 \leq x \infty$. Determine the corresponding orthogonal functions.

   $$\left[\textit{Ans.}\ 1 - x, 1 - 2x + \frac{1}{2}x^2, 1 - 3x + \frac{3}{2}x^2 - \frac{1}{6}x^3 \right]$$

3. Show that the functions $\sin x, \sin 2x, \sin 3x, \ldots$ are orthogonal on the interval $(0, \pi)$.

4. Show that the functions $f_1(x) = 4$ and $f_2(x) = x^3$ are orthogonal on the interval $(-2, 2)$ and determine the constants A and B so that $f_3(x) = 1 + Ax + Bx^2$ is orthogonal to both f_1 and f_2.

5. Show that the set $\left\{ \sqrt{\frac{2}{l}} \cos \frac{n\pi x}{L} \right\}$ $(n = 1, 2, 3, \ldots)$ is orthonormal on the interval $(0, L)$.

6. Show that the functions $\{\cos mx\}$ $(m = 0,1,2,\ldots)$ form orthogonal set of functions on the interval $-\pi \le x \le \pi$. Determine the corresponding orthonormal set of functions.

7. Construct three functions which are orthogonal over $-1 \le x \le 1$ w.r.t. $(1-x^2)^{\frac{-1}{2}}$.

8. Form an orthonormal orthogonal set from the functions $e^{-x}, e^{-2x}, e^{-3x}, e^{-4x}$ on the interval 0 to ∞.

9. Generate an orthonormal set of functions in the interval $(-1,1)$ from the sequence $1, x, x^2, x^3, \ldots$ with respect to the weight function e^{-x}.

10. Is it possible to use Gram-Schmidt procedure to get from $x, 1-x, 3+2x$ an orthogonal set of functions in $(0,1)$? Explain.

11. Given the set of functions $1, x, x^2, x^3, \ldots$, obtain from these a set of functions which are mutually orthonormal in $(-1,1)$ with respect to the weight function x.

5.2 Chebyshev Polynomials

The chebyshev polynomials of the first kind $T_n(x)$, defined on $[-1,1]$ are given by

$$\left.\begin{aligned} T_n(x) &= \cos(x\cos^{-1}x) = \cos n\theta \\ \text{where} \quad \theta &= \cos^{-1}x \text{ or } x = \cos\theta \end{aligned}\right\} \qquad [1]$$

and 'n' is a non-negative integer.

Note: Chebyshev polynomials are also known as Tchebicheft, tchebieheff or Tschebyscheff.

Theorem: $T_n(x)$ and $U_n(x)$ are independent solutions of chebyshev's equation. $(1-x^2)y'' - xy' + n^2y = 0$

Proof: Given $(1-x^2)y'' - xy' + n^2y = 0$ [2]

By data, $T_n(x)$ is a solution of the Equation [2],

we have, $$T_n(x) = \cos(n\cos^{-1}x) \qquad [3]$$

Hence, find $T_n'(x)$, $T_n''(x)$ and substitute from $T_n'(x)$, $T_n''(x)$ and $T_n(x)$ into [2]

$$T_n'(x) = \frac{d}{dx}\{T_n(x)\} = \frac{d}{dx}\left\{\cos(n\cos^{-1}x)\right\}$$

$$= -\sin(n\cos^{-1}x)\cdot n\cdot\frac{-1}{\sqrt{(1-x^2)}}$$

or $$\frac{d}{dx}\{T_n(x)\} = \frac{n}{\sqrt{(1-x^2)}}\sin(n\cos^{-1}x) \quad [4]$$

and $$\frac{d^2}{dx^2}\{T_n(x)\} = \frac{d}{dx}\left[\frac{d}{dx}T_n(x)\right] = n\cdot\frac{d}{dx}\left[(1-x^2)^{\frac{-1}{2}}\cdot\sin(n\cos^{-1}x)\right]$$

$$= n\left[-\frac{1}{2}\left(1-x^2\right)^{-\frac{3}{2}}\cdot(-2x)\cdot\sin(n\cos^{-1}x) + \left(1-x^2\right)^{-\frac{1}{2}}\cos(n\cos^{-1}x)\cdot n\cdot\frac{1}{\sqrt{(1-x^2)}}\right]$$

$$= \frac{nx}{(1-x^2)^{\frac{3}{2}}}\sin(n\cos^{-1}x) - \frac{n^2}{1-x^2}\cos(n\cos^{-1}x)$$

Using [3], [4] and [5], we have

$$(1-x^2)\frac{d^2}{dx^2}T_n(x) - x\cdot\frac{d}{dx}T_n(x) + n^2T_n(x)$$

$$= \frac{nx}{(1-x^2)^{\frac{1}{2}}}\sin(n\cos^{-1}x) - n^2\cos(n\cos^{-1}x)$$

$$-\frac{nx}{(1-x^2)^{\frac{1}{2}}}\sin(n\cos^{-1}x) + n\cos(n\cos^{-1}x) = 0$$

$\Rightarrow$ $T_n(x)$ is a solution of equation [2]

Similarly we can prove that $U_n(x)$ is the solution [2]

where $U_n(x) = \sin(n\cos^{-1}x)$

To show that $T_n(x)$ and $U_n(x)$ are independent solution of [1]. We have, by definition

$$T_n(x) = \cos(n\cos^{-1}x) \text{ and } U_n(x) = \sin(n\cos^{-1}x).$$

$\therefore$ $T_n(1) = \cos(n\cos^{-1}1) = \cos(n\times 0) = 1$

and $U_n(1) = \cos(n\cos^{-1}1) = \cos(n\times 0) = 0.$

Hence $U_n(x)$ cannot be expressed as a constant multiple of $T_n(x)$. This shows that $T_n(x)$ and $U_n(x)$ are independent solutions of [2].

Note: Two solutions $U(x)$ and $V(x)$ are said to be linearly independent if $U(x)$ cannot be expressed as a constant multiple of $V(x)$ i.e., $U(x) = kV(x)$, where k is a constant.

(B) EXAMPLES

1. Find the values of the following.

(a) $T_n(1)$ (b) $T_n(-1)$ (c) $T_{2n}(0)$ (d) $T_{2n+1}(0)$

Solution: We know that

$$T_n(x) = \cos(n\cos^{-1} x) \qquad [1]$$

Put $x = 1$ in [1], we get

(a) $T_n(1) = \cos(n\cos^{-1} 1) = \cos(n \times 0) = 1$

Put $x = -1$, in [1], we get

(b) $T_n(-1) = \cos(n\cos^{-1}(-1)) = \cos(n\pi) = (-1)^n$

(c) Replace x by 0 and n by $2n$ in [1], we get

$$T_{2n}(0) = \cos(2n\cos^{-1}(0)) = \cos\left(2n \times \frac{1}{2}\pi\right) = \cos n\pi = (-1)^n$$

(d) Replace x by 0 and n by $2n+1$ in [1], we get

$$T_{2n+1}(0) = \cos\left[(2n+1)\cos^{-1}(0)\right] = \cos\left[(2n+1)\frac{1}{2}\pi\right] = 0.$$

2. Find the values of the following

(a) $U_n(1)$ (b) $U_n(-1)$ (c) $U_{2n}(0)$ (d) $U_{2n+1}(0)$

Solution: We have

$$U_n(x) = \sin(n\cos^{-1} x) \qquad [1]$$

(a) Put $x = 1$, in [1], we have

$$\boxed{U_n(1) = \sin(n\cos^{-1} 1) = \sin 0 = 0}$$

(b) Put $x = -1$, in [1], we have

$$U_n(-1) = \sin(n\cos^{-1}(-1)) = \sin n\pi = 0,$$

Now putting $x = 0$, in [1], we have

$$\boxed{U_n(0) = \sin(n\cos^{-1} 0) = \sin n\pi}$$

(c) If n is even, say $n = 2m$

$$U_{2m}(0) = \sin m\pi = 0$$

$$\therefore \quad \boxed{U_{2m}(0) = 0}$$

(d) If n is odd, say $n = 2m + 1$

$$U_{2m+1}(0) = \sin(2m+1)\frac{\pi}{2} = \sin\left(m\pi + \frac{\pi}{2}\right)$$

$$= (-1)^m \sin\frac{\pi}{2} = (-1)^n$$

$$\therefore \quad \boxed{U_{2m}(0) = 0}$$

Orthogonal properties of Chebysher polynomials

Examples

1. $$\int_{-1}^{1} \frac{T_m(x)T_n(x)}{\sqrt{(1-x^2)}}dx = \begin{cases} 0, & m \neq n \\ \frac{\pi}{2}, & m = n \neq 0 \\ \pi, & m = n = 0 \end{cases}$$

2. $$\int_{-1}^{1} \frac{U_m(x)U_n(x)}{\sqrt{(1-x^2)}}dx = \begin{cases} 0, & m \neq n \\ \frac{\pi}{2}, & m = n \neq 0 \\ \pi, & m = n = 0 \end{cases}$$

Solution: (1) We know that

$$T_m(x) = \cos(m\cos^{-1} x) \text{ and } T_n(x) = \cos(n\cos^{-1} x) \qquad [1]$$

put $\quad x = \cos\theta;$

$\therefore \quad T_m(\cos\theta) = \cos(m\cos^{-1}\cos\theta) = \cos m\theta$

Similarly, $T_n(\cos\theta) = \cos n\theta$

Let $\quad I = \int_{-1}^{1} \frac{T_m(x)T_n(x)}{\sqrt{(1-x^2)}}dx$

[(put $x = \cos\theta$; $dx = -\sin\theta d\theta$] and using [2] and [3], we have

$$I = \int_{\pi}^{0} \frac{\cos m\theta \cos n\theta}{\sin\theta}(-\sin\theta)d\theta$$

or $\quad I = \int_0^{\pi} \cos m\theta \cos n\theta \, d\theta \qquad [4]$

Case (i): Let $m \neq n$.

$\Rightarrow \quad (m-n) \neq 0,$

$\Rightarrow \quad$ Equation [4] yields

$$I = \frac{1}{2}\int_0^{\pi} 2\cos m\theta \cos n\theta \, d\theta$$

$$= \frac{1}{2}\int_0^{\pi} [\cos(m+n)\theta + \cos(m-n)\theta] d\theta$$

$$= \frac{1}{2}\left[\sin\frac{(m+n)\theta}{m+n} + \sin\frac{(m-n)\theta}{m-n}\right]_0^{\pi} = 0$$

$\therefore \quad \int_{-1}^{1} \frac{T_m(x)T_n(x)}{\sqrt{(1-x^2)}}dx = 0,\quad$ where $m \neq n \qquad$ using [3]

Case (ii): Let $m = n \neq 0$. Then [4] yields

$$I = \int_0^{\pi} \cos^2 m\theta \, d\theta = \int_0^{\pi} \frac{1+\cos 2m\theta}{2m} d\theta$$

$$= \frac{1}{2}\left[\theta + \frac{\sin 2m\theta}{2m}\right]_0^{\pi} = \frac{\pi}{2}$$

$\therefore \quad \int_{-1}^{1} \frac{T_m(x)T_n(x)}{\sqrt{(1-x^2)}}dx = \frac{\pi}{2}\quad$ where $m = n \neq 0 \qquad$ using [3]

Case (iii): Let $m = n = 0$. Then $\cos m\theta = \cos n\theta = 1$. Then [4] yields

$$I = \int_0^{\pi} (1)(1)\, dx = [\theta]_0^{\pi} = \pi.$$

$\therefore$ $\displaystyle\int_{-1}^{1} \frac{T_m(x)T_n(x)}{\sqrt{(1-x^2)}} dx = \pi,$ where $m = n = 0$ using [3]

i.e., $$\int_{-1}^{1} \frac{T_0(x)T_0(x)}{\sqrt{(1-x^2)}} dx = \pi.$$

(2) We have, by definition

$$U_m(x) = \sin(m\cos^{-1} x) \text{ and } U_n(x) = \sin(n\cos^{-1} x) \qquad [1]$$

$$\therefore \quad U_m(\cos\theta = \sin(m\cos^{-1}\theta) = \sin m\theta \qquad [2]$$

Similarly, $U_n(\cos\theta) = \sin n\theta$

Let $$J = \int_{-1}^{1} \frac{U_m(x)U_n(x)}{\sqrt{(1-x^2)}} dx \qquad [3]$$

Putting $x = \cos\theta$ so that $dx = \sin\theta\, d\theta$

and using [2], [3] gives

$$J = \int_{\pi}^{0} \frac{\sin m\theta \sin n\theta}{\sin\theta} (-\sin\theta)\, d\theta$$

or $$J = \int_0^{\pi} \sin m\theta \sin n\theta\, d\theta \qquad [4]$$

Case (i): Let $m \neq n$ so that $(m-n) \neq 0$. Then [4] gives

$$J = \frac{1}{2}\int_0^{\pi} 2\sin m\theta \sin n\theta\, d\theta$$

$$= \frac{1}{2}\int_0^{\pi} [\cos(m-n)\theta - \cos(m+n)\theta]\, d\theta$$

$$= \frac{1}{2}\left[\frac{\sin(m+n)\theta}{m+n} + \frac{\cos(m+n)\theta}{m+n}\right]_0^{\pi} = 0.$$

$\therefore$ $\displaystyle\int_{-1}^{1} \frac{U_m(x)U_n(x)}{\sqrt{(1-x^2)}} dx = 0,$ when $m \neq n$ using [3]

Case (ii): Let $m = n \neq 0$. Then [4] yields

$$J = \int_0^{\pi} \sin^2 m\theta \, d\theta = \int_0^{\pi} \frac{1 - \cos 2m\theta}{2} d\theta$$

$$= \frac{1}{2}\left[\theta - \frac{\sin 2m\theta}{2m}\right]_0^{\pi} = \frac{\pi}{2}$$

$$\therefore \quad \int_{-1}^{1} \frac{U_m(x)U_n(x)}{\sqrt{(1-x^2)}} dx = \frac{\pi}{2}, \quad \text{when } m = n \neq 0.$$

Case (iii): Let $m = n = 0$.

Then $\sin m\theta = \sin n\theta = 0$.

Hence [4] gives

$$J = \int_0^{\pi} (0)(0) \, d\theta = 0$$

$$\therefore \quad \int_{-1}^{1} \frac{U_m(x)U_n(x)}{\sqrt{(1-x^2)}} dx = 0, \quad \text{when } m = n = 0. \qquad \text{using [3]}$$

Recurrence formula for $T_n(x)$ and $U_n(x)$

1. $T_{n+1}(x) - 2xT_n(x) + T_{n-1}(x) = 0$ **2.** $(1-x^2)T_n'(x) = -nxT_n(x) + nT_{n-1}(x)$

3. $U_{n+1}(x) - 2xU_n(x) + U_{n-1}(x) = 0$ **4.** $(1-x^2)U'(x) = -nxU_n(x) + nU_{n-1}(x)$

Proof (1): $T_{n+1}(x) - 2xT_n(x) + T_{n-1}(x) = 0$. By definition of $T_n(x)$, we have

$$T_n(x) = \cos(n \cos^{-1} x) \qquad [1]$$

[put $x = \cos\theta$]

$$\therefore \quad T_n(\cos\theta) = \cos(n\cos^{-1} x) = \cos n\theta, \qquad [2]$$

so that $T_{n+1}(\cos\theta) = \cos(n+1)\theta$, $T_{n-1}(\cos\theta) = \cos(n-1)\theta$ [3]

We are to show that

$$T_{n+1}(x) - 2xT_n(x) + T_{n-1}(x) = 0 \qquad [4]$$

Replacing x by $\cos\theta$ in [4], we must now prove that

$$T_{n+1}(\cos\theta) - 2\cos\theta T_n(\cos\theta) + T_{n-1}(\cos\theta) = 0, \qquad [5]$$

$$\Rightarrow \quad \cos(n+1)\theta - 2\cos\theta\cos n\theta + \cos(n-1)\theta = 0 \qquad \text{using [2] and [3]}$$

$$\Rightarrow \quad \cos(n+1)\theta + \cos(n-1)\theta - 2\cos\theta\cos n\theta = 0 \qquad [6]$$

Now LHS of [6] $= 2\cos n\theta\cos\theta - 2\cos\theta\cos n\theta = 0$
which proves [6] and hence [4] is true.

Proof (2): $(1-x^2)T_n'(x) = -nxT_n(x) + T_{n-1}(x)$

We know that

$$T_n(x) = \cos(n\cos^{-1}x) \qquad [1]$$

$$\therefore \quad T_n'(x) = -\sin(n\cos^{-1}x)\cdot\frac{-n}{\sqrt{(1-x^2}}$$

$$\therefore \quad T_n'(x) = \sin(n\cos^{-1}\cos\theta)\cdot\frac{-n}{\sqrt{(1-\cos^2\theta}}$$

$$\text{or} \quad T_n'(x) = \frac{n\sin n\theta}{\sin\theta} \qquad [2]$$

We are to show that

$$(1-x^2)T_n'(x) = -nxT_n(x) + T_{n-1}(x) \qquad [3]$$

putting $x = \cos\theta$ and using $T_{n-1}(\cos\theta) = \cos(n-1)\theta$

[3] may be written as

$$\sin^2\theta\frac{n\sin n\theta}{\sin\theta} = -n\cos\theta\cos n\theta + n\cos(n-1)\theta$$

$$\Rightarrow \quad \sin\theta\sin n\theta = \cos(n-1)\theta - \cos\theta\cos n\theta \qquad [4]$$

RHS of [4] $= \cos(n\theta-\theta) - \cos\theta\cos n\theta$

$= \cos n\theta\cos\theta + \sin n\theta\sin\theta - \cos\theta\cos n\theta$

$= \sin n\theta\sin\theta = \text{LHS of [4]}$

which proves [4] and hence [3] is true

Proof (3): $U_{n+1}(x) - 2xU_n(x) + U_{n-1}(x) = 0$

We have $U_n(x) = \sin(n\cos^{-1}x)$

$= \sin n\theta$ when $x = \cos\theta$

$\therefore$ putting $x = \cos\theta$

$$U_{n+1}(x) - 2xU_n(x) + U_{n-1}(x)$$

$$= \sin(n+1)\theta - 2\cos\theta \sin n\theta + \sin(n-1)\theta$$

$$= \sin(n+1)\theta - [\sin(n+1)\theta + \sin(n-1)\theta] + \sin(n-1)\theta = 0$$

Proof (4): We have $U_n(x) = \sin(n\cos^{-1} x)$

$$= \sin n\theta \qquad [x = \cos\theta]$$

Also, $U_n'(x) = \dfrac{d}{dx}\left[\sin(n\cos^{-1} x)\right]$

$$= \cos(n\cos^{-1} x) \cdot \frac{-n}{\sqrt{(1-x^2)}}$$

$$= -\frac{n}{\sin\theta} \cdot \cos\theta \qquad [x = \cos\theta]$$

$\therefore$ putting $x = \cos\theta$

$$(1-x^2)U_n'(x) = (1-\cos^2\theta)\frac{-n}{\sin\theta} \cdot \cos n\theta$$

$$= -n\sin\theta \cdot \cos n\theta$$

$$= n(\cos\theta \cdot \sin n\theta - \sin\theta \cos n\theta) - n\cos\theta \sin n\theta$$

$$= nU_{n+1}(x) - nxU_n(x)$$

$$\therefore \qquad (1-x^2)U'(x) = -nxU_n(x) + nU_{n-1}(x)$$

Theorem (1): To prove that

(i) $T_n(x) = \dfrac{1}{2}\left[\left\{x + i\sqrt{(1-x^2)}\right\}^n + \left\{x - i\sqrt{(1-x^2)}\right\}^n\right]$

(ii) $U_n(x) = \dfrac{1}{2}i\left[\left\{x + i\sqrt{(1-x^2)}\right\}^n - \left\{x - i\sqrt{(1-x^2)}\right\}^n\right]$

Proof (i): We have $T_n(x) = \cos(n\cos^{-1}\cos\theta) = \cos n\theta$

(put $x = \cos\theta$)

$$= \frac{1}{2}\left(e^{in\theta} + e^{-in\theta}\right)$$

$$= \frac{1}{2}\left\{\left(e^{i\theta}\right)^n + \left(e^{-i\theta}\right)^n\right\}$$

$$=\frac{1}{2}\left\{(\cos\theta+\mathrm{i}\sin\theta)^n+(\cos\theta-\mathrm{i}\sin\theta)^n\right\}$$

$$=\frac{1}{2}\left[\left\{\cos\theta+\mathrm{i}\sqrt{(1-\cos^2\theta)}\right\}^n+\left\{\cos\theta-\mathrm{i}\sqrt{(1-\cos^2\theta)}\right\}^n\right]$$

$$=\frac{1}{2}\left[\left\{x+\mathrm{i}\sqrt{(1-x^2)}\right\}^n+\left\{x-\mathrm{i}\sqrt{(1-x^2)}\right\}^n\right] \quad (\because\ x=\cos\theta)$$

Proof (ii): We have

$$U_n(x)=\sin(n\cos^{-1}x),\ \text{by definition,} \qquad (\text{put } x=\cos\theta)$$

$$=\sin n\theta=\frac{1}{2\mathrm{i}}\left(e^{\mathrm{i}n\theta}-e^{-\mathrm{i}n\theta}\right)=\frac{1}{2\mathrm{i}}\left\{\left(e^{\mathrm{i}\theta}\right)^n-\left(e^{-\mathrm{i}\theta}\right)^n\right\}$$

$$=\frac{1}{2\mathrm{i}}\left\{(\cos\theta+\mathrm{i}\sin\theta)^n-(\cos\theta-\mathrm{i}\sin\theta)^n\right\}$$

$$=\frac{1}{2\mathrm{i}}\left[\left\{\cos\theta+\mathrm{i}\sqrt{(1-\cos^2\theta)}\right\}^n-\left\{\cos\theta-\mathrm{i}\sqrt{(1-\cos^2\theta)}\right\}^n\right]$$

$$=\frac{1}{2\mathrm{i}}\left[\left\{x+\mathrm{i}\sqrt{(1-x^2)}\right\}^n-\left\{x-\mathrm{i}\sqrt{(1-x^2)}\right\}^n\right] \qquad (\because\ x=\cos\theta)$$

$$=\frac{-\mathrm{i}}{2}\left[\left\{x+\mathrm{i}\sqrt{(1-x^2)}\right\}^n-\left\{x-\mathrm{i}\sqrt{(1-x^2)}\right\}^n\right]$$

Theorem (2): To prove that

(i) $$T_n(x)=\sum_{r=0}^{\left[\frac{n}{2}\right]}(-1)^r\frac{n!}{(2r)!(n-2r)!}\left(1-x^2\right)^r x^{n-2r}$$

(ii) $$U_n(x)=\sum_{r=0}^{\left[\frac{(n-1)}{2}\right]}(-1)^r\frac{n!}{(2r+1)!\,(n-2r-1)!}\left(1-x^2\right)^{r+\frac{1}{2}}x^{x-2r-1}$$

Proof (i): We know that,

$$T_n(x) = \frac{1}{2}\left[\left\{x + i\sqrt{(1-x^2)}\right\}^n + \left\{x - i\sqrt{(1-x^2)}\right\}^n\right]$$

$$= \frac{1}{2}\left[\sum_{s=0}^{n} {}^nc_s x^{n-s}\left\{i\sqrt{(1-x^2)}\right\}^s + \sum_{s=0}^{n} {}^nc_s x^{n-s}\left\{-i\sqrt{(1-x^2)}\right\}^s\right]$$

$$\left\{\begin{aligned} &\because\ (a+b)^n = a^n + {}^nc_1 a^{n-1}b + {}^nc_2 a^{n-2}b^2 + \cdots \\ &\qquad\qquad + {}^nc_s a^{n-s}b^s + \cdots + {}^nc_n b^n \\ &\text{i.e., } (a+b)^n = \sum_{s=0}^{n} {}^nc_s a^{n-s} b^s \end{aligned}\right\}$$

$$\because \quad T_n(x) = \frac{1}{2}\sum_{s=0}^{n} {}^nc_s a^{n-s}\left(1-x^2\right)^{\frac{s}{2}} i^s\{1+(-1)^s\} \qquad [1]$$

$$\text{where } 1+(-1)^s = \begin{cases} 0, & \text{if } s \text{ is odd} \\ 2, & \text{if } s \text{ is even} \end{cases} \qquad [2]$$

Using [2], [1] reduces to

$$T_n(x) = \sum_{s \text{ even}, \le n} {}^nc_s a^{n-s}\left(1-x^2\right)^{\frac{s}{2}} i^s \qquad [3]$$

Since s is even, we take $s = 2r$, where r is an integer.

Hence $s \le n \Rightarrow 2r \le n \Rightarrow r \le \dfrac{n}{2}$.

Now, if n is even, r goes from 0 to $\dfrac{n}{2}$, while if n is odd, r goes from 0 to $\dfrac{(n-1)}{2}$; that is, in all cases, r goes from 0 to $\left[\dfrac{n}{2}\right]$, where

$$\left[\frac{1}{2}n\right] = \begin{cases} \dfrac{n}{2}, & \text{if } n \text{ is even} \\ \dfrac{(n-1)}{2}, & \text{if } n \text{ odd} \end{cases}$$

Hence with $s = 2r$ and the above arguments, [3] reduces to

$$T_n(x) = \sum_{r=0}^{\left[\frac{n}{2}\right]} {}^nc_{2r} x^{n-2r}\left(1-x^2\right)^r i^{2r}$$

$$= \sum_{r=0}^{\left[\frac{n}{2}\right]} \frac{n!}{(n-2r)!\,(2r)!} x^{n-2r} \left(1-x^2\right)^r (-1)^r, \quad \left[\because\ \mathrm{i}^{2r} = \left(\mathrm{i}^2\right)^r = (-1)^r\right]$$

Hence the proof.

Proof (ii): We know that

$$U_n(x) = -\frac{1}{2}\,\mathrm{i}\left[\left\{x + \mathrm{i}\sqrt{(1-x^2)}\right\}^n - \left\{x - \mathrm{i}\sqrt{(1-x^2)}\right\}^n\right]$$

$$= -\frac{1}{2}\,\mathrm{i}\left[\sum_{s=0}^{n} {}^nc_s x^{n-s}\left\{\mathrm{i}\sqrt{(1-x^2)}\right\}^s - \sum_{s=0}^{n} {}^nc_s x^{n-s}\left\{-\mathrm{i}\sqrt{(1-x^2)}\right\}^s\right]$$

$$= -\frac{1}{2}\,\mathrm{i}\sum_{s=0}^{n} {}^nc_s x^{n-s}\left(1-x^2\right)^{\frac{s}{2}} - \mathrm{i}^s\left\{1-(-1)^s\right\} \qquad [4]$$

$$\text{But} \quad 1-(-1)^s = \begin{cases} 0, & \text{if } s \text{ is even} \\ 2, & \text{if } s \text{ is odd} \end{cases} \qquad [5]$$

Using [5], [4] reduces to

$$U_n(x) = -\mathrm{i} \sum_{\substack{s \text{ odd} \\ s \le n}}^{n} {}^nc_s x^{n-s}\left(1-x^2\right)^{\frac{s}{2}} \cdot \mathrm{i}^s \qquad [6]$$

Since s is odd in [6], we take $s = 2r+1$, where r is an integer.

Hence $\quad s \le n \Rightarrow 2r+1 \le n \Rightarrow r \le \dfrac{(n-1)}{2}.$

Now, $\quad$ if n is odd, r goes from 0 to $\dfrac{(n-1)}{2}$,

if n is even, r goes from 0 to $\dfrac{(n-2)}{2}$;

that is in all cases, r goes from 0 to $\left[\dfrac{(n-1)}{2}\right]$,

$$\text{where} \quad \left[\frac{1}{2}(n-1)\right] = \begin{cases} \dfrac{(n-1)}{2}, & \text{if } s \text{ is odd} \\ \dfrac{(n-2)}{2}, & \text{if } s \text{ is even.} \end{cases}$$

Hence with $s = 2r+1$ and the above arguments, [6] reduces to

$$U_n(x) = -\mathrm{i} \sum_{r=0}^{\left[\frac{(n-1)}{2}\right]} {}^nc_{2r+1} x^{n-2r-1} \left(1-x^2\right)^{r+\frac{1}{2}} \mathrm{i}^{2r+1}$$

$$= \sum_{r=0}^{\left[\frac{(n-1)}{2}\right]} (-1)^r \frac{n!}{(2r+1)!\,(n-2r-1)!} \left(1-x^2\right)^{r+\frac{1}{2}} \cdot x^{x-2r-1}$$

$$\left[\because \ \mathrm{i}^{2r+1} = \mathrm{i}^{2r} \cdot \mathrm{i} = \left(\mathrm{i}^2\right)^r \cdot \mathrm{i} = (-1)^r \cdot \mathrm{i} \text{ and } \mathrm{i}^2 = -1\right]$$

Generating function

To prove that

(i) $$\frac{1-t^2}{1-2tx+t^2} = T_0(x) + 2\sum_{n=1}^{\infty} T_n(x) \cdot t^n$$

(ii) $$\frac{\sqrt{(1-x^2)}}{1-2tx+t^2} = \sum_{n=0}^{\infty} U_{n+1}(x) t^n.$$

Proof (i): Putting $x = \cos\theta = \frac{1}{2}(e^{\mathrm{i}\theta} + e^{-\mathrm{i}\theta})$, we have

$$\frac{1-t^2}{1-2tx+t^2} = \frac{1-t^2}{1-t(e^{\mathrm{i}\theta}+e^{-\mathrm{i}\theta})+t^2} = \frac{(1-t^2)}{(1-te^{\mathrm{i}\theta})(1-te^{-\mathrm{i}\theta})}$$

$$= (1-t^2)\,(1-e^{\mathrm{i}\theta}t)^{-1}\,(1-e^{-\mathrm{i}\theta}t)^{-1}$$

$$= (1-t^2) \sum_{n=0}^{\infty} \left(e^{\mathrm{i}\theta}t\right)^r \sum_{n=0}^{\infty} \left(e^{-\mathrm{i}\theta}t\right)^s \qquad \text{(By the binomial series)}$$

$$= (1-t^2) \sum_{n=0}^{\infty} \sum_{n=0}^{\infty} e^{\mathrm{i}(r-s)\theta} t^{r+s}$$

$$= \sum_{n=0}^{\infty} \sum_{n=0}^{\infty} e^{\mathrm{i}(r-s)\theta} t^{r+s} - \sum_{n=0}^{\infty} \sum_{n=0}^{\infty} e^{\mathrm{i}(r-s)\theta} t^{r+s+2} \qquad [1]$$

It should be noted that we need not consider the summation for the coefficient of t^0. Taking $r=0, s=0$ in the first summation in [1], we have coefficient of t^2 in LHS of the required result

$$= e^{i(0-0)\theta} = e^0 = 1 = T_0(x) \qquad [\because T_0(x)=1]$$

Again coefficient of t in [1] is $e^{i(0-1)\theta} + e^{i(1-0)\theta}$

[obtained by putting $r=0$, $s=1$ and $r=1$ and $s=0$ in the first summation]

$$= e^{-i\theta} + e^{i\theta} = 2\cos\theta = 2x = 2T_1(x)$$

In general, the coefficient of t^n in [1] is

$$\sum_{r=0}^{n} e^{i[r-(n-r)]\theta} - \sum_{r=0}^{n-2} e^{i[r-(n-r-2)]\theta}$$

[Obtained by putting $r+s=n$ i.e., $s=n-r$ in the first summation and $r+s+2=n$ i.e., $s=n-r-2$ in the second summation $s \geq 0$ gives $n-r \geq 0$ i.e., $r \leq n$ and $n-r-2 \geq 0$ i.e., $r \leq n-2$]

$$= e^{-in\theta} \sum_{r=0}^{r} e^{i2r\theta} - e^{-i(n-2)\theta} \sum_{r=0}^{n-2} e^{i2r\theta}$$

$$= e^{-in\theta}\left[1 + e^{2i\theta} + e^{4i\theta} + \cdots + \text{ to } (n+1) \text{ terms}\right]$$

$$-e^{-i(n-2)\theta}\left[1 + e^{2i\theta} + e^{4i\theta} + \cdots \text{ to } (n-1) \text{ terms}\right]$$

$$= e^{-in\theta} \cdot \frac{1-\left(e^{2i\theta}\right)^{n+1}}{1-e^{2i\theta}} - e^{-i(n-2)\theta} \cdot \frac{1-\left(e^{2i\theta}\right)^{n-1}}{1-e^{2i\theta}}$$

(The two series within brackets are geometric progression)

$$= \frac{e^{-in\theta} - e^{i(n-2)\theta} - e^{-i(n-2)\theta} + e^{in\theta}}{(1-e^{2i\theta})}$$

$$= \frac{e^{-in\theta}(1-e^{2i\theta}) + e^{in\theta}(1-e^{2i\theta})}{(1-e^{2i\theta})}$$

$$= e^{-in\theta} + e^{in\theta}$$

$$= 2\cos n\theta = 2\cos(n\cos^{-1} x) = 2T_n(x)$$

$\therefore$ From [1], we have

$$\frac{1-t^2}{1-2tx+t^2} = T_0(x) + 2\sum_{n-1}^{\infty} T_n(x)\cdot t^n$$

Proof (ii): Here, we must show that in the expansion of LHS in ascending powers to t, the coefficient of t^n for $n \geq 1$ is $U_{n+1}(x)$.

Putting $x = \cos\theta = \frac{1}{2}(e^{i\theta} + e^{-i\theta})$

LHS is $\dfrac{\sqrt{(1-x^2)}}{1-2xt+t^2} = \dfrac{\sqrt{(1-\cos^2\theta)}}{1-2t\cos\theta+t^2}$

$$= \frac{\sin\theta}{1-2t\cdot\frac{1}{2}(e^{i\theta}+e^{-i\theta})+t^2} = \frac{\sin\theta}{1-e^{i\theta}t-e^{-i\theta}+e^{i\theta}t\times e^{-i\theta}t}$$

$$= \frac{\sin\theta}{(1-e^{i\theta}t)-e^{-i\theta}t(1-e^{i\theta}t)} - \frac{\sin\theta}{(1-e^{i\theta}t)(1-e^{i\theta}t)}$$

$$= \sin\theta(1-e^{i\theta}t)^{-1}(1-e^{-i\theta}t)^{-1}$$

$$= \sin\theta\cdot\sum_{r=0}^{\infty}\left(e^{i\theta}t\right)^r \sum_{s=0}^{\infty}\left(e^{-i\theta}t\right)^s$$

$$= \sin\theta\cdot\sum_{r=0}^{\infty}\sum_{s=0}^{\infty} e^{i(r-s)\theta}t^{r+s} \qquad [2]$$

We get the coefficient of t^n by taking $r+s=n$

(i.e., $s = n-r$, so that $s \geq 0 \Rightarrow n-r \geq 0 \Rightarrow r \leq n$ for the coefficient of t^n)

$\therefore$ the total coefficient of t^n is [2]

$$= \sin\theta\sum_{r=0}^{n} e^{i[r-(n-r)]\theta} = \sin\theta\sum_{r=0}^{n} e^{-in\theta}\cdot e^{2r\theta i}$$

$$= \sin\theta\cdot e^{-in\theta}\cdot\sum_{r=0}^{n} e^{2ir\theta}$$

$$= \sin\theta\cdot e^{-in\theta}\cdot\left[1+e^{2i\theta}+e^{4i\theta}+\cdots \text{ to } (n+1) \text{ terms}\right]$$

$$= \sin\theta \cdot e^{-in\theta} \cdot \frac{1 - e^{2i\theta(n+1)}}{1 - e^{2i\theta}}$$

$$= \frac{e^{i\theta} - e^{-i\theta}}{2i} \cdot \frac{e^{-in\theta}(1 - e^{2i\theta(n+1)})}{-e^{i\theta} \cdot (e^{i\theta} - e^{-i\theta})}$$

$$= \frac{e^{-i(n+1)\theta} \cdot (-1 + e^{2i\theta(n+1)})}{2i}$$

$$= \frac{e^{i(n+1)\theta} - e^{-i(n+1)\theta})}{2i} = \sin(n+1)\theta$$

$$= \sin\left\{(n+1)\cos^{-1} x\right\}$$

$$= U_{n+1}(x)$$

Hence, from [2] we have

$$\frac{\sqrt{(1-x^2)}}{1 - 2tx + t^2} = \sum_{r=0}^{\infty} U_{n+1}(x) \cdot t^n$$

1. Prove that $T_n'(x) = \dfrac{n}{\sqrt{(1-x^2)}} U_n(x)$

Proof: We have, by definitions

$$\left.\begin{cases} & T_n(x) = \cos(n\cos^{-1} x) \\ \text{and} & U_n(x) = \sin(n\cos^{-1} x) \end{cases}\right\} \quad [1]$$

$$\therefore \quad T_n'(x) = \frac{d}{dx}\cos(n\cos^{-1} x)$$

$$= -\sin(n\cos^{-1} x) \cdot \frac{-n}{\sqrt{(1-x^2)}}$$

$$= \frac{n}{\sqrt{(1-x^2)}} \sin(n\cos^{-1} x)$$

$$T_n'(x) = \frac{n}{\sqrt{(1-x^2)}} U_n(x)$$

2. Prove that $T_m\{T_n(x)\} = T_n\{T_m(x)\} = T_{nm}(x)$

Proof: We have, by definition,

$$T_m\{T_n(x)\} = T_m\left[\cos(n\cos^{-1}x)\right]$$

$$= \cos\left[m\cos^{-1}\{\cos(\cos^{-1}x)\}\right],$$

$$= \cos(nm\cos^{-1}x), \qquad [1]$$

Again $T_n\{T_m(x)\} = T_n[\cos(m\cos^{-1}x)]$, by definition

$$= \cos\left[n\cos^{-1}\{\cos(m\cos^{-1}x)\}\right], \text{ by definition again}$$

$$= \cos(nm\cos^{-1}x) \qquad [2]$$

Finally, $T_{mn}(x) = \cos(mn\cos^{-1}x)$ by definition [3]

From [1], [2] and [3], we get the required result.

3. Prove that $\{T_m(x)\}^2 - T_{n+1}(x)T_{n-1}(x) = 1 - x^2$

Proof: We have, by definition,

$$T_n(x) = \cos(n\cos^{-1}x)$$

putting $x = \cos\theta$, we get

$$T_n(\cos\theta) = \cos(n\cos^{-1}\cos\theta) = \cos n\theta \qquad [1]'$$

$$\therefore \quad [T_n(x)]^2 - T_{n+1}(x)T_{n-1}(x)$$

$$= \cos^2 n\theta - \cos(n+1)\theta\cos(n-1)\theta = \cos^2 n\theta - \cos(n\theta+\theta)\cos(n\theta-\theta)$$

$$= \cos^2 n\theta - (\cos^2 n\theta - \sin^2\theta) = \sin^2\theta$$

$$= 1 - \cos^2\theta = 1 - x^2$$

$$\left(\because\ C_{A+B}C_{A-B} = C_A^2 - S_B^2\right)$$

EXERCISES

1. Prove that $2\left[T_n(x)\right]^2 = 1 + T_{2n}(x)$

2. Prove that $T_m(x) - 2x\,T_{m-1}(x) + T_{m-2}(x) = 0$

3. Prove that $\int_{-1}^{+1} x^6\left(1-x^2\right)^{-\frac{1}{2}} T_8(x)dx = 0$

4. Prove that $T_{m+n}(x) + T_{m-n}(x) = 2T_m(x)T_n(x)$.

5. Show that $T_m'(x) = \dfrac{n}{\sqrt{(1-x^2)}}U_n(x)$

6. Show that $\sqrt{(1-x^2)}\,U_n(x) = xT_n(x) - T_{n+1}(x)$

7. Show that $\left\{\dfrac{1}{\sqrt{(1-x^2)}}\right\}U_n(x)$ satisfies the differential equation

$(1-x^2)y'' + 3xy' + (n^2-1)y = 0$

8. Show that Chebyshev's polynomials $T_n(x) = \cos(n\cos^{-1}x)$ are

$$\left[\textit{Solution}:\ (1-x^2)\frac{d^2y}{dx^2} - x\frac{dy}{dx} + x^2y = 0\right]$$

9. Show that the set of Chebyshev polynomials $T_n(x) = \cos(n\cos^{-1}x), (n = 0, 1, 2, \cdots)$ is orthogonal on $(-1, 1)$ w.r.t. the weight function $p(x) = \dfrac{1}{\sqrt{(1-x^2)}}$.

Appendix I

1. Matrices

$$A^{-1} = \frac{1}{|A|}\operatorname{adj} A. \quad (AB)^{-1} = B^{-1}A^{-1}$$

$$(AB)^T = B^T A^T \quad (A^T)^{-1} = (A^{-1})^T$$

2. Ordinary Differential Equations:

First order linear:

$$y' + p(x)y = q(x), \text{ I.F} = e^{\int p(x)dx}$$

$$\text{G.S} : y \cdot (\text{I.F}) = \int (\text{I.F})(Q|x)dx$$

Bessel equation: $x^2y'' + xy' + (x^2 - v^2)y = 0$

Legendre equation: $(1 - x^2)y'' - 2xy' + \lambda y = 0$

θ (degrees)	0	30	45	60	90	180	270	360
θ (radians)	0	$\frac{\pi}{6}$	$\frac{\pi}{4}$	$\frac{\pi}{3}$	$\frac{\pi}{2}$	π	$\frac{\pi}{2}$	2π
sin θ	0	$\frac{1}{2}$	$\frac{1}{\sqrt{2}}$	$\frac{\sqrt{3}}{2}$	1	0	−1	0
cos θ	1	$\frac{\sqrt{3}}{2}$	$\frac{1}{\sqrt{2}}$	$\frac{1}{2}$	0	−1	0	1
tan θ	0	$\frac{1}{\sqrt{3}}$	1	$\sqrt{3}$	∞	0	$-\infty$	0

3. Law of exponents

$$a^m \cdot a^n = a^{m+n}; (ab)^m = a^m b^m, (a^m)^n = a^{mn}, a^{m/n} = \sqrt[n]{a^m}$$

$$\text{if } a \neq 0; \frac{a^m}{a^n} = a^{m-n}, a^0 = 1, a^{-m} = \frac{1}{a^m}$$

4. Difference of like integer power, $n > 1$:

$$a^n - b^n = (a-b)(a^{n-1} + a^{n-2}b + a^{n-3}b^2 + \cdots + ab^{n-2} + b^{n-1})$$

For example,

$$a^3 - b^3 = (a-b)(a^2 + ab + b^2)$$

$$a^4 - b^4 = (a-b)(a^3 + a^2 b + ab^2 + b^2)$$

$$\tanh^{-1} x = \frac{1}{2}\log\frac{1+x}{1-x} = x + \frac{x^3}{3} + \frac{x^5}{5} + \cdots$$

5. *Binomial Series*:

$$(1+x)^n = 1 + nx + \frac{n(n-1)}{1\cdot 2}x^2 + \frac{n(n-1)(n-2)}{1\cdot 2\cdot 3}x^3 + \cdots$$

$$(1+x)^{-1} = 1 - nx + \frac{n(n+1)}{1\cdot 2}x^2 - \frac{n(n+1)(n+2)}{1\cdot 2\cdot 3}x^2 + \cdots$$

$$(1+x)^{-n} = 1 + nx + \frac{n(n+1)}{1\cdot 2}x^2 + \frac{n(n+1)(n+2)}{1\cdot 2\cdot 3}x^3 + \cdots$$

6. Progressions:

Numbers a, $a + d$, $a + 2d$, $\cdots$ are said to be in *Arithmetic progression (A.P.)*

Its *nth term* $T_n = a + \overline{n-1}\,d$ and *sum* $S_n = \frac{n}{2}(2a + \overline{n-1}\,d)$

Numbers a, ar, ar^2, $\cdots$ are said to be in *Geometric Progression (G.P.)*

Its *nth term* $T_n = ar^{n-1}$ and *sum* $S_n = \frac{a(1-r^n)}{1-r}$, $\lim_{n\to\infty} S_n = \frac{a}{1-r}$ when $|r| < 1$,

Numbers $a_1, a_2, a_3 \cdots$ are said to be in *Harmonic progresion (H.P)* if $\frac{1}{a_1}, \frac{1}{a_2}, \frac{1}{a_3}, \cdots$ are in *A.P.*

For any two numbers a and b, their

Arithmetic mean $= \frac{1}{2}(a+b)$,

Geometric mean $= \sqrt{ab}$,

Harmonic mean $= \frac{2ab}{(a+b)}$

For the first n natural numbers 1, 2, 3, ... n,

$$\sum n = \frac{n(n+1)}{2},$$

$$\sum n^2 = \frac{n(n+1)(2n+1)}{6},$$

$$\sum n^3 = \left\{\frac{n(n+1)}{2}\right\}^2$$

7. ***Stirling's approximation*:** When n is large $n! \sim \sqrt{2\pi n}.n^n e^{-n}$

8. Permutation and combination:

$${}^nP_r = \frac{n!}{(n-r)!};$$

$${}^nC_r = \frac{n!}{r!(n-r)!} = \frac{{}^nP_r}{r!}$$

$${}^nC_{n-r} = {}^nC_r, {}^nC_0 = 1 = {}^nC_n$$

$$\sin x = \cos\left(x - \frac{\pi}{2}\right) = \cos\left(\frac{\pi}{2} - x\right)$$

$$\cos x = \sin\left(x + \frac{\pi}{2}\right) = \sin\left(\frac{\pi}{2} - x\right)$$

$$\sin(\pi - x) = \sin x, \cos(\pi - x) = -\cos x$$

$$\cos^2 x = \frac{1}{2}(1 + \cos 2x), \sin^2 x = \frac{1}{2}(1 - \cos 2x)$$

$$\sin x \sin y = \frac{1}{2}[-\cos(x+y) + \cos(x-y)]$$

$$\cos x \cos y = \frac{1}{2}[\cos(x+y) + \cos(x-y)]$$

$$\sin x \cos y = \frac{1}{2}[\sin(x+y) + \sin(x-y)]$$

$$\cos x \sin y = \frac{1}{2}[\sin(x+y) - \sin(x-y)]$$

$$\sin u + \sin v = 2\sin\frac{u+v}{2}\cos\frac{u-v}{2}$$

$$\sin u - \sin v = 2\cos\frac{u+v}{2}\sin\frac{u-v}{2}$$

$$\cos u + \cos v = 2\cos\frac{u+v}{2}\cos\frac{u-v}{2}$$

$$\cos v - \cos u = 2\sin\frac{u+v}{2}\sin\frac{u-v}{2}$$

$$A\cos x + B\sin x = \sqrt{A^2+B^2}\cdot\cos(x\pm\delta)$$

$$\tan\delta = \frac{\sin\delta}{\cos\delta} = \mp\frac{B}{A}$$

9. Tangent, cotangent, secant, cosecant:

$$\tan x = \frac{\sin x}{\cos x},\quad \cot x = \frac{\cos x}{\sin x}$$

$$\sec x = \frac{1}{\cos x},\quad \csc x = \frac{1}{\sin x}$$

$$\tan(x\pm y) = \frac{\tan x \pm \tan y}{1\mp \tan x\tan y}$$

$$\tan 2x = \frac{2\tan x}{1-\tan^2 x}$$

Any *t*-ratio of $(n\cdot 90^0 \pm \theta)$

$= \pm$ same ratio of θ when n is even.

$= \pm$ co-ratio of θ when n is odd.

The sign + or – is to be decided from the quadrant in which $(n\cdot 90^0 \pm\theta)$ lies.

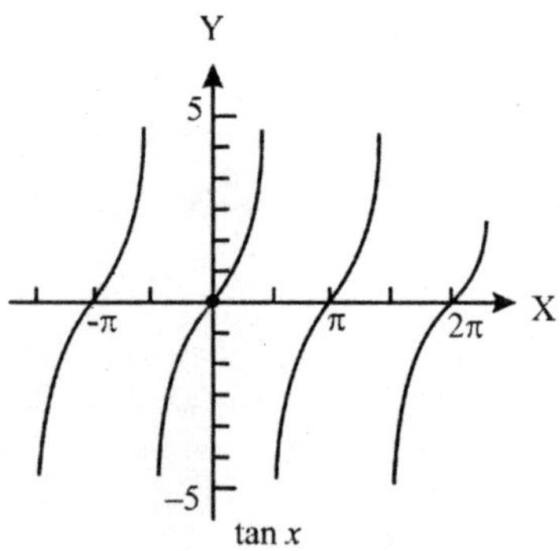

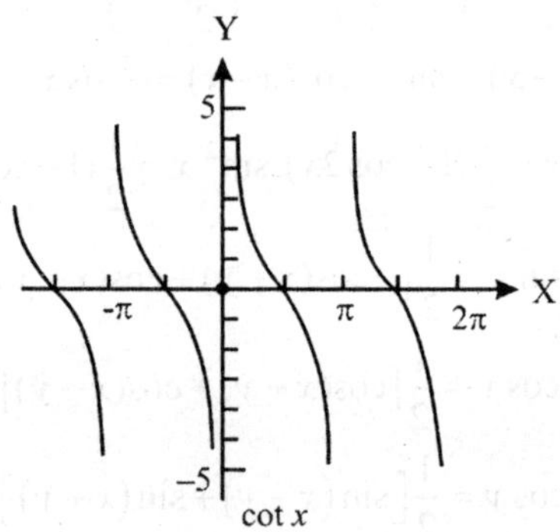

Examples:

$$\sin 570^\circ = \sin(6\times 90^\circ + 30^\circ) = -\sin 30^\circ = -\frac{1}{2};$$

$$\tan 315^\circ = \tan(3\times 90^\circ + 45^\circ) = -\cot 45^\circ = -1$$

In any $\Delta\, ABC$, $\dfrac{a}{\sin A} = \dfrac{b}{\sin B} = \dfrac{c}{\sin C}$ and $\cos C = \dfrac{a^2+b^2-c^2}{2ab}$.

10. Hyperbolic functions:

$$\sinh x = \frac{1}{2}(e^x - e^{-x}), \quad \cosh x = \frac{1}{2}(e^x + e^{-x})$$

$$\tanh x = \frac{\sinh x}{\cosh x}, \qquad \coth x = \frac{\cosh x}{\sinh x}$$

$$\cosh x + \sinh x = e^x, \quad \cosh x - \sinh x = e^{-x}$$

$$\cosh^2 x - \sinh^2 x = 1$$

$$\sinh^2 x = \frac{1}{2}(\cosh 2x - 1); \quad \cosh^2 x = \frac{1}{2}(\cosh 2x + 1)$$

$$\begin{cases} \sinh(x \pm y) = \sinh x \cosh y \pm \cosh x \sinh y \\ \cosh(x \pm y) = \cosh x \cosh y \pm \sinh x \sinh y \end{cases}$$

$$\tanh(x \pm y) = \frac{\tanh x \pm \tanh y}{1 + \tanh x \tanh y}$$

11. Exponential function e^x

$e = 2.71828\ 18284$

$$e^x e^y = e^{x+y}, \ \frac{e^x}{e^y} = e^{x-y}, \quad \left(e^x\right)^y = e^{xy}$$

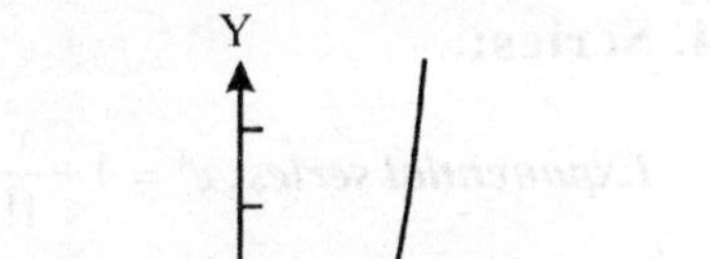

Exponential function e^x

12. Natural logarithm:

In x is the inverse of e^x and has base e

and $e^{\text{In}\, x} = x, \ e^{-\text{In}\, x} = e^{\text{In}\left(\frac{1}{x}\right)} = \frac{1}{x}.$

In $(xy) = \text{In}\, x + \text{In}\, y$

In $\left(\frac{x}{y}\right) = \text{In}\, x - \text{In}\, y$

In $(x^a) = a\, \text{In}\, x$

Logarithm of base ten $\log_{10} x$ or simply $\log x$

(known as common logarithm)

$\log x$ is the inverse of 10^x, and

$10^{\log x} = x, \ 10^{-\log x} = \frac{1}{x}.$

$\log x = M\, \text{In}\, x, \ M = \text{Log}\, e = 0.43429$

$\text{In}\, x = \frac{1}{m} \log x, \ \frac{1}{m} = 2.30258$

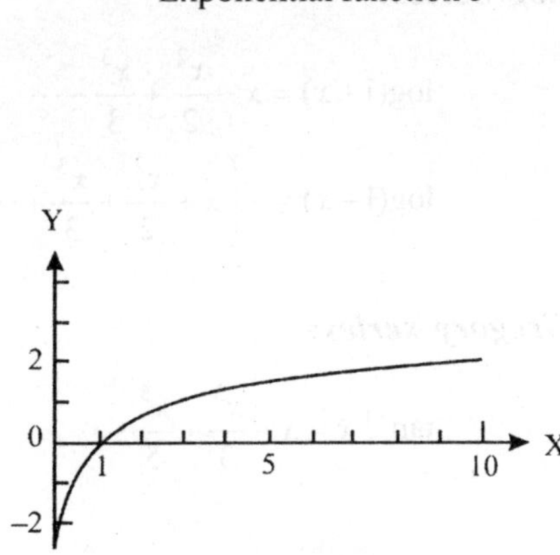

Natural Logarithm In x

13. Sine and Cosine functions:

$\sin x$ is odd, $\sin(-x) = -\sin x$

$\cos x$ is even, $\cos(-x) = \cos x$

Note: Angles are measured in radians in calculus, so that $\sin x$ and $\cos x$ have period 2π.

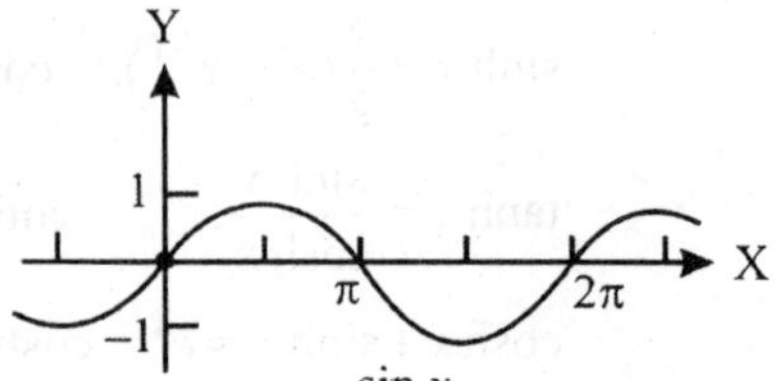

sin x

$$\sin^2 x + \cos^2 x = 1$$

$$\sin(x \pm y) = \sin x \cos y \pm \cos x \sin y$$

$$\cos(x \pm y) = \cos x \cos y \mp \sin x \sin y$$

$$\sin 2x = 2 \sin x \cos x = \frac{2 \tan x}{(1 + \tan^2 x)}$$

$$\cos 2x = (\cos^2 x - \sin^2)x = 1 - 2\sin^2 x$$

$$= 2\cos^2 x - 1 = \frac{1 - \tan^2 x}{1 + \tan^2 x}$$

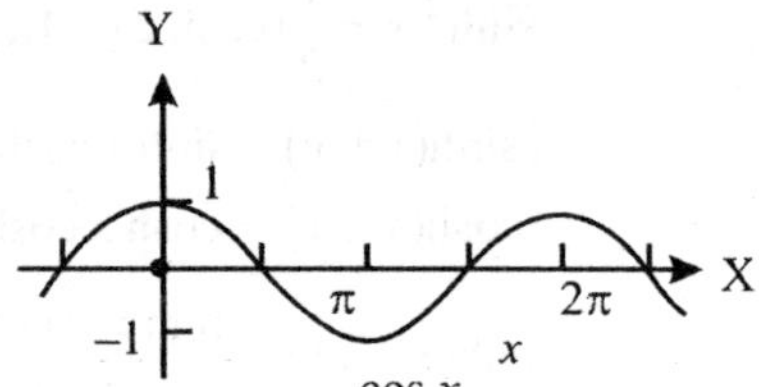

cos x

14. Series:

Exponential series: $e^x = 1 + \frac{x}{1!} + \frac{x^2}{2!} + \frac{x^3}{3!} + \cdots$

sin, cos, sinh, cosh series:

$$\sin x = x - \frac{x^3}{3!} + \frac{x^5}{5!} - \cdots, \quad \cos x = 1 - \frac{x^2}{2!} + \frac{x^4}{4!} - \cdots$$

$$\sinh x = x + \frac{x^3}{3!} + \frac{x^5}{5!} + \cdots, \quad \cosh x = 1 + \frac{x^2}{2!} + \frac{x^4}{4!} - \cdots$$

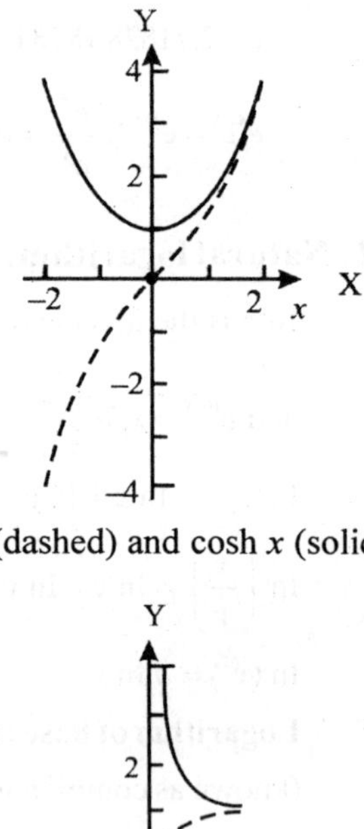

sinh x (dashed) and cosh x (solid)

Log Series:

$$\log(1 + x) = x - \frac{x^2}{2} + \frac{x^3}{3} - \cdots$$

$$\log(1 - x) = -\left(x + \frac{x^2}{2} + \frac{x^3}{3} + \cdots\right)$$

Gregory series:

$$\tan^{-1} x = x - \frac{x^3}{3} + \frac{x^5}{5} - \cdots,$$

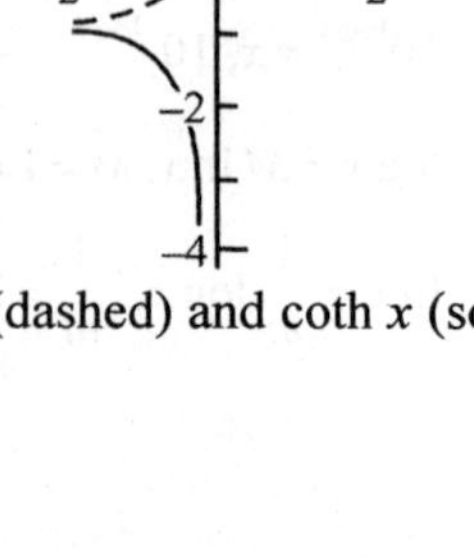

tanh x (dashed) and coth x (solid)

Differentiation

1. $\frac{d}{dx}(uv) = u\frac{du}{dv} + v\frac{du}{dx}$ $\quad\frac{d}{dx}\left(\frac{u}{v}\right) = \frac{vdu/dx - udv/dx}{v^2}$

2. $\frac{du}{dx} = \frac{du}{dy}.\frac{dy}{dx}$ (Chain Rule) $\quad\frac{d}{dx}(ax+b)^n = n(ax+b)^{n-1}.a$

3. $\frac{d}{dx}(e^x) = e^x$ $\quad\frac{d}{dx}a^x = a^x \log_e a$

4. $\frac{d}{dx}(\log_e x) = 1/x$ $\quad\frac{d}{dx}(\log_a x) = \frac{1}{x\log a}$

5. $\frac{d}{dx}(\sin x) = \cos x$ $\quad\frac{d}{dx}(\cos x) = -\sin x$

6. $\frac{d}{dx}(\tan x) = \sec^2 x$ $\quad\frac{d}{dx}(\cot x) = -\text{cosec}^2 x$

7. $\frac{d}{dx}(\sec x) = \sec x \tan x$ $\quad\frac{d}{dx}(\text{cosec}\, x) = -\text{cosec}\, x \cot x$

8. $\frac{d}{dx}(\sin^{-1} x) = \frac{1}{\sqrt{(1-x^2)}}$ $\quad\frac{d}{dx}(\cos^{-1} x) = -\frac{1}{\sqrt{(1-x^2)}}$

9. $\frac{d}{dx}(\tan^{-1} x) = \frac{1}{1+x^2}$ $\quad\frac{d}{dx}(\cot^{-1} x) = \frac{-1}{1+x^2}$

10. $\frac{d}{dx}(\sec^{-1} x) = \frac{1}{x\sqrt{(x^2-1)}}$ $\quad\frac{d}{dx}(\text{cosec}^{-1} x) = \frac{-1}{x\sqrt{x^2-1}}$

11. $\frac{d}{dx}(\sinh x) = \cosh x$ $\quad\frac{d}{dx}(\cosh x) = \sinh x$

12. $D^n(e^{mx}) = m^n e^{mx}$ $\quad D^n(a^{mx}) = m^n(\log a)^n.a^{mx}$

13. $D^n(ax+b)^n = m(m-1)(m-2)...(m-n+1)(ax+b)^{m-n}$

14. $D^n \log(ax+b) = (-1)^{n-1}(n-1)!a^n/(ax+b)^n$

15. $D^n \sin(ax+b) = a^n \sin\left(ax+b+\frac{n\pi}{2}\right)$

16. $D^n \cos(ax+b) = a^n \cos\left(ax+b+\frac{n\pi}{2}\right)$

17. $D^n\{\sin(bx+c)\} = (a^2+b^2)^{\frac{n}{2}} e^{ax} \sin(bx+c+n\tan^{-1}b/a)$

18. $D^n\{\cos(bx+c)\} = (a^2+b^2)^{\frac{n}{2}} e^{ax} \cos(bx+c+n\tan^{-1}b/a)$

19. $D^n(uv) = u_n + {}^nC_1 u_{n-1}v + {}^nC_2 u_{n-2}v_2 + \cdots + {}^nC_r u_{n-r}v_r + \cdots + {}^nC_n v_n$

Appendix II

Results of Some Standard Integrals

1. $\int x^n \, dx = \dfrac{x^{n+1}}{n+1} \qquad (n \neq -1)$

2. $\int e^x \, dx = e^x$

3. $\int \sin x dx = -\cos x$

4. $\int \tan x dx = -\log \cos x$

5. $\int \sec x dx = \log(\sec x + \tan x)$

6. $\int \sec^2 x dx - \tan x$

7. $\int \sinh x dx = \cosh x$

8. $\int \dfrac{dx}{a^2 + x^2} = \dfrac{1}{a} \tan^{-1} \dfrac{x}{a}$

9. $\int \dfrac{dx}{a^2 - x^2} = \dfrac{1}{2a} \log \dfrac{a+x}{a-x}$

10. $\int \dfrac{dx}{x^2 - a^2} = \dfrac{1}{2a} \log \dfrac{x-a}{x+a}$

11. $\int \sqrt{(a^2 - x^2)} dx = \dfrac{x\sqrt{(a^2 - x^2)}}{2} + \dfrac{a^2}{2} \sin^{-1} \dfrac{x}{a}$

12. $\int \sqrt{(a^2 + x^2)} dx = \dfrac{x\sqrt{(a^2 + x^2)}}{2} + \dfrac{a^2}{2} \sinh^{-1} \dfrac{x}{a}$

13. $\int \sqrt{(x^2 - a^2)} dx = \dfrac{x\sqrt{(x^2 - a^2)}}{2} - \dfrac{a^2}{2} \cosh^{-1} \dfrac{x}{a}$

14. $\int \dfrac{1}{x} dx = \log_e x$

15. $\int a^x dx = \dfrac{a^x}{\log_e a}$

16. $\int \cos x dx = \sin x$

17. $\int \cot x dx = \log \sin x$

18. $\int \operatorname{cosec} x dx = \log(\operatorname{cosec} x - \cot x)$

19. $\int \operatorname{cosec}^2 x \, dx = -\cot x$

20. $\int \cosh x = \sinh x$

21. $\int \dfrac{dx}{\sqrt{a^2 - x^2}} = \sin^{-1} \dfrac{x}{a}$

22. $\int \dfrac{dx}{\sqrt{a^2 + x^2}} = \sinh^{-1} \dfrac{x}{a}$

23. $\int \dfrac{dx}{\sqrt{x^2 - a^2}} = \cosh^{-1} \dfrac{x}{a}$

24. $\int e^{ax} \sin bx dx = \frac{e^{ax}}{a^2 + b^2}(a \sin bx - b \cos bx)$

25. $\int_0^\infty e^{-ax} \sin bx dx = \frac{b}{a^2 + b^2}$

26. $\int e^{ax} \cos bx dx = \frac{e^{ax}}{a^2 + b^2}(a \cos bx + b \sin bx)$

27. $\int_0^\infty e^{-ax} \cos bx dx = \frac{a}{a^2 + b^2}$

28. $\int_{-\infty}^\infty e^{-x^2} dx = \sqrt{\pi}$

29. $\int_0^\infty \frac{e^{-ax}}{x} \sin bx dx = \tan^{-1} \frac{b}{c}, c > 0, b > 0$

30. $\int_0^\infty \frac{\sin ax}{x} dx = \frac{\pi}{2}$ if $a > 0$

31. $\int_0^\infty \frac{e^{ax} - e^{-ax}}{e^{-\pi x} - e^{-\pi x}} dx = \frac{1}{2} \tan \frac{a}{2}$

32. $\int_0^\infty \frac{e^{ax} + e^{-ax}}{e^{-\pi x} - e^{-\pi x}} dx = \frac{1}{2} \sec \frac{a}{2}$

33. $\int_0^{\frac{\pi}{2}} \sin^n x dx = \int_0^{\frac{\pi}{2}} \cos^n x dx = \frac{(n-1)(n-3)(n-5)\cdots}{n(n-2)(n-4)\cdots}$

Note: RHS is multiplied by $\frac{\pi}{2}$ when n is even

$$\int_0^{\frac{\pi}{2}} \sin^m x \cos^n x$$

$$= \frac{(m-1)(m-3)\cdots \times (n-1)(n-3)}{(m+n)(m+n-2)(m+n-4)\cdots}$$

Note: RHS is multiplied by $\frac{\pi}{2}$ when both m and n are even

$$\int_{-a}^{a} f(x)dx = \begin{cases} 2\int_0^a f(x)dx, & \text{if } f(x) \text{ is an } \textit{even} \text{ function} \\ 0 & \text{if } f(x) \text{ is an } \textit{odd} \text{ function} \end{cases}$$

$$\int_0^{2a} f(x)dx = \begin{cases} 2\int_0^a f(x)dx, & \text{if } f(2a-x) = f(x) \\ 0 & \text{if } f(2a-x) = -f(x) \end{cases}$$

$$\int udv = uv - \int vdu \qquad \text{(Integration by parts)}$$

$$\int u(x)v(x)dx = uv_1 - u'v_2 + u''v_3 - u'''v_4 + \cdots$$

Note: Superscript denotes differentiation, i.e. u'' denotes differentiation of u twice. Subscript number denotes number of times integration of v, i.e. v_3 denotes integration of v thrice.